Geology of the country around Kingston upon Hull and Brigg

This memoir chronicles the geology of the Kingston upon Hull and Brigg areas of Humberside and north Lincolnshire. The older rocks are known only from deep boreholes, but the youngest Triassic rocks, a full and varied succession of Jurassic strata and more than 300 metres of Cretaceous rocks occur at outcrop, together with a wide range of Quaternary deposits.

The Upper Carboniferous sandstones, mudstones and coals are a continuation of those seen at the surface farther west, as are the Permian rocks, though the latter are much thicker than at outcrop and contain substantial beds of salt and anhydrite. The arid conditions under which the Permian rocks were formed persisted during Triassic times, when largely red sandstones and mudstones were deposited.

The Jurassic and Cretaceous rocks were laid down in seas of varying depth, the former consisting predominantly of mudstones and limestones, but including sandstones and ironstones, one of which, the Frodingham Ironstone, was worked for more than a hundred years. Contemporaneous uplift in the north of the district associated with the Market Weighton Structure was responsible for the thinning in that direction of the Jurassic and Lower Cretaceous rocks, the latter being virtually confined to the area south of Brigg.

The Chalk, which forms the major topographical features of the Yorkshire and Lincolnshire Wolds, is of particular interest because the type areas of its three lowest formations in northern England are found in the district.

The Quaternary deposits provide evidence of two interglacial and at least two glacial episodes as well as a record of dramatic changes of sea level in relatively recent times.

Chapters of the memoir deal with the tectonic history, economic products and hydrogeology of the distrist.

BRITISH GEOLOGICAL SURVEY

G D GAUNT, T P FLETCHER
and C J WOOD

Geology of the country around Kingston upon Hull and Brigg

Memoir for 1:50 000 geological sheets 80 and 89
(England and Wales)

CONTRIBUTORS

Stratigraphy
E G Smith

Economic products and hydrogeology
I N Gale
P M Harris

Geophysics
J D Cornwell
G A Kirby

Palaeontology
B M Cox
H C Ivimey-Cook
A A Morter
N J Riley
G Warrington
I P Wilkinson

LONDON: HMSO 1992

First published 1992

ISBN 0 11 884399 0

Bibliographical reference

GAUNT, G D, FLETCHER, T P, and WOOD, C J. 1992. Geology of the country around Kingston upon Hull and Brigg. *Memoir of the British Geological Survey*, Sheets 80 and 89 (England and Wales).

Authors

T P Fletcher, MSc, PhD
British Geological Survey, Edinburgh

G D Gaunt, BSc, PhD
C J Wood, BSc
formerly British Geological Survey

Contributors

J D Cornwell, MSc, PhD
B M Cox, BSc, PhD
P M Harris, MA, CEng, MIMM
H C Ivimey-Cook, BSc, PhD
G A Kirby, BSc, PhD
N J Riley, BSc, PhD
G Warrington, BSc, PhD
I P Wilkinson, BSc, PhD
British Geological Survey, Keyworth

I N Gale, MSc
British Geological Survey, Wallingford

A A Morter, BSc
E G Smith, BSc, CEng, FIMM
formerly British Geological Survey

Other publications of the Survey dealing with this and adjoining districts

BOOKS

Memoir
Geology of the country around East Retford, Worksop and Gainsborough (Sheet 101), 1973

British Regional Geology
Eastern England from the Tees to The Wash (2nd edition)

Mineral Assessment Reports
No. 22 Scunthorpe, north-west SK 81 1976
No. 29 Scunthorpe, south-west SK 80 1977
No. 33 Gainsborough, north SK 89 1978

MAPS

1:625 000

Geological map of Great Britain, Sheet 2, 3rd edition 1979
Quaternary map of the United Kingdom, South 1977
Bouguer anomaly map of the British Isles, Southern Sheet 1986
Aeromagnetic map of Great Britain, Sheet 2 1965

1:250 000
Humber–Trent 1983
Bouguer gravity anomaly, Humber–Trent 1977
Aeromagnetic anomaly, Humber–Trent 1977

1:50 000 or 1:63 360

71 (Selby) Solid and Drift provisional edition 1973
72 (Beverley) Drift edition reprinted 1968
73 (Hornsea) Drift edition reprinted 1958
79 (Goole) Solid edition 1972, Drift edition 1971
88 (Doncaster) Solid and Drift editions 1969
91 (Great Grimsby) Solid and Drift edition 1990
101 (East Retford) Solid and Drift edition 1969
103 (Louth) Solid and Drift provisional edition 1980

Printed in the UK for HMSO

Dd 240405 C10 06/92

CONTENTS

1 **Chapter 1 Introduction**
Historical 2
Outline of geological history 2

5 **Chapter 2 Carboniferous**
Dinantian (Carboniferous Limestone) 5
Namurian (Millstone Grit) 5
Westphalian (Coal Measures) 8

16 **Chapter 3 Permian and Triassic**
Basal Permian Sands 17
Don Group 17
Aislaby Group 20
Teesside Group 21
Staintondale Group 22
Eskdale Group 22
Sherwood Sandstone Group 23
Mercia Mudstone Group 24
Penarth Group 25
Lias Group 28

29 **Chapter 4 Jurassic**
Lias Group 30
Redbourne Group 44
Ancholme Clay Group 57

71 **Chapter 5 Cretaceous**
Lower Cretaceous 71
Upper Cretaceous (Chalk) 77

102 **Chapter 6 Structure**
Deep structure 102
Carboniferous rocks 104
Permian and Mesozoic rocks 105
Superficial structures 106

109 **Chapter 7 Quaternary**
Pre-Devensian glacial deposits 109
Interglacial deposits 115
Devensian deposits 117
Flandrian deposits 124

128 **Chapter 8 Economic products and hydrogeology**
Building stone 128
Chalk 128
Clay and mudstone 129
Coal 129
Evaporites 129
Hydrocarbons 129
Iron ore 129
Limestone 136
Oil shale 136
Sand and gravel 136
Silica sand 137
Hydrogeology 137

140 **Appendices**
1 List of principal boreholes and wells 140
2 List of Geological Survey photographs 142
3 Notes on the marker horizons in the Northern Province Chalk 143
4 Ancholme Clay Group biostratigraphy 145

149 **References**

158 **Fossil index**

163 **General index**

FIGURES

1 Map showing the location and regional geological setting of the district 1
2 Map showing the principal topographical features of the district 2
3 Borehole sections of the Dinantian and Namurian rocks 6
4 Borehole sections of the Westphalian A rocks 9
5 Borehole sections of the Westphalian B rocks 10
6 Borehole sections of the Westphalian C rocks 11
7 Borehole sections of the Permian rocks 18
8 Isopachytes of the Permian carbonates and evaporites 19
9 Borehole sections of the Mercia Mudstone Group, with isopachytes of the Mercia Mudstone and Sherwood Sandstone groups 24
10 Comparative sections of the Penarth Group 27
11 Distribution and relative abundances of palynomorphs from the Mercia Mudstone Group to basal Lias Group in Blyborough Borehole 28
12 Jurassic stages and principal lithostratigraphical divisions 29
13 Relationship of the Lias of the district to that of adjacent regions 31
14 Isopachytes in metres of a) Scunthorpe Mudstones Formation, b) Coleby Mudstones Formation below top of Marlstone Rock, and c) Coleby Mudstones Formation above Marlstone Rock 31
15 Borehole sections of the Scunthorpe Mudstones Formation 33
16 Sections, mainly from boreholes, of the Coleby Mudstones Formation 37
17 Principal ammonite occurrences in the Lower Jurassic 42
18 Lithostratigraphical classification of the Redbourne Group, with chronostratigraphical correlation 44

19 Sections of the Northampton Sand and Grantham formations 44
20 Sections of the Lincolnshire Limestone Formation 47
21 Sections of the Glentham and Cornbrash formations 53
22 Chronostratigraphical correlation of the Ancholme Clay Group 58
23 Occurrences of selected ammonite taxa in the Ancholme Clay Group of the district 59
24 Sections, mainly from boreholes, of the Kellaways Beds Formation 60
25 Borehole sections of the lower part of the Ancholme Clay Group above the Kellaways Beds Formation 63
26 Sections, mainly from boreholes, of the upper part of the Ancholme Clay Group 64
27 Generalised section of the Lower Cretaceous rocks and the lower part of the Spilsby Sandstone Formation 68
28 Correlation of the Northern Province Chalk, as seen in the district, with the Chalk of the Southern Province by means of marl seams 79
29 Stratigraphy of the Hunstanton Chalk Member (Red Chalk), with ranges of significant macrofossils 81
30 Generalised section of the Ferriby Chalk Formation above the Hunstanton Chalk Member 85
31 Stratigraphy of the Welton and Burnham (pars) Chalk formations, with ranges of significant macrofossils 89
32 Map of the outcrops of the Chalk formations 91
33 Detailed section of the Welton Chalk Formation 92
34 Detailed section of the Burnham Chalk Formation 94–95
35 Geophysical maps and basement interpretation of the district 102
36 Deep geological interpretation of a Bouguer anomaly profile across the Gainsborough Trough and the Askern–Spital Structure 103
37 Structure maps, based largely on seismic evidence, of the Carboniferous and Permian rocks 104
38 Structure maps of the Mesozoic rocks 105
39 Map showing principal Quaternary deposits in the district 110
40 Map showing rockhead contours at and below OD at 5 m intervals 114
41 Sections showing Devensian interbedded glacial and lacustrine deposits in the Humber Gap 120
42 Stone orientations in Devensian till 121
43 Annual production of Frodingham Ironstone 1860–1984 130
44 Frodingham Ironstone reserve areas west of New River Ancholme 132
45 Workings in Claxby Ironstone 134–135
46 Ancholme Clay Group in Nettleton Bottom Borehole: Callovian and Oxfordian sequence 145
47 Ancholme Clay Group in Nettleton Bottom Borehole: Kimmeridgian sequence 146
48 Correlation between Nettleton Bottom Borehole and South Ferriby Pit 147
49 Localities which have yielded biostratigraphical information on the Ancholme Clay Group 148

PLATES

1 Lincolnshire Limestone in Kirton Quarries 49
2 Ancholme Clay Group mudstones overlain by Carstone, Hunstanton Chalk Member and Chalk at South Ferriby Quarry 80
3 Burnham Chalk on Welton Chalk at Burnham Lodge Quarry 96
4 Step-faults in Chalk at Barton upon Humber Quarry 107
5 Quaternary deposits at Red Cliff, North Ferriby 119
6 Abandoned opencast working for Frodingham Ironstone near Bagmoor Farm 130

TABLES

1 Chronostratigraphical range in the Carboniferous of the district 5
2 Subdivision of Permian and Triassic rocks in the district 16
3 Classification of the Northern Province Chalk 78
4 Correlation of Quaternary deposits in the district and the deduced sequence of events 111
5 Detailed analysis of Frodingham Ironstone in a borehole [SE 9197 1311] at Santon 131
6 Frodingham Ironstone reserves 133
7 Typical physico-chemical properties of silica sands, Messingham 137

PREFACE

This memoir describes the geology of the district covered by 1:50 000 New Series sheets 80 (Kingston upon Hull) and 89 (Brigg) of the Geological Map of England and Wales, an area renowned for its mineral resources for 2000 years. The principal towns in the district are Kingston upon Hull, a port since the 12th century, and Scunthorpe, which developed in the 19th century in response to the exploitation of the Frodingham Ironstone. Iron ore had been mined in this district since Roman times, but between 1859 and 1988, when working ceased, around 300 million tonnes of ore were extracted. Although the ore is now regarded as being subeconomic, there is an estimated resource of over 1000 million tonnes left in the ground. Other valuable resources still exploited in the area are chalk and clay for cement-making, chalk for whiting, sand for glass-making, and oil.

The survey of the district was accompanied by the drilling of several fully cored boreholes, enabling a more comprehensive stratigraphical appraisal of the Triassic to Cretaceous succession than was previously possible. In addition, a number of hydrocarbon and coal trial boreholes have helped to elucidate the stratigraphy of the basement Carboniferous rocks. Interpretation of the complex Quaternary succession has been enhanced by a greater understanding of glacial and periglacial processes and by the application of improved dating techniques.

The lithological uniformity of the late Jurassic, mainly mudstone sequence has led to the introduction of a modified lithostratigraphical nomenclature; however, a biostratigraphical appraisal of the sequence in the key borehole at Nettleton Bottom enables correlation with the standard succession in southern Lincolnshire. The type areas of the formations comprising the Chalk Group of the Northern Province occur within the district; the stratigraphy has been refined and additional marker horizons have been named and defined. Geophysical investigations have revealed anomalies suggesting a buried granite on the northern margin of the district and a Carboniferous trough in the south-west.

This new knowledge enables us to better assess the existing resources of the district and will facilitate land-use planning. Additionally, a better understanding of the Quaternary succession provides us with a more complete picture of climatic and relative sea-level change over the past 2 million years; this will help us to forecast the likely impact on Britain of future global change.

Peter J Cook, DSc
Director

British Geological Survey
Keyworth
Nottingham NG12 5GG

23 December 1991

LIST OF SIX-INCH AND 1:10 000 GEOLOGICAL MAPS

The following is a list of six-inch and 1:10 000 geological maps included, wholly or in part, in 1:50 000 sheets 80 (Kingston upon Hull) and 89 (Brigg), with the initials of the surveyors and the dates of survey. The surveyors were R. J. Bull, T. P. Fletcher, G. D. Gaunt, G. H. Rhys, J. G. O. Smart, E. G. Smith and V. Wilson.

Manuscript copies of all complete maps have been deposited for public reference in the library of the British Geological Survey in Keyworth. Incomplete maps are marked with an asterisk.

Map	Name	Surveyors	Date
SE 80 NW	Butterwick	GDG	1964
SE 80 NE	Scunthorpe (south-west)	RJB, TPF	1972–76
SE 80 SW	Owston Ferry	GHR, GDG	1965
SE 80 SE	Messingham	TPF	1972
SE 81 NW	Fockerby	GDG	1965
SE 81 NE	Burton upon Stather	GDG, TPF, VW	1939–76
SE 81 SW	Gunness	GDG	1964
SE 81 SE	Scunthorpe (north-west)	RJB, TPF, GDG	1965–76
SE 82 NW	Gilberdyke	GDG	1968
SE 82 NE	Broomfleet	GDG	1968–69
SE 82 SW	Adlingfleet	GDG	1965–68
SE 82 SE	Alkborough	GDG, TPF	1965–76
* SE 83 SW	Sandholme	GDG	1968
* SE 83 SE	North Cave	GDG	1968
SE 90 NW	Scunthorpe (south-east)	TPF, RJB, VW	1939–73
SE 90 NE	Scawby	TPF	1972
SE 90 SW	Manton	TPF	1972–74
SE 90 SE	Hibaldstow	TPF	1971–72
SE 91 NW	Winterton	GDG, TPF, VW	1940–72
SE 91 NE	Horkstow	EGS, GDG	1968–72
SE 91 SW	Scunthorpe (north-east)	TPF, VW	1940–74
SE 91 SE	Appleby	TPF	1972–75
SE 92 NW	Brough	GDG	1969–70
SE 92 NE	North Ferriby	JGOS	1969
SE 92 SW	Winteringham	GDG, TPF	1972–76
SE 92 SE	South Ferriby	JGOS, GDG	1976–77
* SE 93 SW	South Cave	GDG	1969–71
* SE 93 SE	Little Weighton	JGOS	1970
SK 89 NW	Laughton	GHR	1964–65
SK 89 NE	Scotton	RJB, TPF	1972–76
SK 89 SW	Gainsborough (north)	EGS	1960
* SK 89 SE	Blyton	RJB, EGS	1960–73
SK 99 NW	Kirton in Lindsey	RJB	1972–73
SK 99 NE	Waddingham	RJB, TPF	1971–75
* SK 99 SW	Willoughton	RJB	1973–75
* SK 99 SE	Snitterby	RJB, TPF	1972–75
TA 00 NW	Brigg	TPF	1972–75
TA 00 NE	Barnetby le Wold	TPF, RJB	1974–77
TA 00 SW	North Kelsey	TPF	1975–76
TA 00 SE	Grasby	TPF	1972–75
TA 01 NW	Bonby	EGS	1968
TA 01 NE	Wootton	EGS	1968
TA 01 SW	Elsham	TPF, EGS	1969–76
TA 01 SE	Melton Ross	EGS, TPF	1969–75
TA 02 NW	Hessle	GDG, JGOS	1969–71
TA 02 NE	Kingston upon Hull (south-west)	GDG	1971
TA 02 SW	Barton-upon-Humber	JGOS	1967
TA 02 SE	Barrow upon Humber	EGS, GDG	1977
* TA 03 SW	Willerby	GDG	1971
* TA 03 SE	Kingston upon Hull (north-west)	GDG	1971
TA 10 NW	Great Limber	RJB	1974–77
TA 10 SW	Caistor	RJB, VW	1942–74
TA 11 NW	Thornton Abbey	EGS, GDG	1968–80
TA 11 SW	Kirmington	EGS	1969–74
TA 12 NW	Kingston upon Hull (south-east)	GDG	1971–77
TA 12 SW	Goxhill	GDG, EGS	1977–80
* TA 13 SW	Kingston upon Hull (north-east)	GDG	1971
TF 09 NW	South Kelsey	RJB	1972–75
TF 09 NE	Holton le Moor	RJB	1973–74
* TF 09 SW	Atterby Carr	RJB	1973–75
* TF 09 SE	North Owersby	RJB	1974
TF 19 NW	Nettleton Bottom	RJB	1974–75
* TF 19 SW	Claxby	RJB, VW	1947–75

ACKNOWLEDGEMENTS

The six-inch survey south of the Humber and the resurvey north of the Humber were carried out by Mr R J Bull, Drs T P Fletcher and G D Gaunt and Messrs J G O Smart and E G Smith between 1960 and 1977. A small area surveyed by Mr G H Rhys was included along with some of the reconnaissance survey by Dr V Wilson. The work was accomplished under the supervision of Mr D R A Ponsford and Mr Smith as District Geologists.

The greater part of the present memoir was written by Drs Gaunt and Fletcher, but Mr C J Wood wrote the Chalk section of Chapter 5 besides identifying the Upper Cretaceous fossils. Mr Smith contributed to several parts of the memoir and much of Chapter 8 was written by Messrs P M Harris (economic products) and I N Gale (hydrogeology). Drs J D Cornwell and G A Kirby contributed to Chapter 6 (structure), and palaeontological and biostratigraphical details were contributed by Dr G Warrington (Triassic), Dr H C Ivimey-Cook (Lower and Middle Jurassic), Dr B M Cox (Upper Jurassic) and Mr A A Morter (Lower Cretaceous). Other fossils were identified by Mr I P Wilkinson (Mesozoic microfossils) and Dr N J Riley (Carboniferous). The memoir was edited by Mr W B Evans and Mr Smith.

Grateful acknowledgement is made to numerous organisations and individuals, including landowners, quarry operators and public and local authorities for generous help during the survey. In particular British Petroleum Development (UK) Ltd, British Coal and British Steel (via Dr D Golding and Mr D Elford) have released valuable information in their respective spheres.

NOTES

Throughout this memoir the word 'district' means the area included in 1:50 000 geological sheets 80 (Kingston upon Hull) and 89 (Brigg).

Figures in square brackets are National Grid references within 100 km squares SE, SK, TA and TF.

The authorship of fossil species is given in the fossil index.

CHAPTER 1

Introduction

This memoir describes the geology of the district covered by the Kingston upon Hull (80) and Brigg (89) 1:50 000 geological sheets of England and Wales (Figure 1).

The district was originally surveyed on the one-inch scale to the south of the Humber and on the six-inch scale to the north by A C G Cameron, J R Dakyns, C Fox-Strangways, A J Jukes-Browne, C Reid and W A E Ussher, and the results were published on Old Series one-inch sheets 83 and 86 in 1886 and 1887 respectively. Memoirs for these sheets were published in 1888 and 1890. During the 1939–45 war a six-inch reconnaissance survey of the Frodingham Ironstone and Claxby Ironstone outcrops and adjacent areas was carried out by Dr V Wilson, but the maps were not published. The resurvey north of the Humber and primary six-inch survey south of the Humber were carried out between 1960 and 1977. Separate Solid and Drift editions at the 1:50 000 scale were published for Sheet 89 in 1982 and for Sheet 80 in 1983.

Topographically the district comprises three northward-converging ridges with westward-facing scarps rising above low-lying gently undulating or flat terrain, some of it only 1 to 2 m above OD, and cut across by the Humber Estuary (Figure 2). To the south of the Humber the western ridge, on which Scunthorpe stands, rises to 67 m above OD north

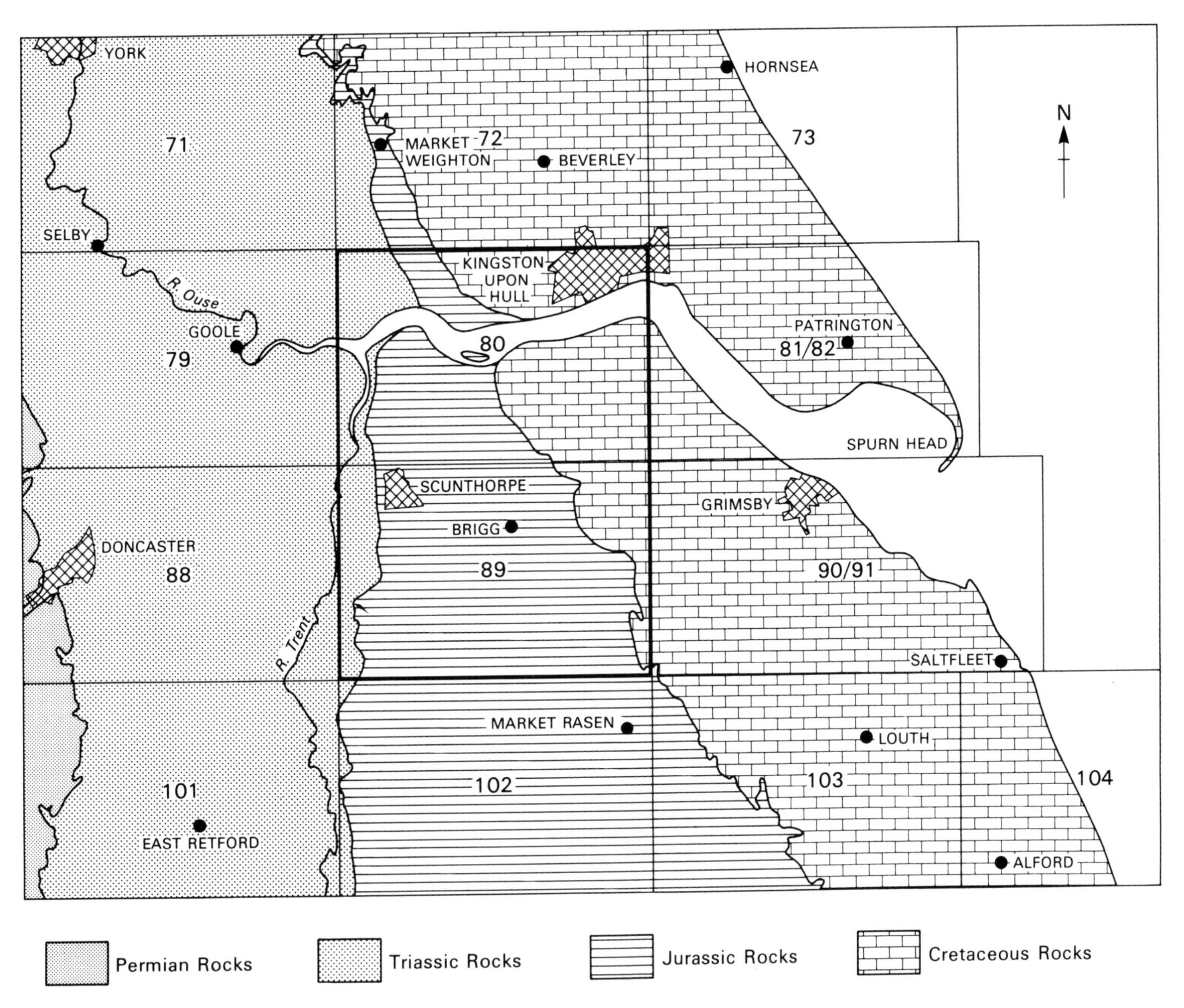

Figure 1 Map showing the location and regional geological setting of the district

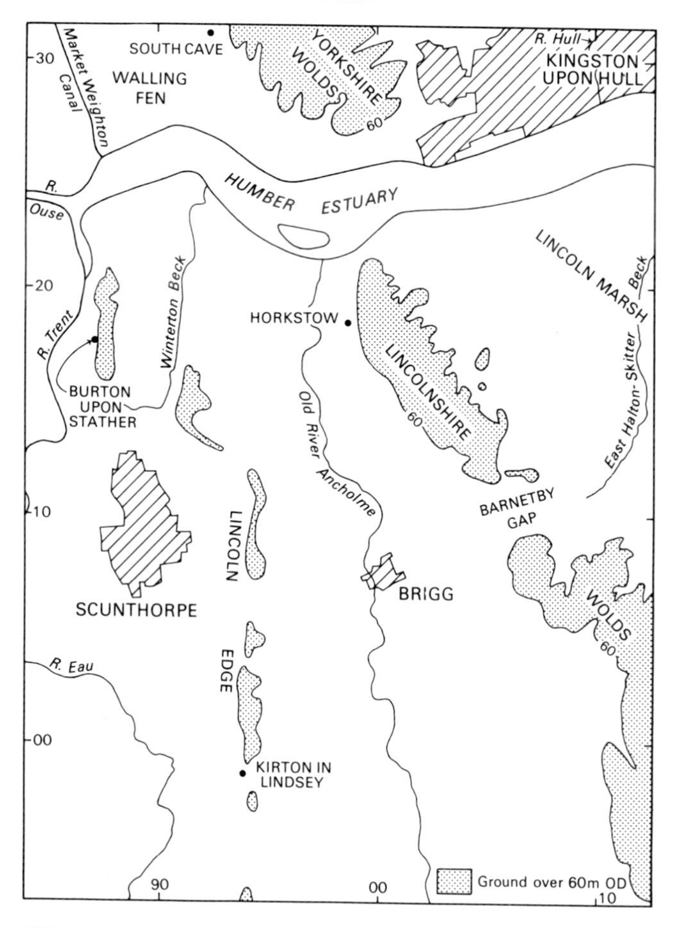

Figure 2 Map showing the principal topographical features of the district

of Burton upon Stather; the central ridge, which as the Lincoln Edge runs from the Humber to beyond Grantham, reaches 77 m above OD north-east of Scunthorpe and 73 m north of Kirton in Lindsey; the eastern ridge is part of the Lincolnshire Wolds, which rise to 100 m near Horkstow and 168 m above OD south of Caistor. North of the Humber, the western and central ridges almost coalesce and form only minor features on the scarp of the eastern ridge, here part of the Yorkshire Wolds which reach nearly 150 m above OD east of South Cave. The low-lying area west of the Yorkshire Wolds forms part of the Vale of York; that to the east, on which Kingston upon Hull is situated, is part of the floodplain of the River Hull. South of the Humber the floodplain of the River Trent lies in the west, the floodplain of the River Ancholme separates the Lincoln Edge from the Lincolnshire Wolds, and the low ground east of the latter continues south-eastwards as the coastal Lincoln Marsh.

The entire district lies within the Humber catchment, and the drainage of one-fifth of England is into the estuary. Spring tides in the Humber rise to 3.6 m above OD and, within the district, the rivers Ouse and Trent are tidal and variably saline, as also were the lower reaches of the Ancholme and Hull before sluices and flood-barriers were built.

HISTORICAL

The earliest traces of man in the district are presumed Lower Palaeolithic flakes discovered at Kirmington. Remains from the later part of the Stone Age and from the Bronze and Iron ages have also been found, the Bronze Age boats from North Ferriby and Brigg deserving special mention. Brough (Petuaria) and Caistor were Roman settlements and the line of Ermine Street can still be seen along the Lincoln Edge. The Humber Estuary gave ready access to Anglian and Danish invaders, who founded most of the towns and villages in the district.

Kingston upon Hull, which originated during the 12th century, served as an important port for the mediaeval wool trade and for the later whaling and fishing industries. Scunthorpe, the only other large town in the district, developed rapidly in the mid-19th century in response to the exploitation of the Frodingham Ironstone. Communications within the district improved considerably with the linking of the Humber Estuary to the canal systems of northern England and, via the River Trent, with the Midlands, and later with the coming of the railways. More recently, motorway links have arrived and the Humber Bridge has been built.

There is archaeological or documentary evidence of exploitation from mediaeval times of most of the mineral resources available at the surface, notably clay, mudstone, limestone and chalk, and there are even traces of old workings in some of the ironstones. In more recent times the principal minerals worked have been ironstone, which reached an annual output of 5.6 million tonnes in the early 1960s, and chalk, clay and mudstone for cement manufacture. Coal has been proved at considerable depth, but has not been mined, and there has been some exploration for deep hydrocarbon sources, with modest success in the Corringham area in the south-west.

OUTLINE OF GEOLOGICAL HISTORY

The only evidence of pre-Carboniferous history is provided by geophysical investigations, which suggest the possible presence of a concealed granite near Market Weighton to the north-west, and of a deep-seated belt of magnetic rocks crossing the district from north-west to south-east. Boreholes have penetrated only as deep as the Lower Carboniferous (Dinantian) rocks. When these rocks were laid down most of the district was covered by a shallow tropical sea, in which the deposition of calcareous sediments generally kept pace with slow subsidence. However, the Gainsborough Trough crossed the extreme south-west of the district, and within it both deposition and subsidence were more rapid. Most of the trough deposits are probably limestone, though geophysical evidence suggests that there could be sandstones at depth and mudstones in the upper parts of the sequence.

In succeeding Namurian times, increasing amounts of mud, silt and sand reached the district, firstly as deltaic and eventually as fluvial sediments, to form the Millstone Grit. Marine deposition was reduced to periodic inundations which produced thin mudstone ‘marine bands’ within the otherwise brackish and freshwater mudstones, siltstones and sandstones.

During the deposition of the higher Millstone Grit strata and the overlying Coal Measures, cyclic sedimentation predominated, producing innumerable repetitions of a sequence beginning with marine mudstone and overlain in turn by fluvial and lacustrine mudstone, siltstone, sandstone, seatearth and coal, the two last indicating terrestrial vegetated conditions. Overall thickness variations suggest that the rate of subsidence, although nowhere rapid, was slightly greater in the north-west.

The Carboniferous rocks were uplifted, faulted and gently folded by the ensuing Hercynian earth movements, and then subjected to prolonged erosion. For much of the succeeding Permian and Triassic periods the district lay in the south-western part of a vast intracontinental basin which stretched eastwards into central Europe and which experienced a hot, dry, low-latitude climate. Initial deposition, probably late in early Permian times, was of wind-blown sands which form a thin veneer unconformably mantling the denuded Carboniferous rocks. In the late Permian the basin was occupied by the shallow inland 'Zechstein Sea', which underwent five major, and many minor cycles of transgression and regression, and which at times during regressions became hypersaline. In the district these cycles produced repeated sequences of dolomitic limestones, dolomites, evaporites (mainly anhydrite but including halite) and clastics (mainly red desert-derived mudstones). By early Triassic times the Zechstein Sea had disappeared, and thick red fluvial sands, the Sherwood Sandstone, spread across the district. Later in the Triassic another hypersaline sea developed to the east and, within the district, the reddish dolomitic and anhydritic fluvial, lagoonal and littoral muds of the Mercia Mudstone were deposited. As the Triassic Period came to a close the basin was invaded by the sea, and the fully marine shallow-water muddy sediments of the Penarth Group were laid down across the district, although there was one temporary partial reversion to more restricted basinal conditions.

Open-sea sedimentation continued in the district through most of the succeeding Jurassic Period, interrupted only by temporary incursions of brackish and freshwater sediments and by transient phases of emergence that were more pronounced towards the north because of differential uplift along the Market Weighton Structure. Deposition during the early Jurassic was mainly of mud and silt in generally quiescent, warm, shallow, marine water, forming the dark mudstones and thin silty limestones of the Lias. At times, phases of extreme shallowing led to the formation, probably to the north and west, of iron-precipitating lagoons, the sediments from which were later redeposited in the district as ironstones. Following extensive uplift towards the end of the early Jurassic, considerable erosion was experienced before the unconformably overlying Redbourne Group was laid down. These sediments are more varied in lithology. The initial marine transgression led to the deposition of the thin ferruginous Northampton Sand. This was subsequently overlain during a temporary regression by the thin brackish-water sands and muds of the Grantham Formation. The succeeding marine calcareous muds with ooliths and silts of the Lincolnshire Limestone accumulated as coastal shoals with some adjacent coral knolls. Following another regression, freshwater and brackish sands and muds spread across the district to form the Glentham Formation. Thin coals and limestones within this sequence reflect intervals of emergence and marine transgression respectively. The succeeding Cornbrash Formation was deposited during another marine transgression, but this thin limestone formation contains evidence of emergence and erosion. The overlying Ancholme Clay Group contains at its base the littoral and sublittoral sands and muds of the Kellaways Beds. Deeper water, more quiescent, open-sea, muddy conditions like those in the early Jurassic were re-established, but there were short-lived periods of shallowing, which gave rise to sandier sequences like the Brantingham and Elsham Sandstone formations.

Towards the end of the Jurassic, intermittent minor earth-movements were renewed. Some of the Ancholme Clay was denuded prior to deposition of the sublittoral Lower Spilsby Sandstone and, following another break in sequence at the Jurassic–Cretaceous boundary, the Upper Spilsby Sandstone was laid down. After a further depositional break, the products of iron-precipitating lagoons were redeposited as the Claxby Ironstone. The succeeding Lower Tealby Clay represents a temporary return to quiescent, open-sea, muddy conditions, before further shallowing and clearing of the water induced the formation of the Tealby Limestone. Renewed deepening led to the deposition of the ferruginous sandy oolitic muds of the Roach Formation. Towards the middle of Cretaceous times there was substantial uplift and erosion. Erosion was more vigorous towards the north, and Lower Cretaceous deposits survive only in the south-east of the district. In the north, erosion removed not only the Lower Cretaceous rocks, but also much of the Ancholme Clay and, still farther north, just beyond the district, the only Jurassic rocks remaining are the lower part of the Lias. When the sea slowly spread over the district again just before the onset of late Cretaceous time, sublittoral reworking produced the thin sandy and locally pebbly Carstone. As the inundation progressed the supply of clastic sediment diminished and deposition became dominantly calcareous. An initial thin layer of reddish calcareous mud, the Hunstanton or Red Chalk, was succeeded by deposition of white Chalk, mainly consisting of minute coccoliths, which continued throughout late Cretaceous time.

From the late Cretaceous onwards until a few hundred thousand years ago, a time span of some 65 million years encompassing the Tertiary Era and more than half of the Quaternary Era, there is no evidence of sedimentary deposition in the district. During this time uplift to the west initiated erosion which eventually exposed Jurassic rocks over much of the district and Triassic rocks in the west. Most theories of preglacial drainage evolution include the premise that the regional river pattern was initiated on a surface that sloped down to the east. On this surface the precursors of the present eastern Pennine rivers established courses that once probably continued eastwards across eastern Yorkshire and northern Lincolnshire. Attempts have been made to correlate both the Humber Gap and various dry gaps across the Yorkshire and Lincolnshire Wolds with the courses of the original rivers. Ultimately, however, the Ouse and the Trent developed as subsequent streams along softer outcrops, diverting much of the original drainage.

This long period of uplift and erosion was terminated late in the Quaternary by the onset of glacial conditions. Ice invaded the district at least twice. On the first occasion it covered the entire district, cutting and filling a major subglacial drainage channel between Kirmington and Immingham, and possibly another in the Humber Gap, which it may even have initiated, and leaving other glacial deposits elsewhere. In one of the subsequent interglacial phases the sea rose to about 22 m above OD, depositing estuarine sediments at Kirmington, and in a later interglacial episode it cut a cliff into the Chalk along the eastern side of the Yorkshire and Lincolnshire Wolds, its base at about 1 to 3 m above OD. During the last glaciation ice invaded only the north-eastern part of the district, but the resulting glacial blockage of the Humber Gap produced the extensive Lake Humber in the Vale of York and the Trent and Ancholme valleys. Thick laminated clays and silts were deposited in the lake, together with beds of sand and gravel around the margin. The accompanying cold and vegetation-poor conditions allowed wind-blown sand to accumulate in several areas. With the amelioration of climate and the postglacial rise of sea level, thick and extensive fluvial and estuarine deposits and peat formed in low-lying areas. The constantly changing course of the Humber channels and the periodic growth and destruction of intertidal banks and islands show that deposition and erosion are still active within the estuary.

CHAPTER 2

Carboniferous

The oldest known rocks in the district, entirely concealed, are of Carboniferous age. All the 14 available boreholes which prove these rocks are situated near the western margin of the district; indeed six of them (Corringham Nos. 3, 5, 7–10) are grouped at one locality in the south-west. Spital Borehole, just south of the district, and North Ewster Borehole, just to the west, have also been considered in this account. Crosby Borehole, drilled between 1907 and 1912, was cored throughout, as were most of the Coal Measures strata in the British Coal's North Ewster and Rock Abbey boreholes. Details from the other boreholes, all drilled by British Petroleum or its predecessors, are derived largely from chipping samples and (except for Spital Borehole) from geophysical logs, but limited lengths of core from these boreholes have yielded valuable fossil evidence. Particulars of Burton on Stather Borehole in this chapter refer to the hydrocarbon borehole, not the water borehole of the same name. Unverified macrofossil identifications, derived mainly from the two British Coal boreholes, are given here in quotes.

The Carboniferous Period lasted approximately from 365 to 290 million years ago (Forster and Warrington, 1985), and the British sequence was deposited in equatorial latitudes (Smith et al., 1981). The chronostratigraphical range proved in the district is summarised in Table 1. In northern England there are several different depositional facies in the lower part of the succession, reflecting the presence of slowly subsiding 'blocks' and more rapidly subsiding 'basins', 'troughs' and 'gulfs'. One of these troughs, the NW–SE orientated Gainsborough Trough,extends across the south-western extremity of the district.

Table 1 Chronostratigraphical range in the Carboniferous of the district (asterisk denotes stages not proved)

Subsystem	Series	Stage
Silesian	Westphalian	Westphalian D
		Westphalian C
		Westphalian B
		Westphalian A
	Namurian	Yeadonian
		Marsdenian
		Kinderscoutian
		Alportian*
		Chokierian*
		Arnsbergian*
		Pendleian*
Dinantian		Brigantian
		Asbian
		Holkerian

DINANTIAN (CARBONIFEROUS LIMESTONE)

By analogy with the surrounding region (e.g. Edwards, 1967, pp.14–23) several hundred metres of limestone and associated rocks probably underlie the district, but only the topmost beds have been proved (Figure 3). The fossils recovered from these suggest a late Dinantian age.

In Butterwick Borehole [SE 8421 0563] the lowest cores comprise 11.27 m of pale brownish grey massive limestone, medium-grained but with some finer grained layers. A few brachiopods, including *Gigantoproductus* sp. and a fragment of the coral *Lithostrotion junceum*, imply a late Asbian or Brigantian age. Chippings from the overlying beds are of cream to white limestone, coarsely crystalline in parts, with traces of grey to black mudstone. A 3.36 m core near the top of the limestones consists of interbedded pale greyish brown, medium-grained limestone and dark grey, thin-bedded, partly argillaceous limestone, with layers of dark grey chert up to 13 mm thick. The paler limestone yielded a few brachiopods, including *Brachythyris* sp. and a smooth-shelled spiriferoid. Dr A R E Strank reports (personal communication) that foraminifera and algae from these beds are 'of Brigantian aspect'.

The limestone at the bottom of Corringham No. 7 Borehole [SK 8963 9300] is pale brown, buff and white, and some chippings suggest the presence of calcite veins. The highest 7 m are probably thinly bedded.

The cored limestone from the bottom of Spital Borehole [SK 9656 9115] is pale fawn to white, compact and porcellanous. According to Dr Strank (personal communication) foraminifera and algae from about 1 m below the top of the limestone suggest a Holkerian or Asbian age; the apparent absence of Brigantian fossils possibly indicates a depositional break or an erosive episode at the top of the limestone.

NAMURIAN (MILLSTONE GRIT)

The Namurian sequences known in the district are shown on Figure 3. Of the Corringham boreholes only No. 7 proved a complete succession, but No. 1 [SK 8940 9276], being mainly cored, is also included. Fossil evidence is confined to the upper strata within the Gainsborough Trough, and even here only the three younger Namurian stages are recognisable. The differences in thickness between sequences within the trough and on the block testify to differential subsidence. With the evidence available, correlation within the trough is difficult, and between trough and block it is virtually impossible except in the uppermost strata.

Beds below the Gracilis Marine Band (pre-Marsdenian)

These beds largely consist of mudstones and siltstones with sandstones of varying thickness. There are a few thin

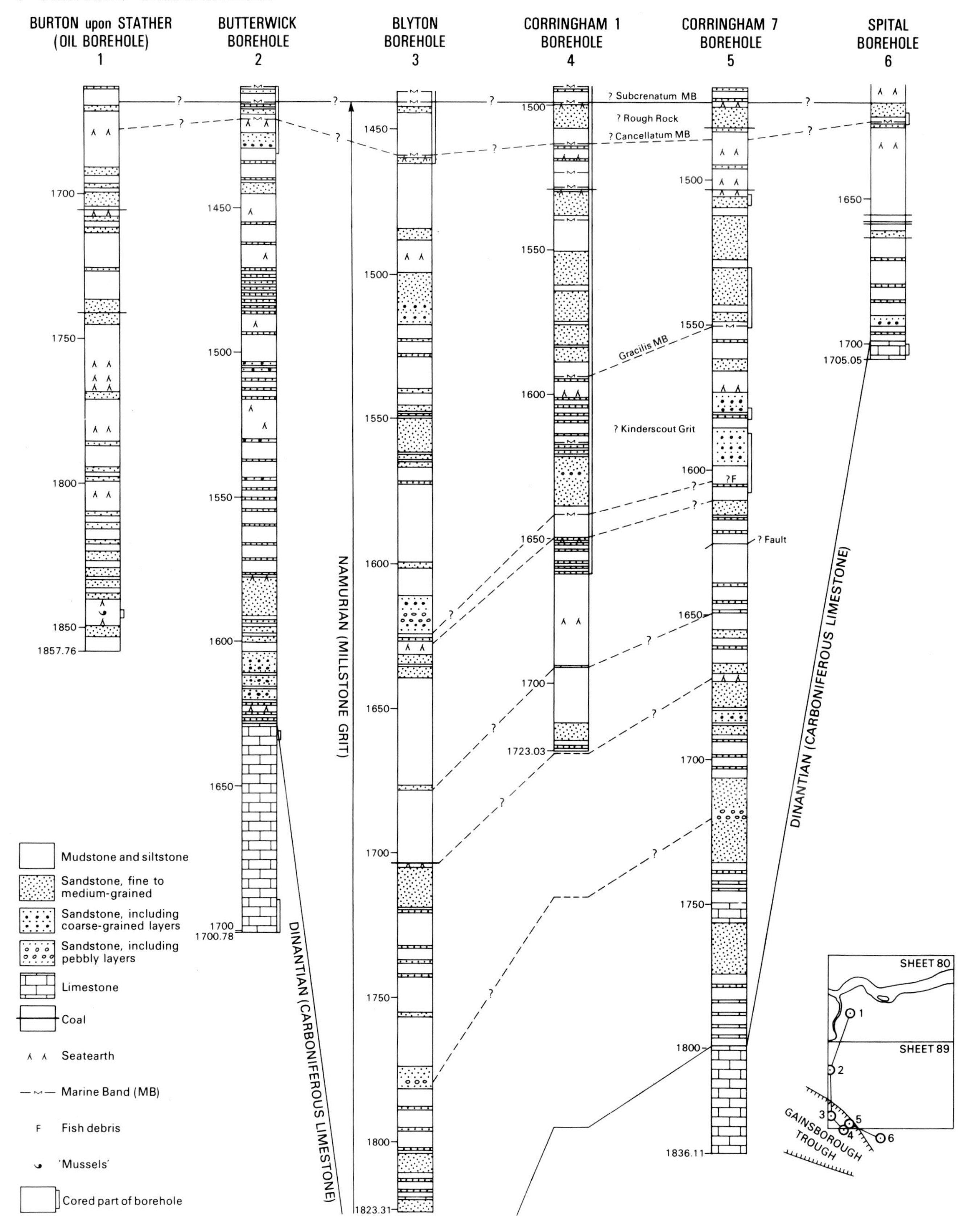

Figure 3 Borehole sections of the Dinantian (Carboniferous Limestone) and Namurian (Millstone Grit) rocks. Depths are in metres

limestones, which occur mainly near the base, and a few thin coals and/or seatearths are recorded at higher levels. In Corringham No. 7 Borehole the lowest 19 m comprise dark grey to black siltstones with thin black and white-mottled sandy limestones. Similar beds from boreholes on the block southwest of the Gainsborough Trough may be partly of late Dinantian age on the basis of gamma-ray correlation with beds in the Ashover area (Smith et al., 1973, pp.12, 15, 17, pl.II). Unfortunately, no gamma-ray log is available for this part of Corringham No. 7.

The succeeding 60 m of strata are dark grey to black pyritic silty mudstones, siltstones and thick fine-grained variably calcareous sandstones, with a few thin grey and white-mottled silty and sandy limestones. Brachiopod spines and crinoid debris were recorded from a limestone at about 1755 m in this borehole. Some of the sandstones are noticeably angular-grained, and in places they contain large green biotite flakes. The lowest 40 m or so of strata in Blyton Borehole [SK 8435 9555] are similar, but more argillaceous, and there is a pebbly layer at the top, similar to that at 1720 m in Corringham No. 7 Borehole. Most of the succeeding 120 to 150 m in Blyton and Corringham No. 7 boreholes, and also the lowest 85 m of strata preserved in Corringham No. 1, are paler and less pyritic than the lower beds. The sandstones are generally coarser; some are calcareous, others have siliceous and silty matrices. Subrounded and rounded grains have been noted in them, as also have feldspar and biotite. A few 'ironstone' layers, apparently nodular, occur in the mudstones and siltstones.

The lowest marine band, at 1641.88 m in Corringham No. 1 Borehole spans at least 0.15 m of dark grey pyritic silty mudstone containing *Lingula* sp., *Orbiculoidea* sp., hindeodellid bars and *Elonichthys* sp. Similar and possibly equivalent mudstone in Corringham No. 7 Borehole yielded only a possible fish fragment. These fossils are at a comparable stratigraphical level to a *Lingula*-bearing layer occurring 3.63 m above a bed containing *Reticuloceras* cf. *reticulatum*, indicative of the Kinderscoutian Stage, in South Leverton No. 1 Borehole [SK 7945 8103] to the south-west (Smith et al., 1973, p.17, plate II). In Burton upon Stather Borehole (oil) [SE 8787 1883], mudstone at 1846 m containing spores and *Curvirimula* sp. denoting a Kinderscoutian or Marsdenian age may be at the same level.

In Corringham Nos. 1 and 7 boreholes there are thick, locally coarse-grained and pebbly feldspathic sandstones containing tourmaline and chlorite in the lower part of the succeeding beds. They are lithologically comparable to the Kinderscout Grit farther west. Plant debris occurs throughout these beds, and there are scattered rootlets and distinct seatearths in the upper part, and thin 'ironstones' near the top. In Corringham No. 1 Borehole a marine band in the middle of the beds comprises at least 1.55 m of dark grey silty mudstone with *Lingula mytilloides* and *Orbiculoidea* cf. *nitida*.

Gracilis Marine Band (Marsdenian)

The base of the Marsdenian Stage is marked by the Gracilis Marine Band, which has been proved in several of the Corringham boreholes. In No. 3 the lowest part of the band comprises at least 0.45 m of calcareous medium to coarse-grained sandstone containing phosphatic debris, *Schizophoria* sp. and fish debris. Elsewhere the main part of the band, up to 3.70 m thick, consists of dark grey micaceous silty mudstone, locally carbonaceous and pyritic, and containing sand laminae, scattered sand grains and clasts of sandstone. Its fauna comprises *Serpuloides* sp., *L. mytilloides*, productoid protegulum, *Caneyella* sp., *Sanguinolites* sp., turreted gastropod, *Anthracoceratites* sp. (abundant), *Bilinguites gracilis*, hindeodellids indet., *Idiognathoides sinuatus*, *Elonichthys* sp. and *Rhadinichthys* sp.

Beds between the Gracilis and Cancellatum marine bands (Marsdenian)

At Corringham these beds are similar to those below the Gracilis Marine Band, being largely arenaceous in the lower part and containing seatearths and thin 'ironstones' near the top. The sandstones are, however, finer grained and more micaceous. One of the seatearths locally carries a thin coal. Three marine bands have been proved in the upper beds. The lowest band (in micaceous siltstone at 1539.39 m in No. 1 Borehole) yielded *Hyalostelia* sp., *L. mytilloides* and fish debris: the middle band (in grey mudstone at 1528.12 m in No. 1 Borehole) yielded *L. mytilloides*, a ribbed shell fragment, and a scolecodont fragment; the highest band (in dark micaceous silty mudstone at 1523.70 m in No. 1 Borehole) yielded *Hyalostelia sp.*, *Serpuloides stubblefieldi*, *L. mytilloides* and a fish fragment. In the absence of goniatites none of these marine bands is positively identifiable, but it seems likely that one of them is the Superbilinguis Marine Band. If so, the thick underlying sandstones presumably represent the Ashover Grit; the Chatsworth Grit, if present, is thin.

The Cancellatum Marine Band (Yeadonian)

The base of the Yeadonian Stage, marked by the Cancellatum Marine Band, cannot be distinguished with certainty amongst several indeterminate marine bands in the district. However, in view of its wide regional occurrence and the common presence of productids and crinoids within it, it is thought to be represented in Corringham No. 1 Borehole by part of 2.44 m of dark grey silty mudstone at 1514.25 m which yielded *Hyalostelia* sp., *Paraconularia* sp., *L. mytilloides*, ?productoid, malacostracan remains (cf. *Pseudogalathea?*), crinoid debris and *Rhabdoderma* sp., and in Corringham No. 2 Borehole by part of 5.65 m of 'dark' micaceous siltstone and mudstone at 1529.64 m containing a productoid fragment, *L. mytilloides*, *O. nitida*, crinoid debris and fish fragments. Identification of the Cancellatum Marine Band elsewhere is more tenuous. It may be represented in Blyton Borehole by part at least of 4.70 m of dark grey micaceous silty mudstone at 1458.62 m, containing *Serpuloides stubblefieldi*, *L. mytilloides*, *Euphemites* sp., *Elonichthys* sp. and *Planolites ophthalmoides*, in Butterwick Borehole by 'dark' micaceous sandy siltstone containing *L. mytilloides* at 1419.72 m, and in Spital Borehole by '*Lingula*'-bearing beds at about 1623 m.

Beds between the Cancellatum and Subcrenatum marine bands (Yeadonian)

These beds cannot be delineated precisely because of uncertainty in the identification of both marine bands, but they are clearly thin. In Corringham No. 1 Borehole they lie within the following sequence:

	Depth m
Undoubted Coal Measures	to 1492.61
Mudstone, dark grey, micaceous, silty; *Hyalostelia* sp. at 1493.06	to 1493.35
Sandstone, grey, micaceous, with mudstone partings	to 1494.10
Mudstone, dark grey, micaceous, silty, with interbedded thin sandstones	to 1495.65
Core not recovered (thought to be mudstone)	to 1498.09
Mudstone, dark grey, silty; *L. mytilloides*, *O.* cf. *nitida* from 1498.40 to 1498.91	to 1499.28
Mudstone, brown, micaceous, silty	to 1499.58
Sandstone, pale grey, fine-grained, with mudstone partings; rootlets near top	to1507.91
Siltstone, grey, micaceous, with interbedded thin sandstones	to 1510.82
Presumed Cancellatum Marine Band (see above)	

The Cumbriense Marine Band may be the fossiliferous mudstone at 1498.91 m, in which case the overlying sandstone is the Rough Rock, only 0.75 m thick, and the Subcrenatum Marine Band is represented only by specimens of *Hyalostelia* sp. Alternatively, the Subcrenatum Marine Band may be at 1498.91 m, in which case the Rough Rock is the substantial sandstone close below, and the Cumbriense Marine Band may rest immediately on, and be indistinguishable from, the Cancellatum Marine Band. The second interpretation is the one preferred here, for it gives the best 'fit' with sequences farther west. Its application to the other boreholes in the district is shown in Figure 3, where the conjectural correlation of Burton upon Stather Borehole is based largely on its lithological similarity to boreholes farther west, notably Crowle No. 1 [SE 7734 1193].

WESTPHALIAN (COAL MEASURES)

The Coal Measures proved in the district are shown on Figures 4, 5 and 6, which cover respectively the Westphalian A, B and C stages. Fossil bands, coals and seatearths allow a fair degree of identification and correlation even though the district lies well east of detailed provings in the Yorkshire Coalfield. Most of the faunas quoted can be matched with those at comparable stratigraphical levels farther west. For the main coal-bearing strata above the Vanderbeckei (Clay Cross) Marine Band only two comprehensively described cored sequences are available, namely those from the British Coal North Ewster and Rock Abbey boreholes. Coal thicknesses are quoted only from the cored sequences. The correlations show that, as Westphalian time progressed, generalised regional subsidence to the north or north-west replaced differential subsidence in the Gainsborough Trough as the dominant cause of thickness variations. Marine band nomenclature is that of Ramsbottom et al. (1978); other lithostratigraphical names are informal.

WESTPHALIAN A (LOWER COAL MEASURES)

Subcrenatum Marine Band

The Subcrenatum (Pot Clay) Marine Band is thought to be represented in Corringham No. 1 Borehole by dark grey silty mudstone containing *L. mytilloides* and *O.* cf. *nitida* at 1498.91 m. In the other Corringham boreholes the equivalent mudstone has yielded *Hyalostelia* sp., possibly *Serpuloides stubblefieldi*, and fish debris. The conjectured correlatives in Blyton and Butterwick boreholes (Figure 4) yielded only *L. mytilloides*.

Subcrenatum Marine Band to Kilburn Coal

These strata contain four marine bands in their lower part. In Corringham No. 1 Borehole all four consist of dark grey silty mudstone. The lowest, at 1493.06 m, contains *Hyalostelia* sp. spicules and equates with one of the marine bands within the 'Soft Bed' sequence of Yorkshire. The marine band, 1.24 m thick at 1487.22 m, contains *S. stubblefieldi*, *L. mytilloides* and fish debris and was tentatively correlated with the Listeri (Alton) Marine Band by Smith et al. (1973, pl.III). It is here thought more likely to be the Honley (or First Smalley) Marine Band, in which case the 1.83 m-thick band with *L. mytilloides* and *Rhadinichthys* sp. at 1481.43 m, correlated by Smith et al. (*ibid.*) with the Forty-Yards Marine Band, must be the Listeri. This latter band is responsible for a distinctive peak on the gamma-log, a feature of the Listeri Marine Band elsewhere. The highest of the four marine bands contains *L. mytilloides* at 1475.59 m. *Rhadinichthys* sp. was found in the succeeding 1.09 m, and pyritised *Carbonicola* aff. *proxima* in 0.81 m of mudstone at 1471.90 m. This 'mussel' indicates the *C. proxima* Fauna characteristic of the beds above the Norton Mussel Band, and so identifies the underlying marine layer as the Amaliae (Norton) Marine Band (Smith et al., 1973, pl.III), resting on a seatearth at the horizon of the Norton Coal. The 'trace' of coal at 1449.63 m in Corringham No. 7 Borehole could therefore well be the Upper Band Coal of the Chesterfield district (Smith et al., 1967, pp.84, 115).

None of the marine bands proved elsewhere in the district in these strata (2.73 m of dark grey silty mudstone at 1493.41 m with sponge spicules, *Serpuloides* sp., *L. mytilloides*, turreted gastropod, *Platysomus* sp. and *Rhadinichthys* sp. in Corringham No. 6 Borehole; 3.83 m of dark grey silty mudstone containing *L. mytilloides* at 1440.18 m in Blyton Borehole; 0.31 m of dark mudstone originally reported to contain '*Lingula*' at 1408.18 m in Butterwick Borehole) can be identified with any certainty. In Burton upon Stather Borehole the medium- to coarse-grained sandstone around 1650 m resembles the Crawshaw Sandstone, but it is more likely to be the Loxley Edge Rock of the Sheffield area (Eden et al., 1957, pp.34–45).

The silty mudstones, siltstones and silty sandstones between the Norton Mussel Band and the Kilburn Coal are appreciably micaceous, and are broad equivalents of the Elland Flags of western Yorkshire and the Wingfield Flags of Derbyshire. In Corringham No. 1 Borehole, a cochliodont tooth, *Rhadinichthys* sp. scales and *Planolites ophthalmoides* were found in the 10 m interval above 1468.65 m. Sandstones are

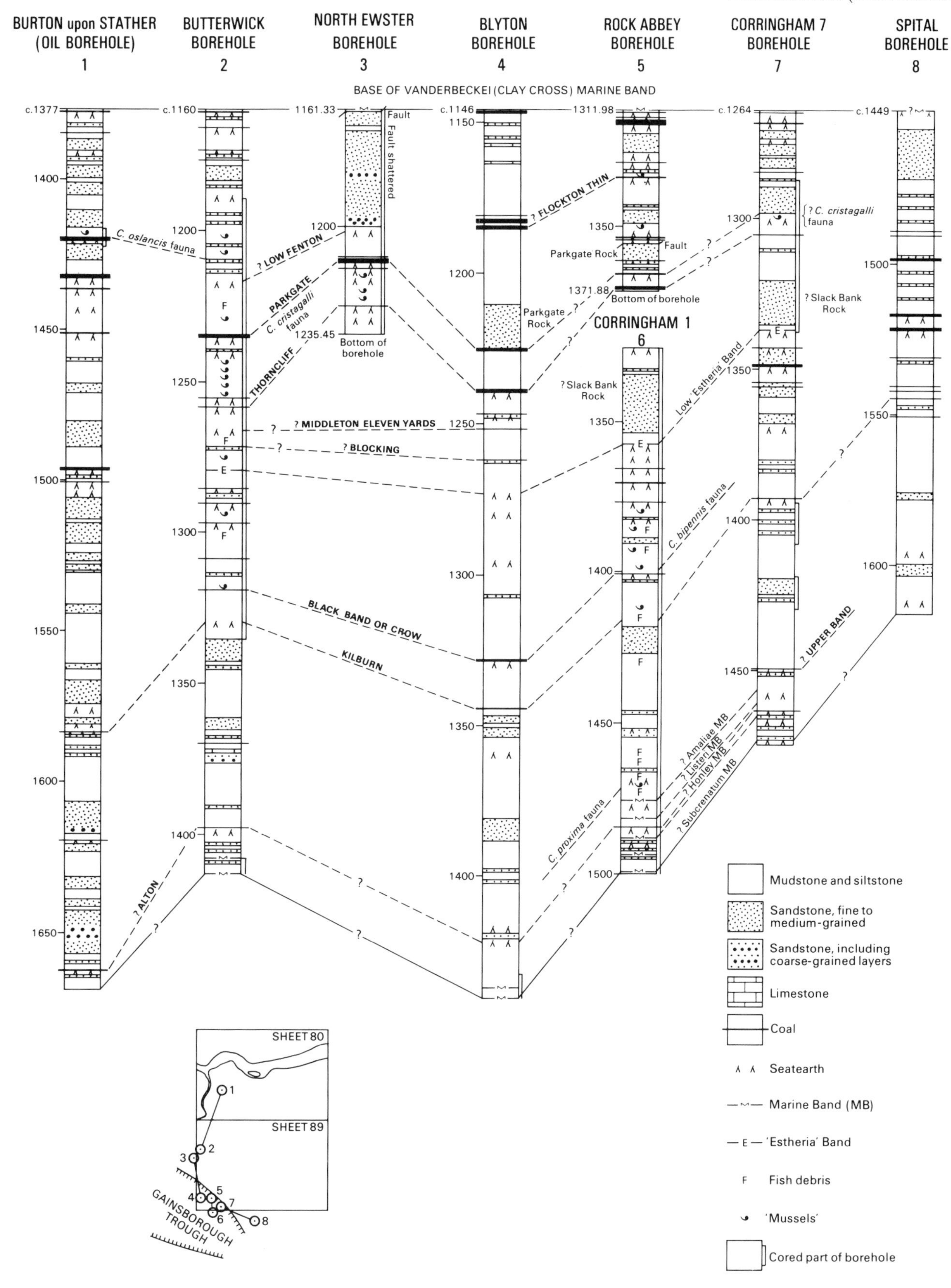

Figure 4 Borehole sections of the Westphalian A rocks (Lower Coal Measures). Depths are in metres

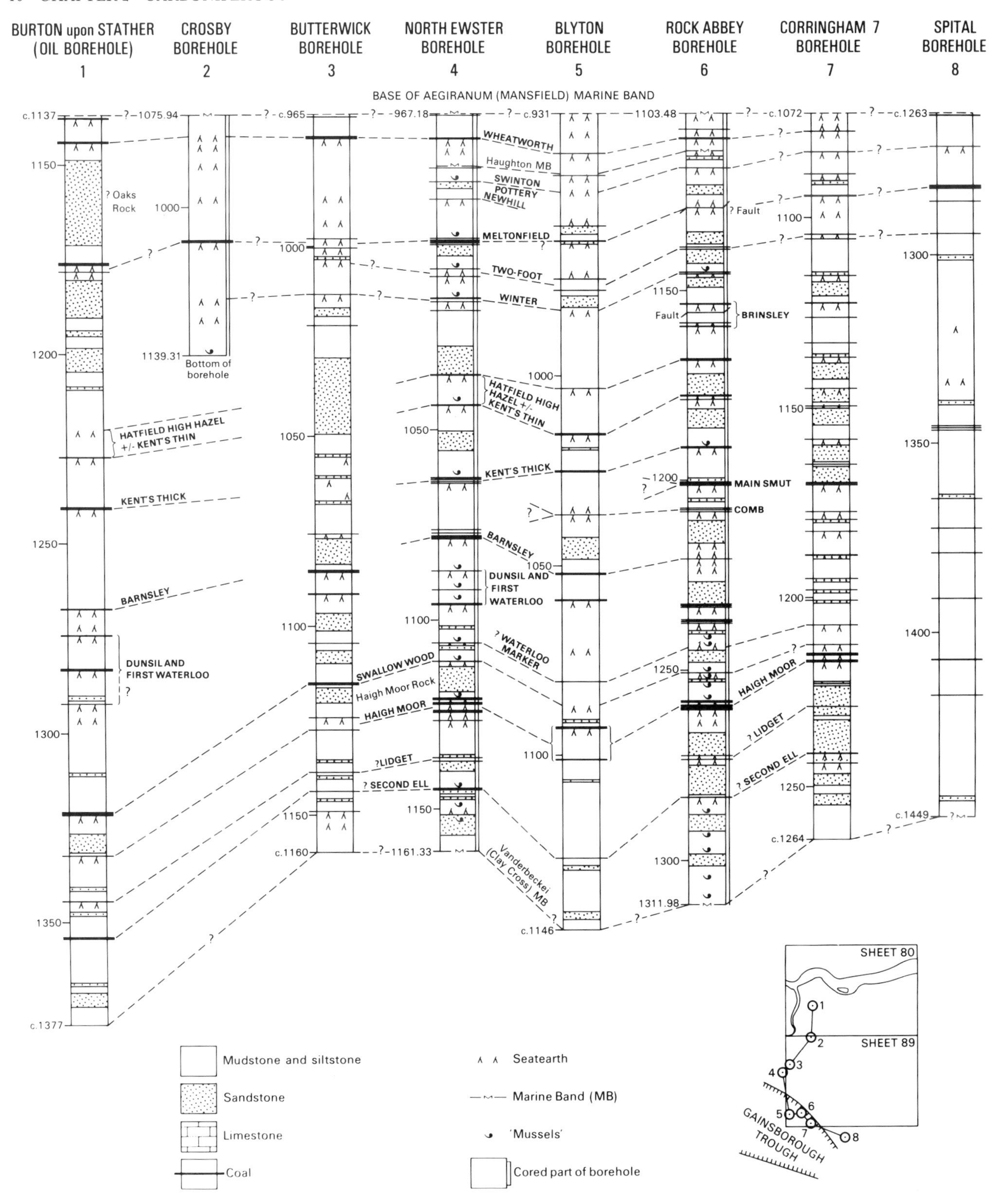

Figure 5 Borehole sections of the Westphalian B rocks (Middle Coal Measures below the Aegiranum Marine Band). Depths are in metres

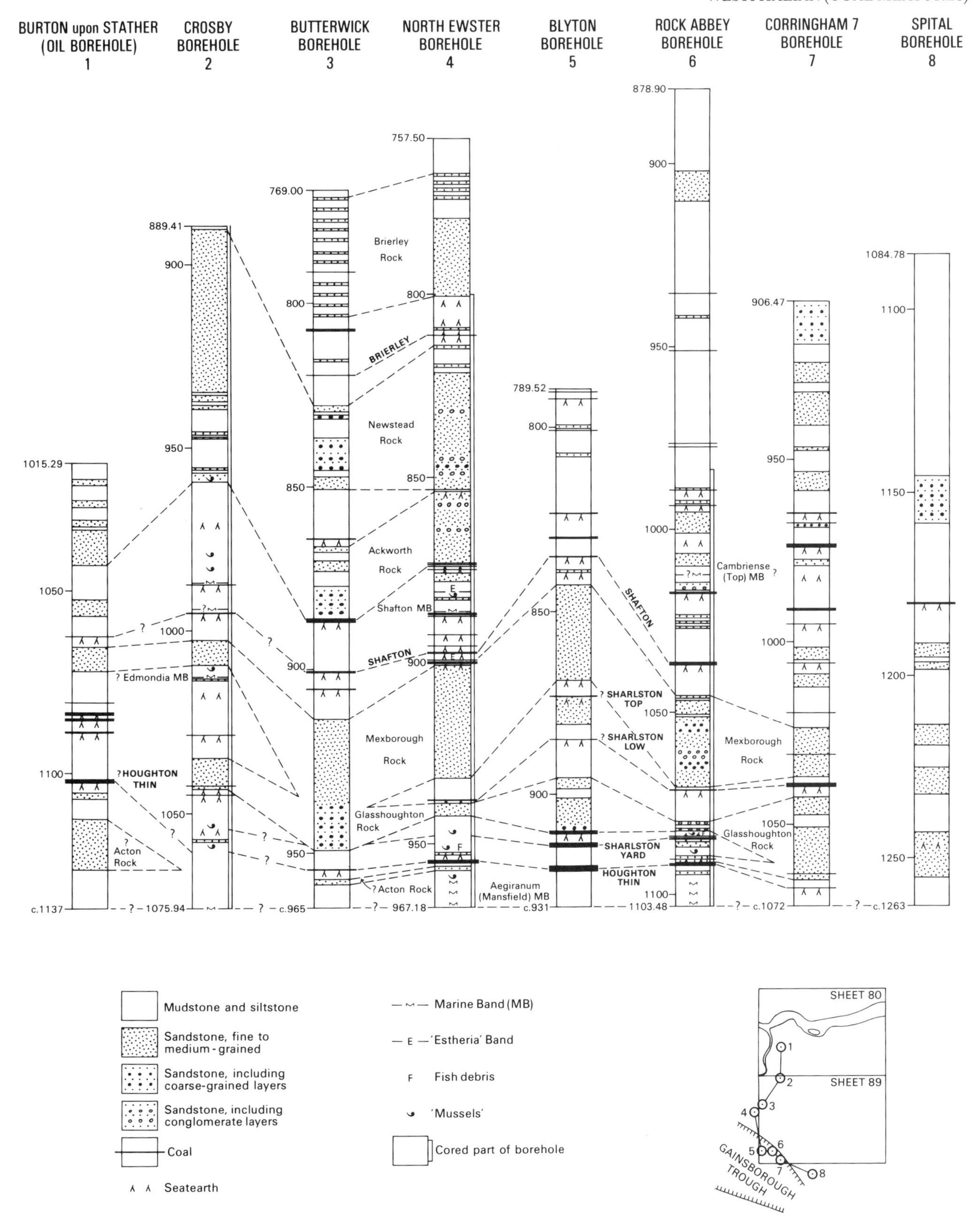

Figure 6 Borehole sections of the Westphalian C rocks (Upper Coal Measures and upper part of Middle Coal Measures). Depths are in metres

concentrated at two levels, which correspond approximately to the Greenmoor (or Brincliffe Edge) Rock and the overlying Grenoside Sandstone of south Yorkshire. The lower sandstones are locally calcareous and contain coarse-grained layers; a seatearth and/or thin coal is widely present within them. The upper sandstones are characterised by large flakes of brown and green mica. No trace has been found of the 'Kilburn Marker' (Strong *in* Falcon and Kent, 1960, p.52), a black chamositic siltstone occurring elsewhere between the two sandstones. Corringham No. 1 Borehole yielded fish debris just below the higher sandstones and a fragment of *Orbiculoidea* cf. *cincta* has been found at the same level in Corringham No. 6. This marine indicator has no known correlative elsewhere. The Kilburn Coal (the Better Bed of west Yorkshire) is possibly up to 0.30 m thick locally; it was not recorded in Corringham No. 1 and Butterwick boreholes.

Kilburn Coal to Low 'Estheria' Band

These beds are largely argillaceous, with several seatearths and/or thin coals. A mussel band above the Kilburn Coal in Corringham Nos. 1 and 5 boreholes yielded *Curvirimula* sp., *Naiadites* sp., cf. *Carbonita* and fish debris. Another mussel band, above a thin coal at 1400.56 m in Corringham No. 1 Borehole (which is presumably either the Black Bed Coal or the Crow Coal of western Yorkshire), contains *Carbonicola* aff. *bipennis*, and so indicates the *C. bipennis* Fauna. In this same borehole the 15 m of strata above 1394.54 m yielded *Spirorbis* sp., *Carbonicola* sp., *Curvirimula* spp. including *C. subovata* and *C. trapeziforma*, cf. *Naiadites*, *Carbonita humilis*, *Geisina arcuata*, *Rhadinichthys* sp. and *Rhabdoderma* sp. In Butterwick Borehole, *Elonichthys* sp. is recorded at 1302.26 m, and indeterminate 'mussel' fragments occur both below and above. The higher coals and seatearths in these beds may equate with the Beeston coals of west Yorkshire, but the Low Silkstone Coal, elsewhere lying immediately below the Low 'Estheria' Band, is represented only by a seatearth.

The Low 'Estheria' Band

This band, proved in most of the Corringham boreholes and in Butterwick Borehole, is up to 0.56 m thick. It consists largely of dark grey to black carbonaceous and locally canneloid mudstone, passing up in places into paler silty mudstone. Its fauna comprises *Euestheria* sp., *G. arcuata*, *Elonichthys* sp., *Rhabdoderma* sp. and *Rhizodopsis* spp. including *R. sauroides*; coprolites are also present.

Low 'Estheria' Band to Vanderbeckei (Clay Cross) Marine Band

These strata contain some thick coals and substantial sandstones, but correlations are speculative. In Butterwick Borehole *C. trapeziforma* occurs just above the Low 'Estheria' Band, and fish debris, including *Rhadinichthys* sp., was found some 5 m higher, just above a thin sandstone. In Blyton Borehole the same thin sandstone carries a thin coal which may be the Blocking (or Top Silkstone) Coal in view of the fish debris above it. In both these boreholes the succeeding seatearth/thin coal is possibly the Middleton Eleven Yards Coal (the Threequarters Coal of Derbyshire). All these strata are cut out at Corringham by a thick sandstone lying at about the level of the Slack Bank (or Silkstone) Rock of western Yorkshire.

The Thorncliffe Coal complex is identifiable by the prominent *C. cristagalli* Fauna above it. In Butterwick Borehole the fauna, ranging through 12.95 m of mudstone, comprises *Spirorbis* sp., *Anthracosphaerium* aff. *dawsoni*, *Anthraconaia fisheri/williamsoni*, *Carbonicola* spp. including *C. cristagalli* and *C. rhomboidalis*, *Naiadites* spp. including *N. flexuosus*, and *Geisina arcuata*. North Ewster Borehole yielded '*Carbonicola*', '*Naiadites*' and '*Geisina*' in the corresponding strata, and in Corringham Nos. 2 and 6 boreholes the fauna comprises *Spirorbis* sp., *Carbonicola* spp. including *C. cristagalli*, *C.* aff. *obtusa* and *C. oslancis*, *N.* aff. *flexuosus*, *Carbonita salteriana*, *G. arcuata* and *Rhabdoderma* sp. The thick overlying coal (1.22 m in Butterwick Borehole and 1.96 m with 'dirt' layers in North Ewster Borehole) is the Parkgate Coal. This seam is cut out in Corringham No. 7 Borehole by the Parkgate Rock, which is well developed in Blyton Borehole but absent farther north. In Butterwick Borehole, *Naiadites* sp. and fish debris including *Platysomus* sp. occur in mudstones above the Parkgate Coal, and the overlying seatearth and its correlative in North Ewster Borehole probably indicate the Low Fenton Coal horizon.

The occurrences in Butterwick Borehole (at and above 1209.04 m) of *C.* cf. *bipennis*, *C. oslancis*, *Naiadites* aff. *subtruncatus* and *Curvirimula* sp. or *Naiadites* sp., and in Burton upon Stather Borehole (above 1416.71 m) of *C.* cf. *oslancis*, *C.* cf. *rhomboidalis*, *N.* cf. *flexuosus* and *G. arcuata* are similar to faunas found between the Deep Hard and Deep Soft coals in the East Retford district (Smith et al., 1973, p.65). In Blyton Borehole the thick seams at about 1185 m may represent the Flockton Thin/Deep Soft coals, but they are not recorded elsewhere in the district.

WESTPHALIAN B (MIDDLE COAL MEASURES, IN PART)

Vanderbeckei (Clay Cross) Marine Band

This band has been identified with certainty only in North Ewster and Rock Abbey boreholes, where it comprises up to 0.62 m of dark grey laminated and slightly silty mudstone containing *Lingula* sp., ?ostracods and fish debris. Specimens of *L. mytilloides* and *Rhadinichthys* sp. were formerly erroneously believed to have come from 0.88 m of strata at 1280.13 m in Corringham No. 7 Borehole, but no core was taken at this depth; they are now thought to have come from Corringham No. 9 Borehole. There are old and inconsistent records of '*Lingula*' at two depths between 1488.95 and 1455.42 m in Spital Borehole, where the presence of *Lingula* sp. and fish debris has recently been confirmed in chipping samples from 1492 to 1484.38 m and at 1456.94 m; the lower occurrence is believed to be due to contamination, but the upper occurrence is thought to be approximately in situ.

Vanderbeckei (Clay Cross) Marine Band to Swallow Wood Coal

These beds are fairly consistent in lithology and thickness, allowing reasonably firm correlation. In Rock Abbey

Borehole pyritised mussels, including '*Naiadites*', were found immediately above the Vanderbeckei Marine Band, and other mussels, including '*Anthracosia*', occur in several layers in the succeeding 25 m of strata. The overlying seam, presumed to be the Second Ell Coal of the East Retford district, is 0.30 m thick in Rock Abbey Borehole and 0.40 m in North Ewster Borehole. At the latter, the overlying mudstone yielded '*Spirorbis*', '*Anthracosia*', '*Naiadites*' and fish scales. The Lidget Coal of the East Retford and Goole–Doncaster districts lies between two thin sandstones which expand locally in the Rock Abbey–Corringham area. In the two cored boreholes, it is split into two thin leaves. A thin buff limestone was reported in Blyton and Corringham No. 7 boreholes in the mudstones below the Haigh Moor Coal. As in many places farther west, the latter is split into two or more seams. In Rock Abbey and North Ewster boreholes three seams range through 2.26 and 3.79 m of strata respectively, their thicknesses in the latter borehole being, in upward order, 0.68 m, 0.44 m (including thin 'dirt' layers), and 0.32 m. The overlying mudstones in these boreholes contain '*Spirorbis*' '*Anthracosia*', '*Naiadites*', ostracods and fish debris. Except in the extreme south-east they are succeeded by sandstone at the level of the Haigh Moor Rock. Unidentified mussels are present locally in the mudstones between this sandstone and the Swallow Wood Coal. This latter seam is 0.28 m thick in Rock Abbey Borehole and 0.12 m in North Ewster Borehole.

Swallow Wood Coal to Barnsley Coal

These beds show considerable lateral variation, and their correlation is speculative. In North Ewster and Rock Abbey boreholes the mudstones above the Swallow Wood Coal yielded '*Spirorbis*', '*Anthracosia*', '*?Naiadites*' and ostracods (including '*Geisina*'). The mudstones above the succeeding seatearth or thin coal, probably the Waterloo Marker, contain '*Spirorbis*', '*?Anthraconaia*', '*Anthracosia*', '*Naiadites*' and fish teeth, scales and spines. In the overlying strata, up to the base of the Barnsley Coal, there are at least three seams and/or seatearths. In North Ewster Borehole the thicknesses of these, in ascending order, are 0.42 m (with a thin 'dirt' layer), 0.22 m (cannel), and 0.02 m. Each has a 'mussel' band above it, and these contain '*Anthracosia*' and '*Naiadites*' in the lowest, '*Anthracosia*' in the middle one, and '*Spirorbis*' and '*Naiadites*' in the highest. The coals presumably include the First Waterloo and Dunsil coals, but exact correlation is impossible.

The Barnsley Coal has been identified in Burton upon Stather and North Ewster boreholes by comparison with sequences farther west. It is 2.01 m thick (including thin 'dirt' layers) in North Ewster Borehole but thins markedly to the south-east, being represented in Rock Abbey Borehole by only 0.19 m of coal within a sequence of closely spaced seatearths containing coaly streaks.

Barnsley Coal to Winter Coal

These strata are subject to appreciable lateral variation, and only broad correlations are possible. The strata between the Barnsley and Kent's Thick coals halve in thickness from Burton upon Stather to North Ewster boreholes, the latter seam being identified in both boreholes by comparison with other holes farther west. In North Ewster Borehole the Kent's Thick Coal comprises four leaves totalling 1.48 m of coal within 2.40 m of strata. Wider splitting is suspected farther south-east, where the Kent's Thick may split into two thin seams in Blyton Borehole and three in Rock Abbey Borehole. In North Ewster and Rock Abbey boreholes '*Spirorbis*', '*Anthracosia*' and fish debris have been found above the Kent's Thick Coal.

Two apparently persistent coals or seatearths in the succeeding 20 to 30 m of strata lie within the Kent's Thin/Hatfield High Hazel interval and vary between 0.39 and 0.47 m thick in North Ewster and Rock Abbey boreholes. '*Spirorbis*' and '*Anthracosia*' occur above the lower of these at North Ewster. Two higher seams are separated by a fault breccia at Rock Abbey. These have been identified by British Coal as leaves of the Brinsley Coal of Derbyshire and are, in ascending order, 1.02 m (with a 0.65 m dirt layer) and 0.60 m thick. Only vague traces of the higher of these seams occur farther north. Although no 'mussels' were found in these strata in North Ewster or Rock Abbey boreholes, pyritised '*Anthracosia*' from the bottom of Crosby Borehole presumably come from the beds above the Hatfield High Hazel. The Winter (or Abdy) Coal is locally split into two leaves, which comprise 0.74 m of coal and 0.26 m of 'dirt' in North Ewster Borehole.

Winter Coal to Aegiranum (Mansfield) Marine Band

These strata are fairly consistent laterally, allowing reasonably firm correlation. In North Ewster and Rock Abbey boreholes '*Spirorbis*' and pyritised '*Anthracosia*' and '*Naiadites*' have been found above the Winter Coal. The thin, and locally split, Two-Foot Coal, with leaves up to 0.29 m thick at Rock Abbey, and fish debris between them at North Ewster, is identifiable by comparison with boreholes farther west, where the overlying Maltby (Two-Foot) Marine Band is widely present. This marine band has not been found in the district, but the overlying mussel band is represented by '*Spirorbis*', '*Anthracosia*' and '*Naiadites*' in North Ewster Borehole. Northward thickening of the overlying sandstone may have cut out the Two-Foot Coal in Burton upon Stather Borehole. In North Ewster Borehole the Meltonfield Coal comprises six leaves totalling 0.63 m of coal (the thickest leaf being 0.41 m) in 1.57 m of carbonaceous mudstones and seatearths; '*Anthracosia*' is recorded above it. In this borehole, the Newhill Coal (the Clowne Coal of Derbyshire) is 0.09 m thick, but it is not traceable across the district, and neither the Manton 'Estheria' Band nor the Clowne Marine Band, which farther west occur respectively below and above the Newhill Coal, have been found. The Swinton Pottery Coal, 0.10 and 0.13 m thick in North Ewster and Rock Abbey boreholes respectively, is overlain by mudstone containing '*Anthracosia*', '*Naiadites*', '*Carbonita*' and fish scales. The overlying Haughton Marine Band comprises up to 1.70 m of dark grey to black, variably silty and partly canneloid mudstone containing foraminifera, '*Serpuloides*', '*Lingula*', fish debris and '*Planolites*'. The Sutton Marine Band has not been found, but the Wheatworth Coal is widely traceable, being 0.80 and 0.40 m thick in North Ewster and Rock Abbey boreholes respectively. The tentative identification of the Wheatworth Coal in Burton upon Stather Borehole implies that the underlying thick sandstone there is the Oaks Rock of west Yorkshire, which has cut out most of the strata above the Meltonfield Coal.

WESTPHALIAN C AND D (MIDDLE COAL MEASURES, IN PART, AND UPPER COAL MEASURES)

Aegiranum (Mansfield) Marine Band

This marine band marks the base of the Westphalian C Stage, and is 7.68 and 6.89 m thick in North Ewster and Rock Abbey boreholes respectively. In the latter, a 0.25 m calcareous siltstone containing cone-in-cone structures may be the 'Mansfield Cank' (Dunham *in* Edwards and Stubblefield, 1948, pp.251–253). The recorded fauna of the band comprises foraminifera, '*Lingula*', '*?Dunbarella*', '*?Edmondia*', mussels, gastropods, fish debris and '*Planolites ophthalmoidies*'. In Crosby Borehole 'blue shale and ironstone' containing '*Lingula mytilloides*', '*?Productus*' and goniatites at 1075.94 m is presumed to be this marine band, and its conjectured position in Butterwick, Blyton and Corringham No. 7 boreholes coincides with a gamma-ray peak.

Aegiranum (Mansfield) Marine Band to Shafton Marine Band

In North Ewster Borehole, bivalves including '*Naiadites*' occur close above the Aegiranum Marine Band. The succeeding thin sandstone is at the level of the Acton Rock farther west; it appears, from the log of Burton upon Stather Borehole, to thicken greatly towards the north-western part of the district. In North Ewster and Rock Abbey boreholes, the Houghton Thin (or Second Wales) Coal is 0.60 and 0.48 m thick respectively. The Sharlston Yard Coal at Rock Abbey comprises three thin leaves totalling 0.22 m of coal within 0.65 m of mudstone, and '*Spirorbis*', mussels (including '*Naiadites*') and '*?Strepsodus*' and other fish debris occur above. This coal was not proved at North Ewster, but the overlying faunal layer is represented by mussels including '*Anthracosia*'.

The Glasshoughton Rock varies considerably in thickness and lithology (Figure 6). Locally it unites with the underlying thin sandstone, and also in places with the Mexborough Rock above. In Rock Abbey Borehole it is represented by an extremely thin, fine- to coarse-grained conglomeratic sandstone. The presumed Sharlston Low Coal in North Ewster and Rock Abbey boreholes is 0.24 and 0.23 m thick respectively. If this identification is correct, a pronounced gamma-ray peak above the coal at Rock Abbey may indicate the tonstein found at this level in the East Retford district and elsewhere (Smith et al., 1973, p.89), and the thin coal capping the overlying sandstone would be the Sharlston Top, presumably cut out elsewhere by the Mexborough Rock. In Crosby Borehole a 3.96 m band of 'Blue Bind (Marine)' at 1013.31 m, containing '*Lingula*' in the lower part and '*Carbonicola*' above, may be the Edmondia Marine Band.

Except in the north, the Mexborough Rock is thick, locally coarse grained, pebbly and conglomeratic. In Butterwick Borehole it cuts down to rest on the Glasshoughton Rock. The presence of '*Estheria*' above the Mexborough Rock in North Ewster Borehole accords with similar records at this level elsewhere in the region; it also identifies the 0.32 m coal in this borehole as the Shafton Coal. In Rock Abbey Borehole the presumed Shafton Coal is 0.73 m thick, including 'dirt' layers.

Shafton Marine Band to Cambriense (Top) Marine Band

In North Ewster Borehole the Shafton Marine Band comprises 2.30 m of grey, variably silty mudstone containing foraminifera, '*Lingula*', '*Orbiculoidea*', '*Myalina*', pyritised '*Carbonita*', fish debris, '*Planolites*' and, in the upper part, cf. '*Anthracosia pruvosti*' and '*Estheria*'. Slightly higher in the same borehole there is a mussel band containing '*Naiadites*' and '*Geisina*', with '*Estheria*' immediately above. These faunas compare well with those from the equivalent strata farther west and south-west. The 18 m or so above the presumed Shafton Coal in Rock Abbey Borehole, within which the Shafton Marine Band would be expected to lie, are almost all siltstones with thin sandstones. Somewhat higher, however, a solitary *?Tomaculum* at 1013.29 m indicates the position of a marine band, presumably the Cambriense Marine Band. The records of marine fossils from two beds of 'Rotten blue shale' in Crosby Borehole (3.96 m thick at 987.55 m and 3.65 m thick at 993.34 m) are suspect. The higher band, which is said to have yielded '*Lingula mytilus*', '*Myalina compressa*', '*Beyrichia*' and, at the top, '*Carbonicola*' and '*Naiadites*' with '*Anthraconaia modiolaris*' some 8 m higher, is likely to be the Shafton Marine Band; the lower band probably represents cavings from it. In Burton upon Stather Borehole a gamma-ray peak close below the base of the Ackworth Rock at 1043 m may be produced by the Cambriense Marine Band. This band has probably been cut out by the Ackworth Rock in North Ewster and Butterwick boreholes.

Cambriense (Top) Marine Band to Brierley Coal

These beds are the Ackworth Division of Goossens and Smith (1973). They can be delineated only in North Ewster and the more northerly boreholes, where the two dominant components, the Ackworth Rock and the Newstead Rock, are well developed. In Crosby Borehole the lowest bed of the presumed Ackworth Rock is a thin feldspathic sandstone at 959.51 m containing fragments of cannel coal, '*Anthracomya phillipsii*' and '*Beyrichia*'. *Anthraconauta phillipsii* is the most characteristic mussel above the Cambriense Marine Band, although it is rarely found within or below the Ackworth Rock. Both the Ackworth and Newstead rocks consist largely of medium-grained sandstones with, in places and mainly in the upper part, interbedded siltstones and ironstone lenses. Coarse-grained sandstones and some conglomeratic layers containing rolled fragments of discoloured ironstone and of siltstone are locally present, and parts of both sandstones are patchily coloured red and purple. The thin coal at 864.11 m in Butterwick Borehole and a thin carbonaceous mudstone containing coaly streaks and resting on a seatearth at

854.08 m in North Ewster Borehole are the Scofton Coal or a subjacent unnamed seam. In the latter borehole they separate the Ackworth Rock from the Newstead Rock.

The Brierley Coal and the beds above

These strata are recognisable only in North Ewster and Butterwick boreholes and, except possibly for the highest few metres in the former, are referable to the Brierley Division of Goossens and Smith (1973).

In North Ewster Borehole the Brierley Coal is presumed to be the 0.10 m seam at 811.17 m because the dark grey coal-free mudstone-seatearth immediately above it is a characteristic feature of this coal elsewhere. Although the overlying Brierley Rock is well developed in this borehole, the equivalent strata in the nearby Butterwick Borehole are described as 'variegated marls' with thin siltstones and fine- to medium-grained sandstones. This suggests a rapid lateral passage, something to which the Brierley Rock is prone. A gamma-ray peak in the overlying mudstones at 764.80 m in North Ewster Borehole may indicate the position of the mussel and *Euestheria*-bearing Fourth Cherry Tree Marker at the base of the Hemsworth Division of Goossens and Smith (1973).

The highest Carboniferous strata, lying immediately beneath the sub-Permian unconformity, commonly exhibit staining, the mudstones and siltstones being red, purple, yellow and 'variegated', and the sandstones red, pink, orange and brown. Within the district, only four boreholes provide relevant information. In Burton upon Stather and Butterwick boreholes red colours extend down 18 and 22 m respectively below the Permian, and 'variegated' and partly yellow strata are recorded for another 40 and 70 m respectively. In Burton upon Stather Borehole the 4.27 m of strata immediately beneath the Permian rocks are 'muddy grey siltstones' possibly owing their colour to reduction by the subsequent late Permian Zechstein Sea. In Blyton Borehole the topmost 12 m are stained, as are the highest 46 m in Corringham No. 7 Borehole.

CHAPTER 3

Permian and Triassic

Rocks of Permian and Triassic age are described together here because they have some features in common and because the position of their mutual boundary is uncertain. They underlie the entire district, dipping eastwards beneath Jurassic rocks. The youngest Triassic strata crop out in the extreme west but are largely obscured by Quaternary deposits. Most of the information on the Permian and Triassic comes from deep boreholes (Crosby, Risby, South Cliffe and Spital) within and just outside the district, of which only the first was cored. In other boreholes the only information available is that provided by chipping samples and, in some cases, geophysical logs. Interpretation of some of the latter is based on unpublished work by Dr D W Holliday.

The lithostratigraphical subdivisions of the Permian[1] and Triassic rocks recognised within the district are set out in Table 2. Permian rocks rest unconformably on the Carboniferous, and the Triassic includes those basal strata of the Lias Group that lie below the lowest occurrence of the ammonite *Psiloceras*.

On palynological evidence the lowest marine strata (in the Don Group) are placed in the Upper Permian, miospore assemblages from the same group in the adjoining Goole/Doncaster district being indicative of a late Permian (Kazanian – Tartarian age (Gaunt, in preparation). The exact position of the boundary between the Permian and Triassic is unknown. In eastern England it is thought to lie at the base of, or within the Saliferous Marl (Warrington et al., 1980, pp.9 – 10) but, because the junction between the Saliferous Marl and the overlying Sherwood Sandstone Group is markedly diachronous, it is possible that in places it is within the lower part of the latter. Tethyan ammonites are used to define stages within the Triassic, but as these are absent in most of western Europe, including Britain, palynomorphs (principally miospores) are used to recognise the Triassic stages in this district.

The Permian and Triassic periods, spanning some 40 and 45 million years respectively, together lasted from about 290 to about 205 million years ago (Forster and Warrington,

Table 2 Subdivision of Permian and Triassic rocks in the district

Group	Formation	Approximate thickness (m)	German equivalents
Lias (basal beds)	Basal beds of Scunthorpe Mudstones	(4 – 7)	Unterer Lias (basal beds)
Penarth	Lilstock Formation	(4 – 8)	Keuper
	Westbury Formation	(5 – 12)	
Mercia Mudstone		(243 – 271)	Muschelkalk
Sherwood Sandstone		(300 – 465)	Bunter
Eskdale (EZ5)	Saliferous Marl	(25 – 56)	Zechsteinletten
	Top Anhydrite	(0 – 1)	Grenzanhydrit
	Sleights Siltstone*	(6 – 13)	
Staintondale (EZ4)	Upper Anhydrite	(4 – 7)	Pegmatitanhydrit
	Carnallitic Marl	(9 – 28)	Roter Salzton
Teesside (EZ3)	Boulby Halite	(0 – 18)	Leine Halit
	Billingham Main Anhydrite	(0 – 5)	Hauptanhydrit
	Upper Magnesian Limestone	(18 – 54)	Plattendolomit
Aislaby (EZ2)	Fordon Evaporites	(0 – 49)	Stassfurt Salze
	Kirkham Abbey Formation	(0 – 114)	Hauptdolomit
	Middle Marl*	(0 – 61)	
Don (EZ1)	Hayton Anhydrite	(0 – 133)	Werraanhydrit
	Lower Magnesian Limestone	(4 – 139)	Werradolomit and Zechsteinkalk
	Marl Slate	(0 – 3)	Kupferschiefer
	Basal Permian Sands	(1 – 34)	Rotliegendes (in part)

* These mainly argillaceous formations are characteristic of the margins of the Zechstein basin and do not properly fall within the cyclic concept

1 The nomenclature of the Upper Permian lithostratigraphical divisions in north-east England has been revised recently (Smith et al., 1986) but the old names are retained in this account.

1985). During this time, that part of the earth's crust on which the British Permian and Triassic rocks were being formed was drifting northwards from a few degrees to just over 30° north of the equator (Smith et al., 1981). By early Permian times the hot humid swampy environment of the late Carboniferous had disappeared. Hercynian uplift, folding and faulting, and differential subsidence associated with the opening up of a seaway in the 'north Atlantic', had produced a substantially denuded upland landscape separating intracontinental basins.

The three principal components of the Permian and Triassic rocks of the district, namely reddish clastics, dolomitic carbonates and evaporites, clearly reflect deposition around the margin of a fairly arid intracontinental basin situated in hot low latitudes. This, the Southern North Sea Basin, stretched from eastern England across the North Sea into central Europe, and subsidence continued within it for much of Permian and Triassic time. Depending on the rate of subsidence and on the periodic but transient establishment of restricted oceanic connections, this basin was variously the focus of aeolian, fluvial, lacustrine and shallow marine sedimentation. The Don, Aislaby, Teesside, Staintondale and Eskdale groups were formed in and around the basin when it was occupied by the Zechstein Sea, a shallow epeiric sea which persisted for most of late Permian time. This sea was subject to five major (and many minor) cycles of transgression and regression, which produced repeated sequences of carbonates, evaporites and clastics. In eastern England these five cycles are designated EZ1 – EZ5 (Smith, 1970). They equate with the Z1 – Z5 cycles of Germany (although there are differences of interpretation) and correspond broadly to the lowest five named groups in Table 2.

The Kingston upon Hull – Brigg district lies near the south-western margin of the Southern North Sea Basin, which was flanked to the west by uplands in the region of the present Pennines, and to the south by the so-called London – Brabant Massif.

BASAL PERMIAN SANDS

Resting everywhere unconformably on Carboniferous rocks, the Basal Permian Sands extend northwards into County Durham, eastwards under the North Sea, and westwards, impersistently in places, to outcrop. They consist of loose quartz sand or friable, largely incohesive sandstone. Although most of their features, such as rounded sand grains and dune bedding, imply an aeolian origin, some aspects suggest localised subaqueous reworking (Versey, 1925; Pryor, 1971). Farther south they are associated with, or pass laterally into rock-pediment breccias. The Basal Permian Sands are assumed to be a westerly continuation of the basal beds of the German Rotliegendes, and may therefore be of early Permian age. However, some subaqueous reworking probably occurred during the initial transgression of the Zechstein Sea in late Permian times.

Within the district the sands are dark to pale grey and white, and fine to coarse grained. Individual grains, which in Nettleton Borehole include chert, range from subangular to rounded, some of the latter being of the polished 'millet-seed' variety that indicates an aeolian phase in their history. However, localised clayey matrices and partings or thin beds of white, grey and red silty clay occur in some of the eastern boreholes and suggest that in places the higher beds were subaqueous. In a few borehole logs the sands are described as pyritic. They appear not to be more than 12 m thick (Figure 7), except in the south-east. Here, in Nettleton Borehole, they are 34 m thick, and include 10 m of red to pale yellow-brown sandy and apparently dolomitic 'clay shale' in the middle of the sequence. This record helps to validate the identification by Edwards (1951, p.271) of 32 m of Basal Permian Sands in Spital Borehole, farther west, for 8 m of 'fawn limestone' were recorded there in the middle of sands which, from their pink, purple and hematitic appearance, might otherwise have been regarded as weathered Coal Measures. It is possible, however, that cavings coincidentally contaminated both boreholes at the same stratigraphical level.

DON GROUP

Marl Slate

The Marl Slate formed in the initial Zechstein transgression, its marked sapropelic and bituminous nature suggesting reducing conditions and restricted circulation. Although previously thought to be a shallow lagoonal deposit, the presence of thin mantles of Marl Slate over high ridges of Basal Permian Sands suggests depths locally in excess of 60 m and implies a substantial and rapid initial transgression (Smith, 1979). The high metallic concentrations present in County Durham and Germany may be due to phytoplankton activity or to high salinity.

Within the district (Figure 7) the Marl Slate consists of up to 3 m of darkish grey, grey-brown, bluish and black mudstone and shale. It is commonly fissile, locally silty, pyritic, bituminous and carbonaceous, and in a few places also apparently slightly calcareous or dolomitic. In the south it appears to be impersistent, for it was not detected in Nettleton and Spital boreholes, and if present in Blyton Borehole is difficult to distinguish from the overlying Lower Marl. In the East Retford district, farther south-west, some interbanding of the Marl Slate with the underlying Basal Permian Sands, and with the overlying Lower Magnesian Limestone, was noted by Smith et al. (1973, p.113).

Lower Magnesian Limestone

The Lower Magnesian Limestone is highly variable in thickness and lithology, the result of deposition in estuarine, littoral, open-shelf, reef, basin-slope and basin environments (Smith, 1974) during the maximum transgressive phase of EZ1.

An argillaceous member, the Lower Marl, is recognised at the base of the Lower Magnesian Limestone in the East Retford district (Smith et al., 1973, pp.112 – 120) and adjacent areas. It continues into the southern part of the present district (Figure 7), where it consists of olive-grey to red, variably calcareous and dolomitic, micaceous mudstone and siltstone with thin interbedded dolomite. Traces of anhydrite and halite were recorded within it in Nettleton Borehole.

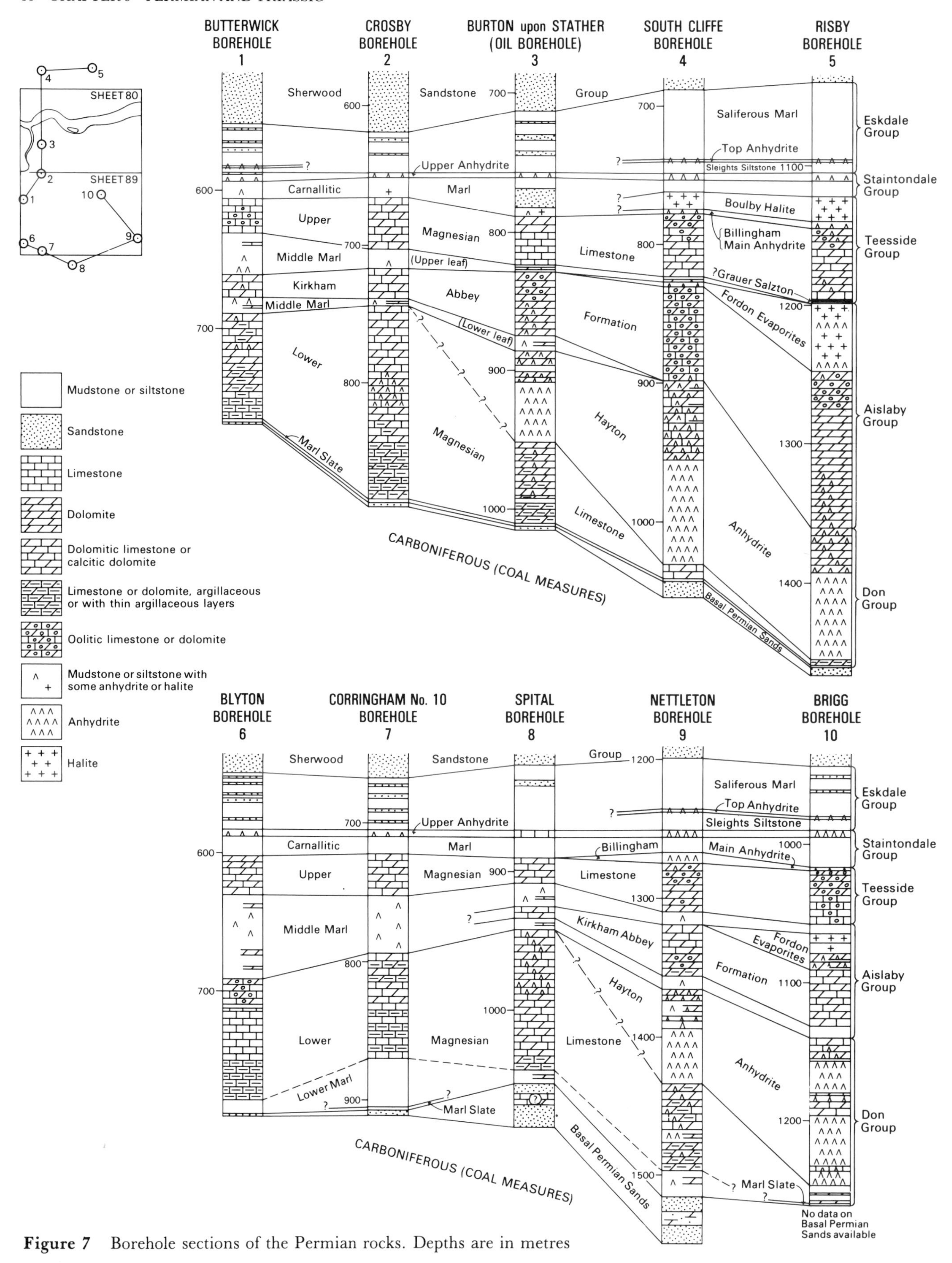

Figure 7 Borehole sections of the Permian rocks. Depths are in metres

Above the Lower Marl (or, in the north-west, above the Marl Slate) are pale to dark grey and locally reddish brown, silty, finely crystalline, dolomitic limestones, containing laminae and thin layers of red-brown to grey calcareous dolomitic mudstone and siltstone; a few scattered quartz grains have also been noted. These argillaceous carbonates and the subjacent Lower Marl are interpreted as an estuarine facies, formed in a wide embayment, the Nottingham Bight (Smith, 1970), with a source of terrigenous detritus to the south-west.

In the southern and western parts of the district, the overlying carbonates are mainly pale to medium grey and grey-brown dolomitic limestones which apparently become more dolomitic towards the top. They are largely finely crystalline except near their top in the south-west, where some of the highest beds are oolitic in Blyton Borehole, and coarsely crystalline (suggesting recrystallised oolites) in Corringham No. 10 and Butterwick boreholes. The presence of oolitic limestone at this level invites comparison with similar rocks occurring in the upper division of the Lower Magnesian Limestone at outcrop farther west. Except in the extreme south-west, appreciable traces of anhydrite occur about half way up the crystalline limestones and may possibly represent a sabkha indicative of the Hampole Discontinuity (Smith, 1968). Traces of anhydrite are also recorded at higher levels in these carbonates, which represent an open-shelf facies built out from the south-west above and beyond the estuarine facies; they form a thick platform in the south-west, with a steep sinuous edge facing north-east towards the centre of the basin (Figure 8a). North-eastwards beyond this shelf edge the entire Lower Magnesian Limestone thins appreciably and consists of pale to dark grey and brown, fine-grained, silty, calcitic dolomite, interbedded with grey to red-brown, calcareous and dolomitic, bituminous, micaceous and locally pyritic mudstone and siltstone; in the extreme north there is also an appreciable amount of anhydrite. These strata represent a basin-slope facies formed in deeper-water euxinic conditions.

Hayton Anhydrite

Like its analogues elsewhere, the Hayton Anhydrite is thickest on the basin slope of the underlying carbonates, and thins both towards the shelf and towards the basin centre. At the western end of the horizontal section B′–B″ on the Kingston upon Hull (80) Geological Sheet, the Hayton Anhydrite is shown as being 60–70 m thick. Reinterpretation of boreholes during preparation of this memoir, however, suggests that the Hayton Anhydrite thins markedly westwards along the most westerly 6–10 km of this section and is absent at its western end (cf. Figure 8b).

Two subdivisions are discernible, the lower and thicker being mainly anhydrite, and the upper containing considerable amounts of dolomite. The origin of these evaporites is uncertain. In Germany the equivalent thick marginal Werraanhydrit was thought to be an offshore deposit formed in water that was not necessarily shallow, but some English interpretations (Taylor and Colter, 1975; Taylor, 1980; Smith, 1980a) suggest deposition in shallow intertidal to supratidal sabkha conditions around the margins of a hypersaline sea.

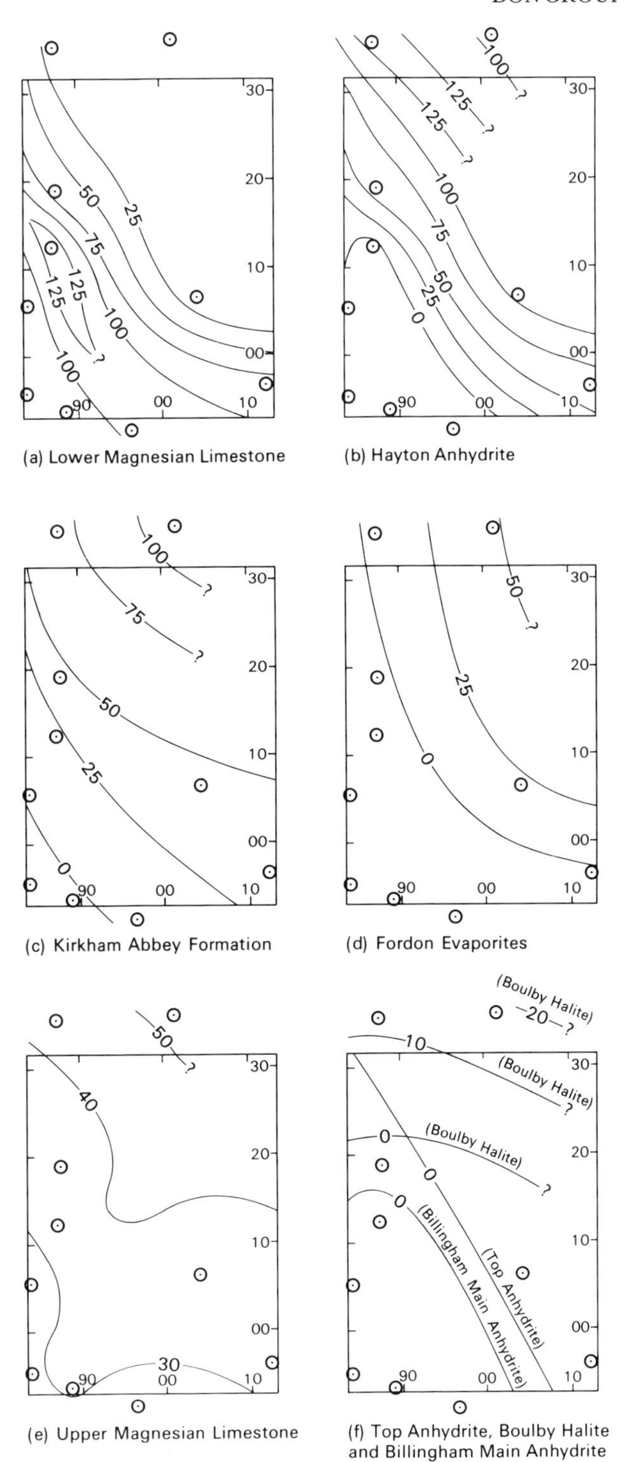

Figure 8 Isopachytes in metres of the Permian carbonates and evaporites. The borehole sites are as shown in Figure 7

Both subdivisions thin out south-westwards although this trend is reversed in the extreme north-east of the district (Figures 7, 8b). The lower subdivision consists largely of translucent to white, mottled pale grey and grey-brown massive anhydrite, much of which is finely crystalline, although descriptions such as 'saccharoidal' and 'sucrosic' suggest coarser textures in places. Pale brown to grey, finely crystalline, locally bituminous, slightly silty, calcitic dolomites are also present. In the absence of cores, the relationships between anhydrite and dolomite are uncertain. Some logs suggest an intimate mixture but in others, notably that of Brigg Borehole, the dolomite clearly forms discrete beds or masses. Traces of red and grey dolomitic mudstones are also present, probably representing laminae or stylolitic layers.

The most common lithology in the upper subdivision is pale buff to grey, finely crystalline, variably silty, calcitic dolomite. Anhydrite or gypsum is also abundant, as inclusions, veins and beds. There is, in addition, interbedded red and grey variably silty and dolomitic mudstone, particularly towards the south-west.

The prevailing view (Taylor and Colter, 1975; Smith, 1980a) is that, in marginal Zechstein areas where it is thickest, the Hayton Anhydrite and its analogues are younger than the Lower Magnesian Limestone and its equivalents (Figures 7, 8a and b). It presumes a major lowering of Zechstein sea level and a marked increase in salinity prior to deposition of the Hayton Anhydrite, followed by considerable sea-level variation consequent upon oceanic recharge but with continuing high salinity, even allowing for the 'permeable barrier' hypothesis put forward by Taylor (1980). The juxtaposition of carbonate and anhydrite within the district (Figure 7), however, suggests that some of the lower part of the Hayton Anhydrite may interdigitate with, and so be coeval with, the middle, locally anhydritic, part of the Lower Magnesian Limestone. It is also possible that some of the slightly anhydritic dolomites in the highest part of the Lower Magnesian Limestone may be the south-westerly continuations of the upper subdivision of the Hayton Anhydrite. If the upper subdivision of the Hayton Anhydrite oversteps the lower, as postulated by Taylor and Colter (1975, fig. 2A) and Taylor (1980, fig. 4), it would be extremely difficult to separate it from the underlying Lower Magnesian Limestone. Farther south-west, as off the Norfolk coast (Taylor, 1980, p.96), the identity of the upper subdivision may be 'lost in red shales and mudstones, probably equivalent to the Middle Marls of north-eastern England'. If these possibilities were to prove correct, there is less need to invoke marked sea-level variations during the EZ1 cycle.

AISLABY GROUP

Middle Marl

At outcrop, to the west of the district, the Aislaby Group is represented almost entirely by the Middle Marl. It consists of red-brown and grey-green mudstone and siltstone, locally dolomitic and sandy. Ripple marks and cross-bedding imply that this terrigenous detritus, presumably derived from the west, was deposited in anastomosing streams on intertidal flats and in shallow lagoons or playas on or fringing a wide flat coastal plain (Smith, 1974).

In the south-west of the district (Figure 7) the Middle Marl is up to 61 m thick and contains thin layers of dolomitic limestone and traces of anhydrite. To the north and east it is split into two leaves by the Kirkham Abbey Formation and Fordon Evaporites. Neither leaf persists to the extreme north-east of the district. The lower leaf is sandy in the west, in Butterwick Borehole; thin white to fawn dolomitic limestones become more prominent farther east and north, as does anhydrite. The upper leaf is generally sandy and contains thin dolomitic limestones and both layered and nodular anhydrite.

It is apparent that the EZ2 transgression did not reach the south-western limit of the district, except possibly to form transient fringing lagoons. In contrast, the absence of terrigenous sediments in the extreme north-east suggests either that this area was subject to continuous carbonate-evaporite deposition before and after the EZ2 transgression or that undetected nonsequences are present there.

Kirkham Abbey Formation

The Kirkham Abbey Formation of Yorkshire and the north-eastern Midlands does not extend to outcrop. It forms a thick carbonate shelf in the eastern part of the region, and thins markedly farther east into a widespread basinal facies. This shallow-water shelf facies represents the maximum transgressive phase of the EZ2 cycle. According to Smith (1980a, pp.19–21) 'it seems likely that salinity in the shelf sea was only slightly higher than in normal sea water', but it was appreciably higher in the nearshore lagoons and in the groundwater.

Within the district the formation thins from the north-east and is not recognisable in the extreme south-west (Figure 8c). It is composed predominantly of white to pale buff calcitic dolomite (Figure 7) of open-shelf facies. In the north and east it is appreciably oolitic, and pisoliths are recorded in the upper beds in Risby Borehole [TA 0106 3578]. Abundant algal traces, including oncoliths and stromatolitic banding, have also been noted in the northern boreholes. Where oolitic and algal traces are not distinct, various cellular, granular and, less commonly, platy textures have been recorded. The matrix is mainly finely crystalline dolomite but there are also traces of clear to white anhydrite, particularly in the lower beds, and anhydrite has also been noted in 'shrinkage' vugs, syneresis cracks, and inclusions. Traces of carbonaceous (?sapropelic) laminae were found in Risby Borehole. Farther south-west the formation becomes more finely crystalline and near its south-western limit consists of dolomitic mudstone with some interlaminated red and brown anhydritic mudstone. It is slightly more anhydritic in western boreholes such as Burton upon Stather (oil) and Butterwick, and traces of pseudomorphs after halite were noted in the former. These lithological variations suggest a south-westward passage from subtidal conditions to intertidal mud flats. In Risby Borehole the thickness (which is close to the known maximum), the presence of pisoliths (noted by Taylor and Colter (1975, p.254) as occurring near the outer shelf edge) and traces of carbonaceous laminae all suggest that the transition to a basin-slope facies occurs only a short distance to the north-east.

Fordon Evaporites

The Fordon Evaporites are lithologically the most variable, and volumetrically the largest of the 'Zechstein' formations.

Unlike earlier formations they are thickest (in places more than 400 m) in the middle of the basin, though here, mainly under the North Sea, local thicknesses are distorted by halokinesis. Where undisturbed, the thick middle part of the formation consists mainly of halite, with anhydrite dominating the thin lower and upper parts. The origin of these basinal evaporites is uncertain. Smith (1980a, pp.21 – 22) favoured deposition 'in a range of predominantly shallow-water and peritidal environments during a phase of sea-level recovery following a sharp sea-level decline at the end of carbonate deposition'; but Colter and Reed (1980), who demonstrated an inherent foresetting arrangement, invoked a deep-water origin. Whatever the circumstances, these evaporites eventually filled in most of the Southern North Sea Basin, reducing it to a virtually flat plain around a small relict Zechstein Sea. Along their margins the evaporites are largely contained within the basinal slope of the underlying EZ2 carbonates, but they transgress thinly on to the carbonate shelf surrounding this slope.

This marginal zone extends into the north-eastern half of the district (Figure 8d). There, in Risby Borehole (Figure 7), the lowest 8 m of the Fordon Evaporites are greyish white to pink mottled, massive anhydrite containing traces of vaguely pelletal, finely crystalline anhydritic dolomite and thin layers of red silty mudstone. The rest of the sequence, 41 m thick, consists of halite, with layers, up to 3 m thick, of pale grey massive anhydrite, thinner layers or laminae of red and greenish grey dolomitic mudstone, and traces of buff to brown finely crystalline, locally bituminous, dolomitic limestone. There is a marked westward thinning from Risby of the whole formation to a mere 3 m of white to pale grey massive anhydrite with some anhydritic dolomite in South Cliffe Borehole. The thinning is more gradual along the eastern margin of the district, for Brigg Borehole proved 26 m of evaporites. Here the lowest 11 m comprise finely crystalline to 'sucrosic', calcitic, dolomitic anhydrite; the upper part passes gradually, apparently by interdigitation, into interbedded grey to buff 'pellety' silty mudstone (with some siltstone) and halite, which is topped by 7 m of massive halite. The three sequences suggest deposition early in the EZ2 regression from localised sabkhas and evaporating lagoons or playas. There was also an input of terrigenous sediment, heralding the subsequent deposition of the upper leaf of the Middle Marl, and considerable postdepositional mineral changes due to the action of hypersaline groundwater. This interpretation implies that the thin basal and marginal anhydritic sequences within the district equate with the lowest subdivision (unit 1) of Colter and Reed (1980), and also that there is no representation within the district of their highest subdivision (unit 8), the supposedly transgressive anhydritic layer which these authors suggest is represented by a possible solution residue resulting from the EZ3 transgression.

TEESSIDE GROUP

Upper Magnesian Limestone

The Upper Magnesian Limestone represents the open-shelf facies of the EZ3 cycle. The transgression reached almost as far as that of EZ1, but its shelf carbonates were built out farther towards the basin centre than those of the two earlier cycles. Conditions across this wide shelf were largely quiescent, producing carbonate muds, but oolites formed locally, probably nearshore, in moderate-energy shoals. During the transgressive phase salinity in the shallow shelf waters may not have been abnormally high.

In Risby Borehole (Figure 7) there is a thin basal layer of buff to black, finely crystalline, dolomitic limestone containing grey to black mudstone partings and anhydrite inclusions, which produces a high gamma-ray response. These characteristics suggest equivalence with the Grauer Salzton of Germany, a thin illitic shale which occupies a comparable stratigraphical position. The lowest beds of the overlying carbonates in Risby Borehole and the basal carbonates in South Cliffe and Nettleton boreholes are slightly argillaceous and silty, with partings and thin beds of grey and, in the north, red calcareous mudstones, inclusions of anhydrite, and minute amounts of carbonaceous debris, some identifiable as plant remains. The rest of the Upper Magnesian Limestone in the district, which thins gently but irregularly south-westwards (Figure 8e), consists of white to grey and pale brown carbonate. In places, most of the lower part consists of dolomitic limestone, but dolomite becomes increasingly common upwards. Small amounts of anhydrite are present near the top in the more northern and eastern boreholes.

Much of the carbonate is finely crystalline but, except in the south-west, the upper part is also variably oolitic or contains textures described as coarse grained, granular, cellular and porous, which may represent recrystallised ooliths. Pisoliths were noted in the higher beds in Brigg Borehole. Algal remains, including oncoliths, are widespread, although not abundant, in the higher beds. Faunal remains include 'skeletal' debris and indeterminate bivalves (in Brigg Borehole) and ostracods (in Burton upon Stather Borehole). The incidence of ooliths, algal traces and anhydrite in the highest beds suggests shallowing at the onset of a regressive phase.

Billingham Main Anydrite

The Billingham Main Anhydrite reflects a regressive phase in the EZ3 cycle. Within the district it is present only in the north and east (Figure 8f), thinning out south-westwards from a maximum of 5 m. It comprises transparent to white and pale grey, partly massive and partly fibrous anhydrite containing traces of anhydritic calcitic dolomite. Grey dolomitic silty mudstone partings were noted in Nettleton Borehole. In Brigg Borehole about 5 m of silty mudstone containing at least one thin dolomite layer intervene between the anhydrite and the underlying Upper Magnesian Limestone.

Boulby Halite

There are no clearly defined shelf or basinal facies of the Boulby Halite, which constitutes the main evaporitic part of the EZ3 cycle. Like the Fordon Evaporites, it thickens appreciably away from marginal areas. Allowing for halokinesis, it is up to 120 m thick offshore (Smith and Crosby, 1979) and about 90 m in the Whitby area.

The Boulby Halite has not been proved in the district, though it is believed to occur in the extreme north (Figure 8f) because of its presence in Risby and South Cliffe boreholes (Figure 7), where it is 18 and 12 m thick respectively. In Burton upon Stather Borehole 'minor salt' was recorded from mudstones resting on the Upper Magnesian Limestone, and in Crosby Borehole the equivalent strata were logged as 'Marl and Salt', but whether these represent the 'feather edge' of the Boulby Halite or are merely isolated occurrences of halite within the Carnallitic Marl is uncertain.

STAINTONDALE GROUP

Carnallitic Marl

The Carnallitic Marl is the lowest leaf of the Upper Marl of peripheral western areas. It consists largely of terrigenous sediment but, unlike the Middle Marl, it continues basinwards, thickening to 30 m in that direction. Sedimentary structures suggest shallow subaqueous deposition from anastomosing streams and in residual playas on a wide flat plain. There were intervals of desiccation, and the interstitial and nodular evaporitic minerals within the sediment may well have crystallised from hypersaline groundwater.

Within the district (Figure 7) the Carnallitic Marl largely comprises reddish brown and greenish grey, locally fissile and micaceous, silty mudstone with some siltstone. Green reduction haloes are widely present, as are nodules and inclusions of anhydrite and gypsum and traces of dolomite and calcite. Traces of 'salt' were recorded in the basal beds in Risby Borehole and in comparable strata (see above) in Burton upon Stather and Crosby boreholes. A 15 m red and grey sandstone containing 'marl' bands and gypsum traces is recorded from the middle of the sequence at Burton upon Stather. The Carnallitic Marl here is 28 m thick, and is 22 m thick in Brigg Borehole; elsewhere it is 9 to 15 m. The two thickest sequences lie just beyond the outer limits of the Billingham Main Anhydrite and the Boulby Halite, and may incorporate residues from solution of these evaporites, although in Burton upon Stather Borehole the sandstone, possibly a channel-fill deposit, has contributed appreciably to the expansion. Sandstones increase in thickness and number beyond the district to the south and west and all the components of the Upper Marl at outcrop eventually pass into sandstone (Smith et al., 1973, p.157 *et seq.*).

Upper Anhydrite

The thin EZ4 Upper Anhydrite has ripple structures, small-scale cross-bedding and magnesitic laminae, which have led Smith (1974, p.138; 1980a, p.26) to envisage deposition in an extensive but shallow hypersaline sea, followed by considerable post-depositional mineralogical change. It extends across the entire district (Figure 7), with a thickness of 4 to 7 m, and consists largely of clear to white, pale grey and locally pink, finely to coarsely crystalline anhydrite containing some gypsum and, in Risby Borehole, traces of red and green mottled gypsiferous mudstone. In Butterwick Borehole the highest 2 m or so comprise grey to fawn finely crystalline dolomitic limestone. In Spital Borehole no anhydrite was recorded, and the formation is evidently represented there by interbedded fawn and brown limestone, a marginal facies according to Smith (1980a, p.26).

ESKDALE GROUP

Sleights Siltstone

The Sleights Siltstone is the middle leaf of the Upper Marl of the peripheral western areas. Like the lower leaf (the Carnallitic Marl), it is a terrigenous, part-fluvial and part-lagoonal deposit that becomes finer grained and more evaporitic towards the basin. Within the district (Figure 7) it is 6 to 13 m thick, with a slight north-eastward thickening, and consists mainly of reddish brown to greenish grey, locally calcareous and dolomitic, silty mudstone and siltstone containing reduction haloes and in places some finely divided carbonaceous debris. White and (rarely) pink fibrous gypsum is widespread, occurring in disseminated form and as nodules and veins. A few small pseudomorphs after halite were noted in Nettleton Borehole.

Top Anhydrite

EZ5 carbonates are unknown in the Zechstein basin, where the Top Anhydrite is merely the most extensive of several evaporitic layers within the dominantly terrigenous Eskdale Group. These evaporites are thought to have formed during transient expansions of the shallow residual Zechstein Sea, which by this time was little more than a central playa (Smith. 1980a, p.28).

The Top Anhydrite appears, largely on the evidence of geophysical logs, to extend into the northern and eastern parts of the district (Figure 7) as about 1 m of anhydrite. In some boreholes, notably Butterwick, only traces of white, pink and pale grey anhydrite or gypsum are recorded at this level.

Saliferous Marl

The Saliferous Marl is the upper leaf of the Upper Marl of peripheral western areas. Like the other leaves it is a dominantly argillaceous formation, consisting mainly of reddish mudstone and siltstone with thin impersistent evaporites. Small-scale cross bedding, ripple structures and evidence of periodic or localised desiccation imply fluvial and shallow-water deposition (the latter in residual playas) on a vast plain that stretched across the Southern North Sea Basin. Thin sandstones are locally present near the top of the formation so that in places there is a transition into the overlying Sherwood Sandstone Group. These sandstones become thicker, more numerous, and occur at increasingly lower levels south-westwards. In the East Retford district (Smith et al., 1973, pp.157–161) the entire Upper Marl passes by interdigitation into the Sherwood Sandstone Group, the base of which is, therefore, demonstrably diachronous. Towards the north-east the lower sandstones pass laterally into argillaceous beds and, as a consequence, the offshore strata (the Bunter Shale Formation), comparable to the Saliferous Marl (Rhys, 1974), thicken considerably in this direction. The slightly more silty Bröckelschiefer Member lies at the base of the Bunter Shale, and in Germany the base of the Triassic System is placed arbitrarily at its base. Correlation across the North Sea suggests, therefore, that the base of the Triassic, as defined in Germany, may lie in eastern England at or close to the base of the Saliferous Marl; where the latter is attenuated or absent the system boundary may be within the lower part of the Sherwood Sandstone Group.

Within the district the Saliferous Marl (Figure 7) consists mainly of reddish brown, and to a lesser extent greyish green, slightly micaceous, locally slightly calcareous, variably silty mudstone and siltstone. In Brigg Borehole it contains reduction haloes, small angular fragments of lignite and traces of pyrite; elsewhere small anhydrite nodules and layers with veins of white gypsum and/or anhydrite are widespread. Traces of grey to black micaceous shale and pseudomorphs after halite were noted in Risby Borehole. Siltstones, some of which are slightly dolomitic, become more common towards the south-west, and thin sandstones also develop in this direction. The formation is 56 m thick in Risby Borehole and thins irregularly south-westwards.

SHERWOOD SANDSTONE GROUP

In eastern England the Sherwood Sandstone Group includes rocks formerly called Lower Mottled Sandstone and Bunter Pebble Beds in southern areas, and Bunter Sandstone and Keuper Sandstone farther north. In Germany, where the terms originated, 'Bunter' and 'Keuper', although originally lithostratigraphical, have come to have chronostratigraphical implications. Palynomorphs (mainly miospores) have shown, however, that in Britain the lithostratigraphical and chronostratigraphical boundaries do not coincide. In consequence, British names incorporating the terms Bunter and Keuper have been replaced by others defined lithostratigraphically (Warrington et al., 1980). In Nottinghamshire the Sherwood Sandstone Group has been divided into the Lenton Sandstone Formation (formerly Lower Mottled Sandstone) and the succeeding Nottingham Castle Formation (formerly Bunter Pebble Beds), but farther north, where pebbles are virtually absent, the group remains undivided.

The Sherwood Sandstone Group of eastern England consists of mainly red-brown, fine- to medium-grained sandstone with sporadic thin lenses of red-brown to grey-green mudstone and layers of rolled mudstone fragments; rounded quartzitic pebbles are common in southern areas. Both parallel-bedding and cross-bedding are present. The sandstone is interpreted (Warrington, 1974) as a sequence of fluvial sediments deposited along the western margin of the intracontinental Southern North Sea Basin. Hot and generally arid conditions are implied by the reddish colours and localised occurrence of wind-rounded grains, ventifacts and desiccation cracks. The cross-bedding in the East Retford district dips mainly to the north-east (Smith et al., 1973, p.172, fig. 29), suggesting fan-like spreads derived largely from the south-west.

Within the district the characteristics of the Sherwood Sandstone Group are intermediate between those of the Nottinghamshire sequence, as represented in the East Retford district (Smith et al., 1973, pp.161–177), and those of the Yorkshire and offshore regions. In the extreme south it is possible to recognise representatives of the Lenton Sandstone and Nottingham Castle formations, but their vertical and lateral limits are indistinct, and no formal subdivision is attempted. An overall thinning is apparent from more than 450 m in the north-east to less than 300 m in the south-west (Figure 9).

The lowest sandstones, 18 to 44 m thick, are variably coloured red, brown, orange, grey, buff and white, and are micaceous, clayey, silty, well sorted and fine grained. The grains are mainly subangular to subrounded, but some layers are rich in rounded grains. Much of this sequence is cemented by a white slightly calcareous and locally gypsiferous clay matrix. The appreciably argillaceous nature of these sandstones is enhanced by numerous interbedded thin red silty mudstones which are in places gypsiferous and which contain green reduction haloes, minute carbonaceous debris and, in one such bed in Brigg Borehole, a fragment of lignite.

The succeeding main part of the Sherwood Sandstone thickens northwards from about 230 m to nearly 400 m and is distinguished from the underlying beds by being dominantly red, less argillaceous, generally coarser grained, and locally pebbly. At or just above the base there is a pebble layer in Nettleton and Crosby boreholes, and a distinctly coarse-grained sandstone in Corringham No. 10 and Butterwick boreholes. In general the sandstones vary from fine- to coarse-grained and from well to poorly sorted, and the grains range from subangular to rounded. A white slightly calcareous clay matrix is locally present but much of the rock has been described as uncemented, unconsolidated, friable, porous and permeable. Pebbles are reported from most of the south-western boreholes, but they are clearly not abundant; some records suggest that they are scattered throughout the middle part of the group, but others imply concentrations at or near the base and at a level about half way up. Some thin red and green-mottled silty mudstone is present, mainly in the higher beds, and some beds are reported to be calcareous, dolomitic or gypsiferous. In South Cliffe Borehole interbedded red-brown and green silty mudstones containing gypsum, possibly in veins, occur about 90 m above the base of the group, but there is little or no trace of equivalent strata in adjacent boreholes. However, scarce traces of gypsum and anhydrite are reported elsewhere in the sandstone at various levels.

In the uppermost part of the Sherwood Sandstone Group there is an upward passage into paler, finer-grained and more argillaceous sandstones which, although their base is indistinct, appear to be thicker in the north (30 to 45 m) than in the south (5 to 15 m). Comparable strata have been noted farther to the north-east (Geiger and Hopping, 1968). These sandstones are mainly white, grey and buff, fine to medium grained, micaceous and silty, loosely cemented, friable and largely free of pebbles. The grains range from subangular to rounded. Numerous interbedded, thin, red and subsidiary grey silty mudstones and siltstones are also present. In some

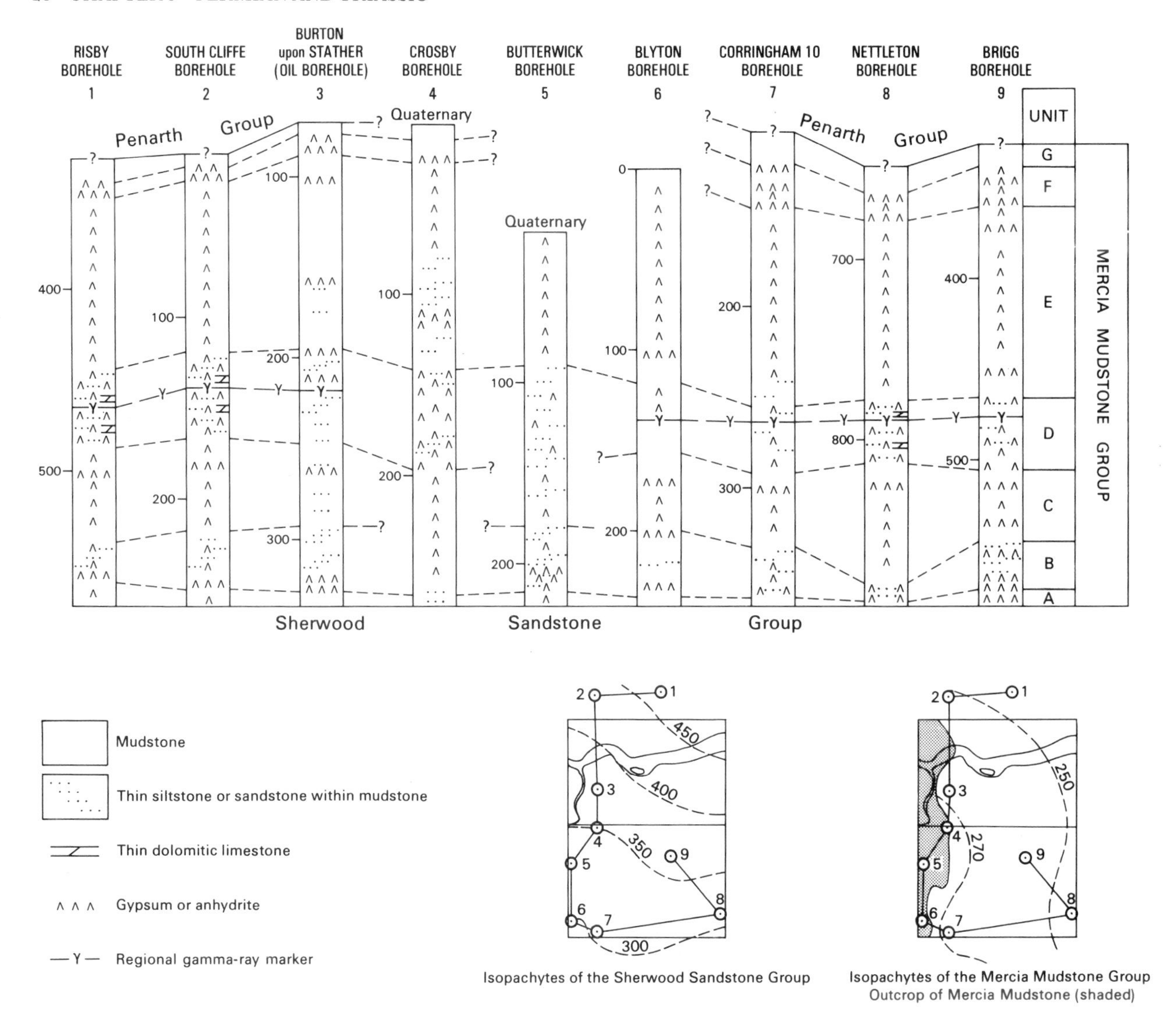

Figure 9 Borehole sections of the Mercia Mudstone Group, with isopachytes of the Mercia Mudstone and Sherwood Sandstone groups. Depths are in metres, as are the isopachytes

of the southern and western boreholes there are suggestions that the topmost sandstone is coarse grained, poorly sorted, and even pebbly—features that may indicate a wind-winnowed *remanié* layer. It is possible that this layer marks the Hardegsen Disconformity of mid-Scythian age within the Buntsandstein of Germany, for the top of the Sherwood Sandstone in eastern England is thought to approximate to this break (Geiger and Hopping, 1968), and a nonsequence, marked by an erosion surface, is thus not unexpected at this level.

MERCIA MUDSTONE GROUP

The Mercia Mudstone Group predominantly comprises red argillaceous rocks that occur between the Sherwood Sandstone Group and the Penarth Group. Palynomorphs, mainly miospores, show that in eastern England the lower part of the Mercia Mudstone is of late Scythian, Anisian and early Ladinian age, and so is coeval with the Upper Buntsandstein to Lower Keuper sequence of Germany. The name 'Keuper Marl' formerly given to much of the Mercia Mudstone Group in Britain is, therefore, inappropriate. Some parts of the group have been locally subdivided, for example around Nottingham (Elliott, 1961) and in the East Retford district (Smith et al., 1973).

The Mercia Mudstone consists of reddish brown and, to a lesser extent, greenish grey, locally dolomitic and anhydritic or gypsiferous mudstones with some siltstones. It represents coastal-plain fluvial, littoral, lagoonal (playa) and shallow marine sediments deposited around the western margins of the Southern North Sea Basin during a period when the basin contained a hypersaline sea with periodically restricted oceanic connections. The argillaceous nature of most of the sediments indicates a lowering of relief of the surrounding land, but the prevailing red colour reflects the continuing hot intracontinental environment. The gypsiferous deposits may have formed as sabkhas or in shallow lagoons. Indications of marine influence include sparse *Lingula* recorded near Eakring by Rose and Kent (1955). In Germany the transgressive phases reached a zenith with deposition of the dolomitic and halitic Muschelkalk, and an analogous facies is traceable across the North Sea to Tetney Lock Borehole near Grimsby (Geiger and Hopping, 1968).

There is little variation in the thickness of the Mercia Mudstone within the district (Figure 9). None of the lithostratigraphical subdivisions of adjacent areas can be identified with certainty, and for descriptive purposes an informal subdivision into seven units (A to G in ascending order) has been made. These units are traceable from borehole to borehole largely by geophysical logs, a technique demonstrated by Balchin and Ridd (1970).

Unit A consists of red-brown, locally slightly dolomitic, silty mudstone with grey-green mottling, bleached haloes, and a small amount of nodular anhydrite. These basal beds produce an exceptionally high gamma-ray response (as noted by Balchin and Ridd, 1970, p.94) and thicken from between 3 and 6 m in the south-west to 13 m in the north-east. They may equate with the thin red mudstone at the base of the 'Green Beds' *sensu lato* of Smith and Warrington (1971) to the south-west, and possibly also with the Seaton Carew Formation of Smith (1980b) to the north.

Unit B is an interbedded sequence of red and pale grey mottled silty mudstone, micaceous siltstone, fine-grained sandstone and 'marly' to massive anhydrite and gypsum, the evaporites apparently occurring mainly in the lower beds. The unit is only 10 m thick in Nettleton Borehole, but elsewhere it is 23 to 37 m thick, with a slight eastward thinning. The presence of evaporites invites comparison with the Esk Evaporite Formation of Warrington et al. (1980, p.57) to the north and the Röt Halite Member near the base of the Dowsing Dolomitic Formation (Rhys, 1974) offshore. The unit may have formed in a transient south-westward extension of a hypersaline playa with a sandy regressive littoral fringe.

Unit C is a sequence of red-brown silty mudstones with grey-green mottling, bleached haloes, a few thin siltstones, and possibly some sandy layers. Small amounts of anhydrite and gypsum are present as nodules and thin layers. Some of the latter can be correlated between boreholes by geophysical logs, one in the upper beds forming a useful marker. The unit is 30 to 52 m thick, with an irregular south-westward thinning.

Unit D consists largely of red-brown and grey-green silty mudstone with some grey-green micaceous, slightly calcareous and/or dolomitic siltstones and silty fine-grained sandstones. Gypsum and anhydrite occur throughout, but more commonly in the upper beds. To the north and east thin layers of silty dolomitic limestone and/or dolomite are also present. The 'regional gamma-ray marker' of Balchin and Ridd (1970) occurs at or just above the middle of the unit, which exhibits a slight and irregular eastward thinning from 48 to 35 m. The carbonates in the north-east suggest a transient marine incursion from that direction, with a sandy littoral fringe to the west and local sabkha development. The unit is at the same general stratigraphical level as the Muschelkalk Halite Member in the upper part of the Dowsing Dolomitic Formation (Rhys, 1974) to the east; it may pass south-westwards into the Clarborough Beds of Smith and Warrington (1971).

Unit E is 95 to 114 m thick, thinning slightly to the east. It consists of mottled partly silty mudstone with bleached haloes, a few thin siltstones and nodules and layers of anhydrite and gypsum; some of the last can be correlated between boreholes. Grey-green colours become more common towards the top, as also does the incidence of gypsum, particularly in the highest 10 m or so. The highest 5.76 m of these beds were cored in Cockle Pits Borehole; they comprise pale grey and red-brown mottled mudstone containing anhydrite as isolated nodules up to 0.12 m across and nodular layers up to 0.05 m thick, and gypsum in cross-cutting veins up to 0.05 m wide.

Unit F comprises nodular and layered anhydrite and veinous gypsum in a matrix of greenish grey mudstone. It is 5 to 25 m thick, the minimum thickness being recorded in Cockle Pits Borehole, where about 70 per cent of the unit consists of gypsum. The anhydrite and gypsum here occurs as scattered to closely packed nodules and thin cross-cutting veins. This unit probably correlates with the Newark Gypsum to the south and with part of the Keuper Anhydritic Member in the middle of the Triton Anhydritic Formation of Rhys (1974) offshore. Sparse miospore assemblages containing *Alisporites* (21.02 m below base of Penarth Group) and *Ovalipollis pseudoalatus* (14.84 and 11.91 m below base of Penarth Group) have been recovered from this unit in Cockle Pits Borehole.

Unit G, 7 to 18 m thick, consists predominantly of grey-green mudstone. It equates with the former Tea Green Marl, which is now included in the Blue Anchor Formation. A sparse association of *Classopollis torosus*, ?*Vesicaspora fuscus* and indeterminate bisaccates was recovered from this unit at 159 m in Blyborough Borehole (Figure 11).

Units E to G form rockhead in the west of the district, but they are almost entirely concealed by Quaternary deposits. Blue-green mudstone is visible under alluvium in the bed of the River Eau [SE 8793 0257] north of Scotterthorpe, in a nearby irrigation pond [SE 8790 0278], and as spoil from another pond [SE 8790 0259]. The blue-green colours imply the presence of Unit G. Slightly silty mudstones, 2.4 m thick, mainly pale red-brown with blue-grey mottling but with some thin blue-grey layers, are visible at the foot of The Cliff [SE 8679 1968] north of Burton upon Stather, and weathered traces of similar mudstone occur in a few other localities for several hundred metres to the north and south. The colours of these mudstones are suggestive of Unit E.

PENARTH GROUP

The Penarth Group, formerly called the Rhaetic, contains, in ascending order, the Westbury and Lilstock formations. Both are largely argillaceous. The former is marine and dark grey, having many features in common with the Lias above. The Lilstock Formation varies from greenish grey to reddish brown, and resulted from more restricted depositional conditions, somewhat similar to those affecting the uppermost parts of the Mercia Mudstone. The Penarth Group thus reflects the transition, with a temporary regressive phase, from an intracontinental basin environment to fully marine conditions. A Rhaetian age is indicated by the fauna and palynomorphs. The latter comprise miospores and organic-walled microplankton that increase in diversity upwards through the succession (Figure 11 and Lott and Warrington, 1988). The palynomorph assemblages are comparable with those recorded from the group farther south (Orbell, 1973; Morbey, 1975; Warrington, 1974, 1977, 1978, 1982, 1987).

Westbury Formation

Dark grey to black fissile mudstones typify the Westbury Formation, but a few thin sandstones occur, mainly at and near the base, and in parts of Nottinghamshire and Lincolnshire one or more thin beds or nodular layers of limestone are present locally in the upper part. In general the formation rests nonsequentially on Mercia Mudstone in eastern England, with local evidence of erosion. The formation contains a marine fauna consisting mainly of bivalves, notably *Protocardia rhaetica* and *Rhaetavicula contorta*; the old name of Contorta Shales originated from the abundance and restricted stratigraphical range of the latter. Most of the shells occur as single-species concentrations on distinct bedding planes, and many are pyritised. The sandstones locally contain vertebrate debris, mainly small bone fragments, fish teeth and scales. A rapid increase in dinoflagellates (Ivimey-Cook, p.157 *in* Warrington, 1974), periodic nondeposition, a sharp lowering of salinity and reworking have all been suggested as explanations of these concentrations of organic remains. A clearer water shelly fauna characterised by *Chlamys valoniensis* is present in the limestones. The fauna and dark colour indicate that a marine transgression rapidly established a shallow-water environment over a wide region, but with reducing conditions close below the sediment surface. The direction of the transgression is uncertain, but was probably from the south-west and possibly also from the south-east (Ziegler, 1982, p.59).

Within the district there is some evidence of a depositional break at the base of the formation. In Blyborough Borehole the base is irregular and the thin basal sandy layer with bone fragments extends down as burrow fills into the uppermost Mercia Mudstone. In Cockle Pits Borehole (Gaunt et al., 1980, p.3) dark grey silty mudstone containing fish scales occupies inclined cracks extending down 0.08 m into the Mercia Mudstone. The basal bed of the Westbury Formation is here a pale grey, strongly cemented, ill-sorted, fine-to medium-grained sandstone 0.05 m thick, containing subangular to rounded quartz grains, irregular masses of pyrite, fish debris and indeterminate shell casts. Elsewhere within the district the basal beds appear to be mudstones.

Except near the base in Cockle Pits Borehole, where there are thin pale to medium grey mudstones, the bulk of the formation is composed of dark grey to black, variably silty and micaceous mudstone. Some of the silt is concentrated in laminae which render the mudstones fissile; certain of the shelly layers have the same effect. Pyrite is fairly common in association with shelly layers and as small irregular masses. A few thin calcite 'beef' layers and calcareous and sideritic nodules have been noted in places. Thin impersistent sandstones within the mudstones are mainly fine-grained and silty, but some are slightly calcareous; ripple structures, pyrite, bioturbation and vertebrate debris have been noted in some boreholes. Sand is also present in the mudstones as wispy lenses accentuating small-scale ripples and cross-bedding, and as horizontal sinuous trails and transverse burrow fills. Thin limestones were recorded by Kent (1953) from the upper part of the Westbury Formation in shot holes near Scotter and Pilham (Figure 10); a limestone at Pilham yielded *?C. valoniensis*. Cockle Pits and Blyborough boreholes proved thin calcareous siltstones at comparable levels, the siltstone in the former yielding *C. valoniensis*. In both these boreholes the shelly fauna in the mudstones becomes more diverse upwards.

From the district as a whole the fauna includes *Cardina regularis*, *Chlamys valoniensis*, *Eotrapezium concentricum*, *Lyriomyophoria postera*, *Protocardia rhaetica*, *R. contorta*, *Tutcheria cloacina*, '*Natica*' *oppellii* and *Gyrolepis alberti*. Miospores, principally *Classopollis torosus*, *Ovalipollis pseudoalatus*, *Ricciisporites tuberculatus* and disacciatriletes including *Alisporites* dominate assemblages from the Westbury Formation (Figure 11). Organic-walled microplankton comprise *Rhaetogonyaulax rhaetica*, the acanthomorph acritarch *Micrhystridium* and the herkomorph *Cymatiosphaera*. Remains of foraminifera occur in the upper part of the formation in Blyborough Borehole.

Known thicknesses for the formation (Figure 10) range from 4.97 to 12.19 m. In more eastern boreholes the base and top of the formation cannot be accurately determined, but geophysical logs suggest that in Nettleton Borehole the total thickness is about 11.6 m and in Risby Borehole between 6.7 and 10.4 m. There is no detectable regional trend in thickness variations.

During the survey only a few poor exposures of the Westbury Formation were seen, all in the south-west. Dark grey fissile mudstone was seen in a ditch [SK 8604 9457] along the Blyton to Pilham Road, and dug out at Blyton School [SK 8535 9492]. Similar mudstone containing micaceous silty laminae and thin 'beef' layers was noted in a dyke [SE 8525 0030] west of Scotter, and fragments of thin-bedded flaggy micaceous sandstone containing fish scales and spines have been ploughed up at a locality [SE 8668 0184] north-west of Scotter. Dark grey fossiliferous mudstones are exposed in the River Eau [SE 8795 0245] to the north-east of Scotterthorpe.

Lilstock Formation

The Lilstock Formation in this district and farther north includes only the Cotham Member; the succeeding Langport Member has not been proved.

Most of the Cotham Member consists of calcareous blocky mudstone which ranges in colour from bluish and greenish grey to (less commonly) reddish brown; much of it has a typically smooth or 'soapy' texture. Fine-grained sandstones and dolomites are present in places. In some areas there is evidence of nonsequence and erosion at the base of the member, and less commonly within it. Except for scattered *Euestheria* and a few ostracods, the indigenous macrofauna is sparse. The general characteristics suggest sedimentation in a microplankton-rich lagoonal environment. Such conditions followed a marine regression and represent a partial reversion to the restricted conditions that prevailed during the deposition of the last phase of the Mercia Mudstone. An

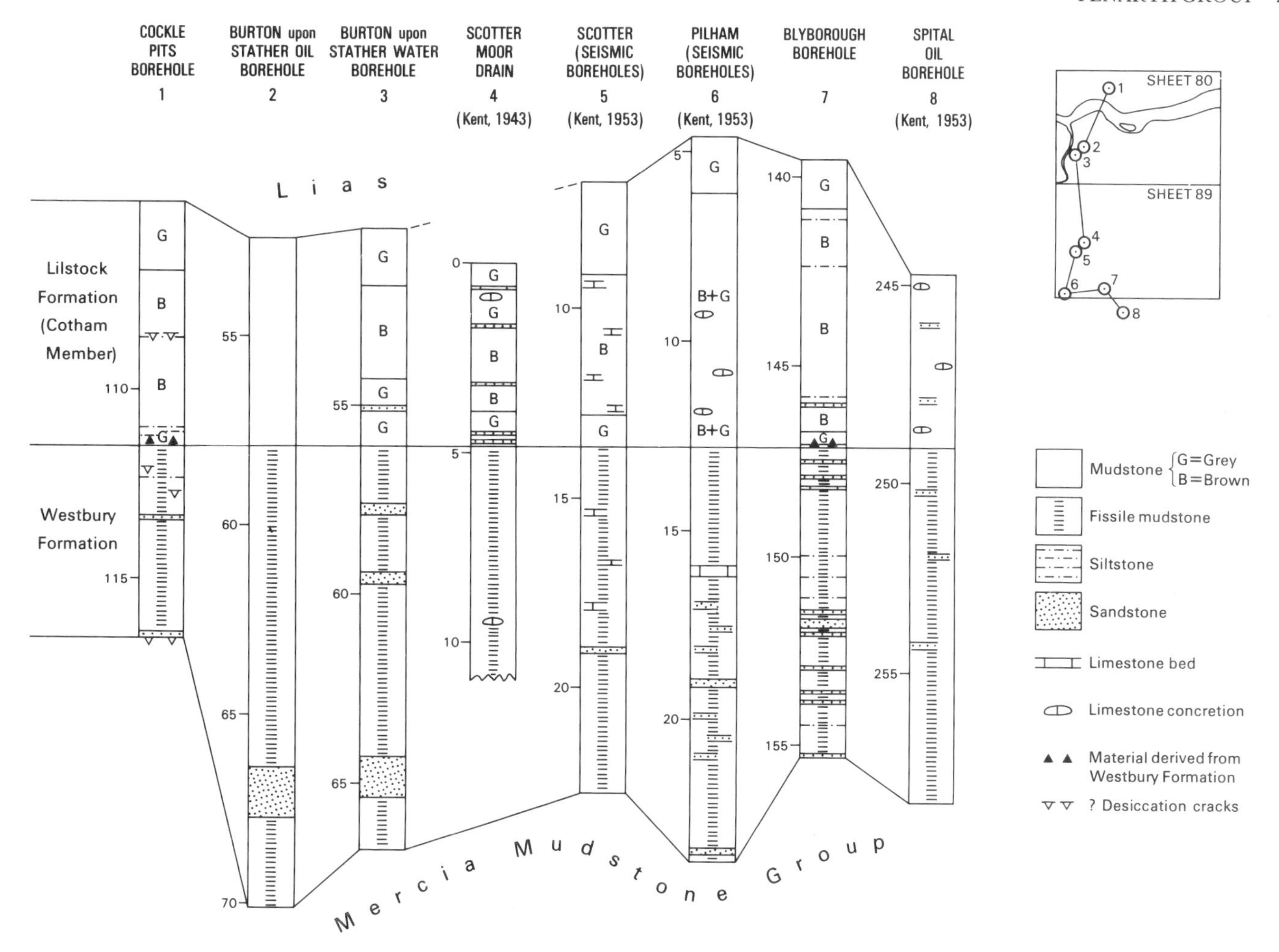

Figure 10 Comparative sections of the Penarth Group. Depths are in metres

input of terrigenous mud, conceivably wind-borne, is suggested by the presence of reddish brown mudstone, although Kent (1970) suggested that this coloration may be the result of erosion of red strata pushed up by rising salt diapirs in the area now covered by the North Sea.

Most of the complete borehole sections (Figure 10) show a tripartite colour arrangement. Near the base and in the upper part, the mudstones are pale bluish or greenish grey, but in the thick middle part they are mainly pale reddish brown. The colour boundaries are generally indistinct, being either interbedded or mottled passages. Laminae and very thin beds and lenses of pale grey to white, locally slightly calcareous siltstone and fine-grained sandstone occur within the mudstone. Some contain fining-upward units and ripple structures, and in places these coarser layers accentuate burrow fills and small-scale convolutions, the latter suggestive of slump structures. Nodular layers of silty limestone are recorded from some localities. Pyrite occurs widely as small crystal clusters and along sinuous trails. Material derived from the Westbury Formation, including dark grey mudstone fragments, bivalves and fish debris, is found in the basal layers of the Cotham Member in Cockle Pits Borehole (Gaunt et al., 1980, p.4), and similar material occurs in the basal beds in Blyborough Borehole, signifying an erosive episode at the contact. At Cockle Pits vertical cracks, suggestive of desiccation, extend down from the top of a hard calcareous silty layer about half way up the Cotham Member.

Extensive assemblages of miospores and organic-walled microplankton have been identified in Blyborough and Cockle Pits boreholes (Figure 11 and Lott and Warrington, 1988). The lower assemblages from the member are mostly dominated by organic-walled microplankton and the higher ones by miospores. The principal Westbury Formation miospores continue into the Cotham Member, though *Ovalipollis pseudoalatus* is less abundant; *Deltoidospora* and *Convolutispora microrugulata* are significant components of the assemblages from the upper part of the member. The rich organic-walled microplankton associations from the lower part are dominated by dinoflagellate cysts, principally *Rhaetogonyaulax rhaetica* but including *Dapcodinium priscum* and, in Blyborough Borehole, *Sverdrupiella mutabilis*. The last-named was described from Late Triassic marine deposits in the Sverdrup Basin, arctic Canada (Bujak and Fisher, 1976) and has been reported from Late Triassic sequences in the north-west Europe—arctic Norway region, though without details (Morby and Dunay, 1978). Its occurrence in Blyborough Borehole (Figure 11) is the first record of this form in Brit-

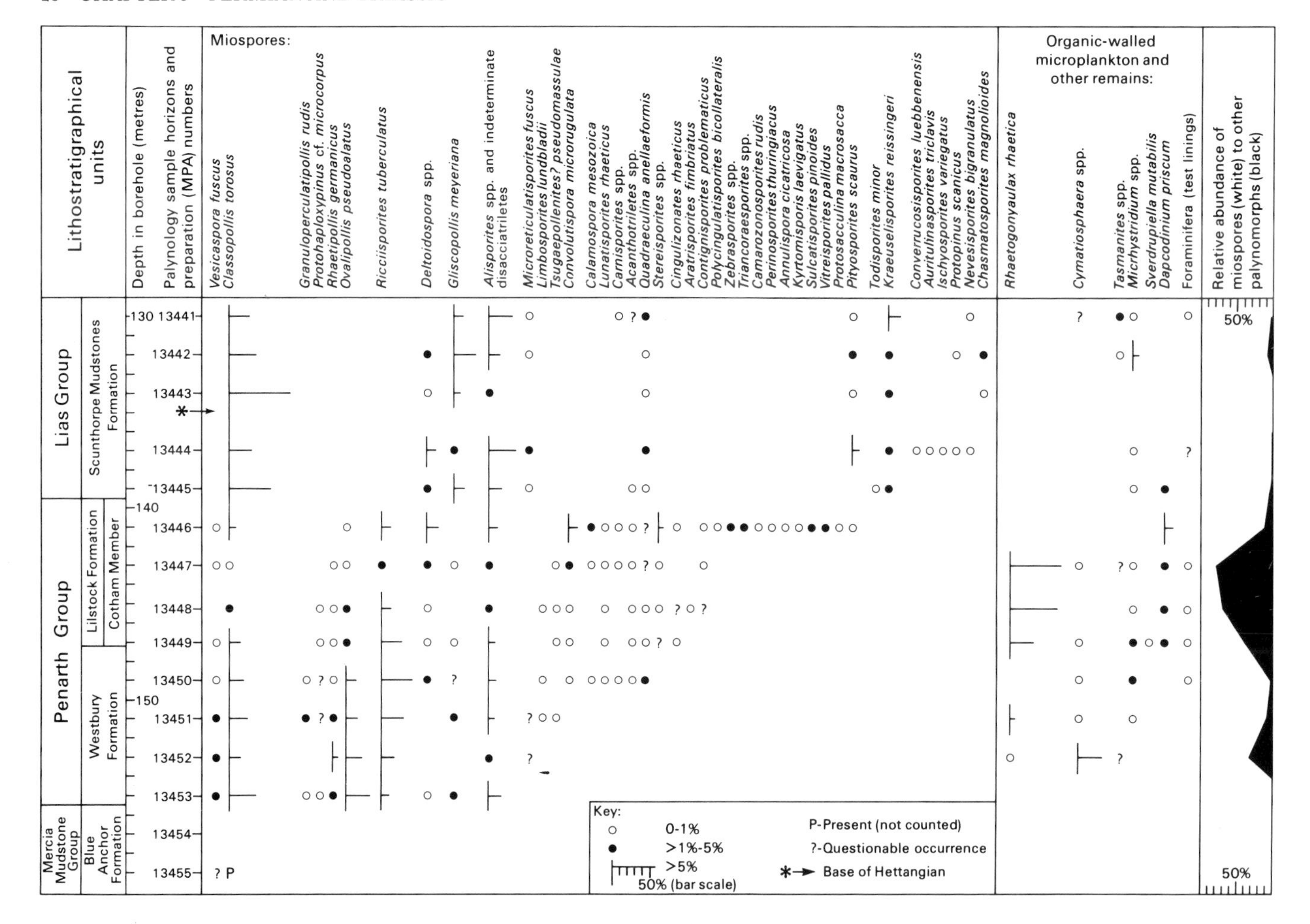

Figure 11 Distribution and relative abundances of palynomorphs from the highest part of the Mercia Mudstone Group to the basal part of the Lias Group in Blyborough Borehole. Relative abundances are based upon counts of 200 specimens

ain. Acritarchs, including *Micrhystridium* and *Cymatiosphaera*, occur sporadically in these associations, and the polygonomorph *Veryhachium* is recorded in Cockle Pits Borehole. The occurrence of foraminifera in the lower part of the member in Blyborough Borehole coincides with the richest organic-walled microplankton associations (Figure 11). Organic-walled microplankton are comparatively sparse in the upper part of the member and comprise *Dapcodinium priscum*, with *Micrhystridium* and *Tasmanites* in Cockle Pits Borehole.

There is no regional thickness trend, and the minor variations are probably due to local erosion at the base of the member, and possibly also at the top.

During the survey pale brown slightly calcareous mudstone was seen in a ditch [SK 8610 9441] along the Blyton to Pilham road, the only exposure of the Cotham Member noted in the district.

LIAS GROUP

Those Lias strata that lie below the lowest occurrence of the ammonite *Psiloceras* fall within the Triassic System (Cope et al., 1980a, pp.18–19; Warrington et al., 1980, p.10, fig. 1). For convenience these strata are described with the rest of the Lias in Chapter 4.

CHAPTER 4

Jurassic

Jurassic rocks are present at the surface or beneath Quaternary deposits across more than half of the district. Their outcrop widens from only 4.5 km in the extreme north to more than 27 km, almost the entire width of the district, in the south. Their dip is generally eastwards, and in the east they are concealed beneath Cretaceous rocks. Of the many exposures examined during the survey most were small and ephemeral. The only large exposures were opencast mines in the Frodingham Ironstone, now becoming degraded and backfilled, limestone quarries around Kirton in Lindsey, which are of limited stratigraphical range, and 'clay' pits for the cement industry near the Humber, in which much of the sequence is distorted by superficial structures. The systematic stratigraphy described below is, consequently, based mainly on the evidence of cored boreholes, notably several drilled for BGS during the survey.

The Jurassic succession comprises four primary lithostratigraphical divisions (Figure 12). The lowest is the Lias Group, which consists largely of mudstones with subordinate limestones and siltstones and three thin 'ironstones'. Although the basal strata are of Triassic age, their description is included in this chapter for convenience. The Redbourne Group, which rests on the Lias with marked unconformity, comprises limestones, sandstones and mudstones. It is overlain, in some places with slight unconformity, by the Ancholme Clay Group, which consists predominantly of mudstones. Another marked unconformity separates these mudstones from the highest unit, the Spilsby Sandstone Formation which, within the district, is largely Jurassic in age, although in at least one locality it extends up into the Cretaceous (as it does widely farther to the south-east); for convenience, however, the whole of the formation is described in this chapter.

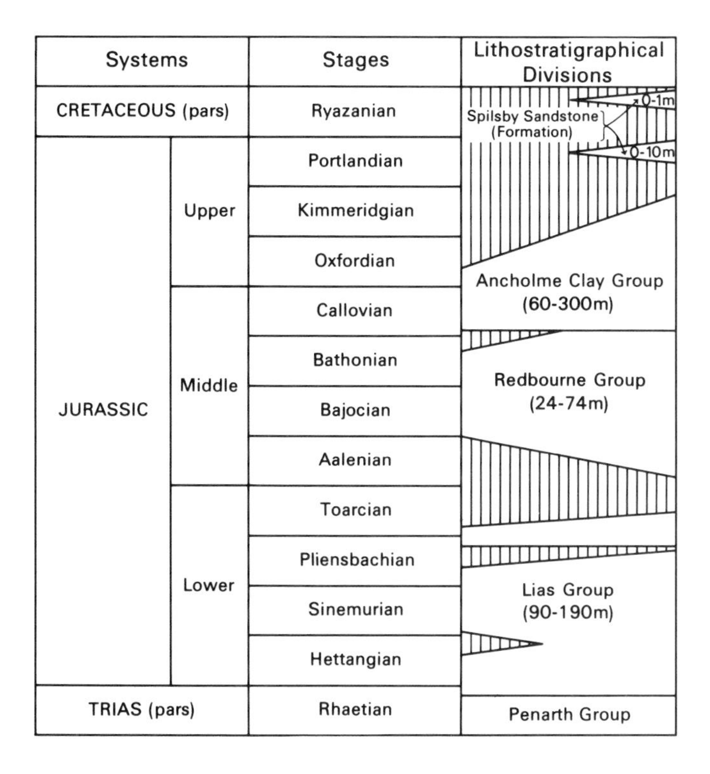

Figure 12 Jurassic stages and principal lithostratigraphical divisions

Biostratigraphical zonation of the Jurassic is based primarily on ammonites, although other fossils, notably foraminifera and ostracods, are also used. The ammonites are generally considered to have evolved sufficiently rapidly for the life span of each species to have been approximately synchronous over wide regions. Hence 'biozones can be defined in common with chronozones at points in rock' (Cope et al., 1980a, p.4), and also used chronostratigraphically to determine the limits of stages (Figure 12). Many of the Callovian and Upper Jurassic zones and subzones have formal definitions and type sections and, consequently, can be used as standard chronostratigraphical units; they are conventionally printed in 'Roman' type with capitals for the index taxon. In the pre-Callovian stages few formal definitions have been completed, so the zones and subzones are here all treated as biostratigraphical units with the names of their nominal taxa in italics.

The Jurassic Period lasted approximately from 205 to 135 million years ago (Hallam et al., 1985), and the British Jurassic sediments were deposited in latitudes between about 30° and 40° north (Smith et al., 1981). During latest Triassic times intermittent northward transgressions from the Tethyan Sea flooded the old intracontinental Southern North Sea Basin and adjacent areas, producing open, generally shallow, warm, temperate, marine conditions across much of north-western Europe by the early Jurassic. Most of England lay within this inundated region, although an appreciably reduced London–Brabant Massif and an area approximately coincident with the present Pennines probably remained emergent. Periodic and localised tectonic activity, for example in the Market Weighton area, produced small areas of shallowing and emergence. The mainly argillaceous Lias was deposited in this sea.

In mid Jurassic times the central part of the present North Sea, from Scotland to Denmark, was emergent, resulting in widespread erosion, and the northern part of eastern England became part of a large embayment with shallow open sea to the south and land to the west, north and north-east. Deposition in this embayment reflected the interplay between land and sea, with mainly calcareous marine sediments alternating and interdigitating with brackish and non-marine arenaceous and silty sediments. The Redbourne Group was deposited in the southern part of this embayment, where marine influence was dominant.

In Callovian times the last vestiges of terrestrial influence were terminated by a major eustatic rise of sea level, and throughout the late Jurassic most of eastern England again became part of an extensive shallow open sea in which the thick and largely argillaceous Ancholme Clay Group was deposited. This sea had both northerly and southerly connections, as indicated by Boreal and Tethyan taxa in its fauna. The open-marine environment was brought to an end in Portlandian times by widespread uplift, and considerable erosion followed. Before the close of the Jurassic Period, however, slight differential subsidence allowed a Boreal marine transgression to invade parts of eastern England from the east and coastal sands, the Spilsby Sandstone, were deposited.

The Jurassic succession in the district clearly reflects the changing depositional environments summarised above. It also shows, by lithological and thickness variations and depositional breaks, the more localised effect of the Market Weighton Structure, some 10 to 20 km to the north (p.106), which acted as a hinge between subsiding areas to its north and south and which at times was subject to uplift. From its earliest effects in the Hettangian (Gaunt et al., 1980, pp.26–28) until the Callovian, it separated the 'East Midlands Shelf' (which includes the present district) from the more rapidly subsiding 'Cleveland Basin' to the north. Later in the Jurassic, the 'East Midlands' area subsided faster as inversion movements began that eventually converted the Cleveland Basin into a broad arch (Kent, 1980a, p.516). In early Cretaceous times, major uplift centred on the Market Weighton Structure was responsible for the erosion of much of the Jurassic succession over and adjacent to it.

LIAS GROUP

The term 'Lias' has varied in meaning and definition in the past, but has been applied in one sense or another to approximately the same strata throughout Britain for well over a century. In this account 'Lias' is used as a lithostratigraphical term of group status for the predominantly argillaceous marine strata that rest on the slightly eroded top of the Penarth Group and extend up to the unconformable base of the largely calcareous and arenaceous Redbourne Group.

Lithostratigraphical subdivisions of the Lias into 'lower', 'middle', and 'upper' have varied with time, region and author, and have been inextricably confused with palaeontological subdivisions. In their place two formations have been defined for the district and adjacent areas, in ascending order the Scunthorpe Mudstones and the Coleby Mudstones. The former contain numerous thin limestones and calcareous siltstones that form prominent topographical features across much of the district, and they also include the Frodingham Ironstone Member at the top of the formation. The Coleby Mudstones generally lack limestones, but contain two ferruginous members, the Pecten Ironstone and the Marlstone Rock. Figure 13 shows the lithostratigraphical correlation of the Lias in the district with adjacent regions.

Biostratigraphical zonation of the Lias is based on ammonites, and enough of these have been found within the district to provide a fair degree of correlation (Figure 13). The base of the *planorbis* Zone, and of the Hettangian Stage and the Jurassic System, is now defined at the first appearance of ammonites of the genus *Psiloceras* (Cope et al., 1980a, pp.18–19). As a result the lowest Lias strata in the district (those below the lowest *Psiloceras*) are referred to the Rhaetian Stage of the Triassic (Gaunt et al., 1980, p.4). The highest Lias in the district belongs to the *falciferum* Subzone, in the Lower Toarcian Substage. Except for the lowest beds, therefore, the Lias is entirely of early Jurassic age.

The Lias consists largely of pale to dark grey kaolinite-illite mudstone, generally slightly calcareous and commonly silty. It contains numerous small-scale graded cycles (mainly fining-upward) in which the lime content and particle size vary considerably, in places to the extent that thin limestones and siltstones (and less commonly fine-grained sandstones) are present. Bioturbation has blurred many lithological contacts, but the bases of many of the coarser-grained beds rest on slightly eroded surfaces. Some beds are rich enough in ferruginous minerals, commonly in oolith form, to be classed as ironstone. In the weathering zone lime has been removed from the mudstones and redeposited as calcite cement in the interstices of the coarser-grained or shell-rich beds. This process has accentuated the original differences in hardness within the succession and has led to the development of pronounced topographical features. The existence of these features has resulted in oversimplified assessments of the succession, expressed in names like 'Angulata Clays' and 'Granby Limestones', which largely obscure the true nature of the unweathered rocks.

Deposition of the Lias followed the transient emergent phase responsible for slight erosion of the underlying Penarth Group. The district lay on the northern part of the slowly subsiding 'East Midlands Shelf' with the more rapidly subsiding 'Cleveland Basin' to the north and the 'Sole Pit Trough' to the east, all forming part of the epeiric 'Southern North Sea Basin' (Kent, 1980a). The Lias within the district is thicker to the south and south-east (Figure 14), reflecting greater subsidence rates in these directions. The locus of greatest deposition appears to have been farther south between Lincoln and Great Ponton in south-western Lincolnshire, beyond which the Lias thins towards the 'London–Brabant Massif'. The northern edge of the East Midlands Shelf lay near Market Weighton, 10 to 15 km north of the district. The Market Weighton Structure (see p.106) experienced periodic positive movements which produced depositional thinning and many condensed sequences, non-sequences and slight unconformities that increase in number and magnitude northwards. The westward component of thinning in the Lias suggests that land may have existed in the Pennine region, probably at times with a spur extending into the Market Weighton area. The iron mineral berthierine and its derivatives, that make up the ferruginous strata, are believed to have been formed in lagoons adjacent to a well-vegetated lateritic landmass lying to the west (Fletcher and Knox, in preparation). On a global scale the early Jurassic was characterised by rising sea levels interrupted by only minor (and transient) lowerings (Hallam, 1981). Evidence from the Liassic rocks of the district suggests that localised tectonic movements, largely emanating from the Market Weighton area, imposed minor variations on this sea-level history. Lias sedimentation terminated during the Toarcian and the whole region suffered uplift and erosion (again more extensive near Market Weighton), to

Figure 13 Relationship of the Lias of the district to that of adjacent regions

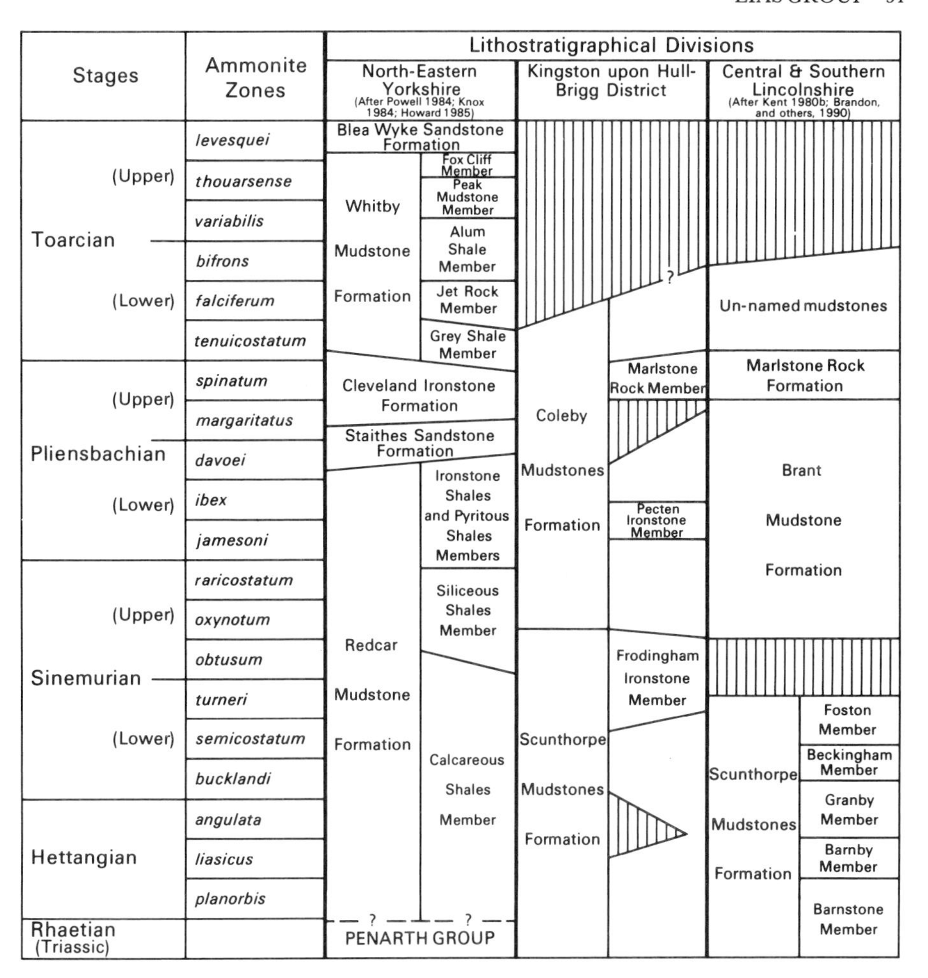

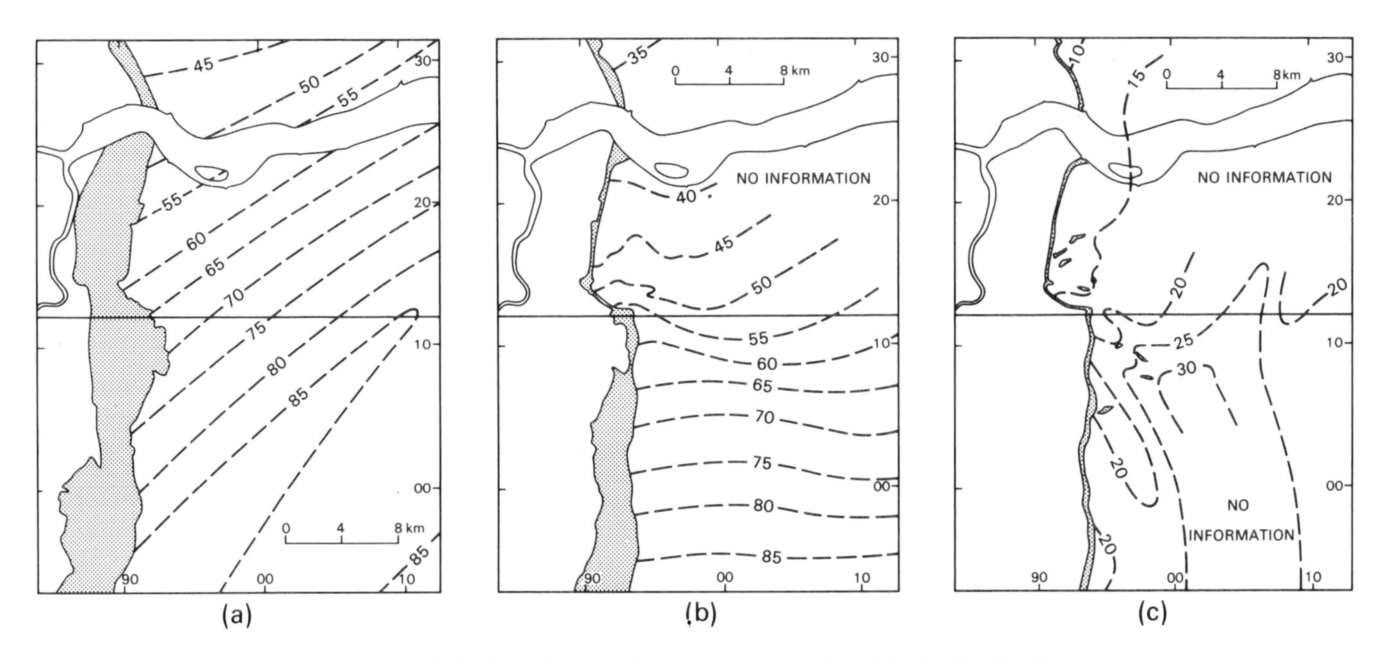

Figure 14 Isopachytes in metres of a) Scunthorpe Mudstones Formation, b) Coleby Mudstones Formation below top of Marlstone Rock, and c) Coleby Mudstones Formation above Marlstone Rock. Outcrop of each sequence is stippled

the extent that the succeeding Redbourne Group rests unconformably on rocks of the *falciferum* Subzone in the southern part of the district and the *semicelatum* Subzone in the north (in South Cave Borehole).

Scunthorpe Mudstones Formation

The Scunthorpe Mudstones Formation (Figure 15) is characterised by grey, variably calcareous, silty mudstone containing numerous thin limestones and calcareous siltstones, particularly near the base and in the upper half. The succession becomes increasingly ferruginous towards the top, a trend that culminates in the local development of the Frodingham Ironstone Member. No permanent exposure of the contact with the Penarth Group is available and so the base of the formation is defined by reference to Blyborough Borehole where, at a depth of 139.50 m, grey shell-layered calcareous mudstone rests with a sharp base on the Cotham Member of the Penarth Group. Elsewhere, for example in Cockle Pits Borehole, the basal bed of the formation is a thin bioclastic limestone. The formation is thickest, almost 90 m, in the southern part of the district (Figure 14a) and thins to less than 45 m in the north-west. It ranges in age from late Rhaetian to late Sinemurian.

Beds below the Frodingham Ironstone Member

Except in degraded railway cuttings (Dudley, 1942), the beds below the Frodingham Ironstone are largely unexposed, and descriptions rely on boreholes and a few temporary exposures (notably those for the M180 Motorway south-west of Bottesford). Blyborough Borehole provides the most complete and detailed section. Much of the succession, especially in its higher part, consists of fining-upward cycles. In a complete cycle three lithologies can be distinguished. The basal strata are calcareous siltstones (locally fine-grained sandstones) with shell debris at the base. They grade up quite sharply into grey, blocky, slightly calcareous and micaceous, silty, kaolinitic mudstones with subordinate chlorite and traces of smectite; these mudstones are generally the major component of the cycle. Both the basal strata of a cycle and the silty mudstones can be strongly bioturbated, with large *Diplocraterion* and *Rhizocorallium* prominent. The silty mudstones grade upwards into much darker, finer-grained laminated mudstones containing *Chondrites* and pyritised 'trails'. Both calcitic and sideritic concretions are present in the higher parts of some cycles, the result of diagenesis prior to compaction. In many cycles the coarser basal layers and/or the uppermost laminated mudstones are thin or absent.

The Scunthorpe Mudstones below the Frodingham Ironstone can be subdivided into a lower unit containing only a few limestones, and an upper one in which limestones are thicker, coarser grained and more numerous. Most of the limestones appear to result from calcite diagenesis at the bases of those cycles where shell-debris layers are sufficiently thick for this process to be effective. In Nettleton Bottom Borehole, however, the most easterly and deeply buried sequence examined, limestones are rare, particularly in the higher part of the unit (Figure 15), and the basal cyclic layers are calcareous siltstones with some fine-grained sandstones. This down-dip change may be because of an east–west variation in original depositional conditions, or because calcitic diagenesis has occurred near outcrop in relatively recent times under the influence of meteoric water.

The mudstones in the lower unit are finely laminated and only slightly bioturbated; they include dark carbonaceous paper shales and contain rare slump and ripple-drift structures. The mudstones in the upper unit contain numerous graded silty layers, and individual beds are thicker and more bioturbated than in the lower unit. They also contain discoidal calcitic concretions and small sideritic nodules. Deposition of the upper unit may have taken place in more agitated and better oxygenated water than that of the lower unit. There is an increase in faunal diversity in the upper unit, with several genera that had been absent in the Hettangian appearing (or reappearing) in the Lower Sinemurian.

The lower unit is 33.90 m thick in Blyborough Borehole, thinning northwards to 21.66 m in Cockle Pits Borehole. The lowest beds, approximately 4 m thick in both boreholes, are medium to dark grey calcareous mudstones rendered fissile by numerous laminae and thin beds of siltstone that give them a striped appearance. They contain a few thin sparry limestones recrystallised from shelly lag layers rich in echinoid spines. In Blyborough Borehole a 0.06 m concretionary porcellanous limestone lies 0.24 m above the base, and is similar to one that occurs at the base of the Lias farther south (Swinnerton and Kent, 1976, p.22). The fossils in the basal beds are mainly bivalves with Rhaetian affinities, such as *Cardinia regularis, 'Gervillia' praecursor* and *Protocardia rhaetica*.

The immediately overlying strata are lithologically similar, except that interbedded sparry limestones account for more than half the thickness, which varies from 0.87 m in Blyborough Borehole to just over 3 m in Cockle Pits Borehole. This northward thickening is contrary to the northward thinning of almost all other parts of the Lias in the district, and appears to reflect a short-lived trend that extended north of Market Weighton where the equivalent beds are even thicker (Gaunt et al., 1980). The hard sparry limestones, formerly called 'Hydraulic Limestones' (Wilson, 1948), form a low feature in places, as near Musgrave's Farm [SE 8754 0505]. The bivalve fauna, collected both from sparry limestones and interbedded mudstones, contains fairly rugged shallow-water genera such as *Cardinia, Liostrea, Modiolus* and *Plagiostoma* which, with echinoid and crinoid debris, indicates well-oxygenated stable sea-bottom conditions. These taxa are not diagnostic of age, but the lowest ammonites (*Psiloceras* in Cockle Pits Borehole and *P. planorbis* in Blyborough Borehole) occur 0.12 and 0.27 m respectively below the top of this interbedded sequence, and mark the base of the Jurassic System.

The succeeding strata, which thin northwards from 29 m in Blyborough Borehole to 14.5 m in Cockle Pits Borehole, are mainly smooth, medium grey, blocky mudstones, although some darker 'striped' fissile mudstones are interbedded in the lower part. Rippled silty laminae are present, as are thin siltstones, and a few irregular shell-lag layers have been diagenetically altered to limestone. Large calcareous concretions and small sideritic nodules are common in the blocky mudstones. Some beds contain numerous pyritised horizontal burrows, and a few are slightly disturbed by

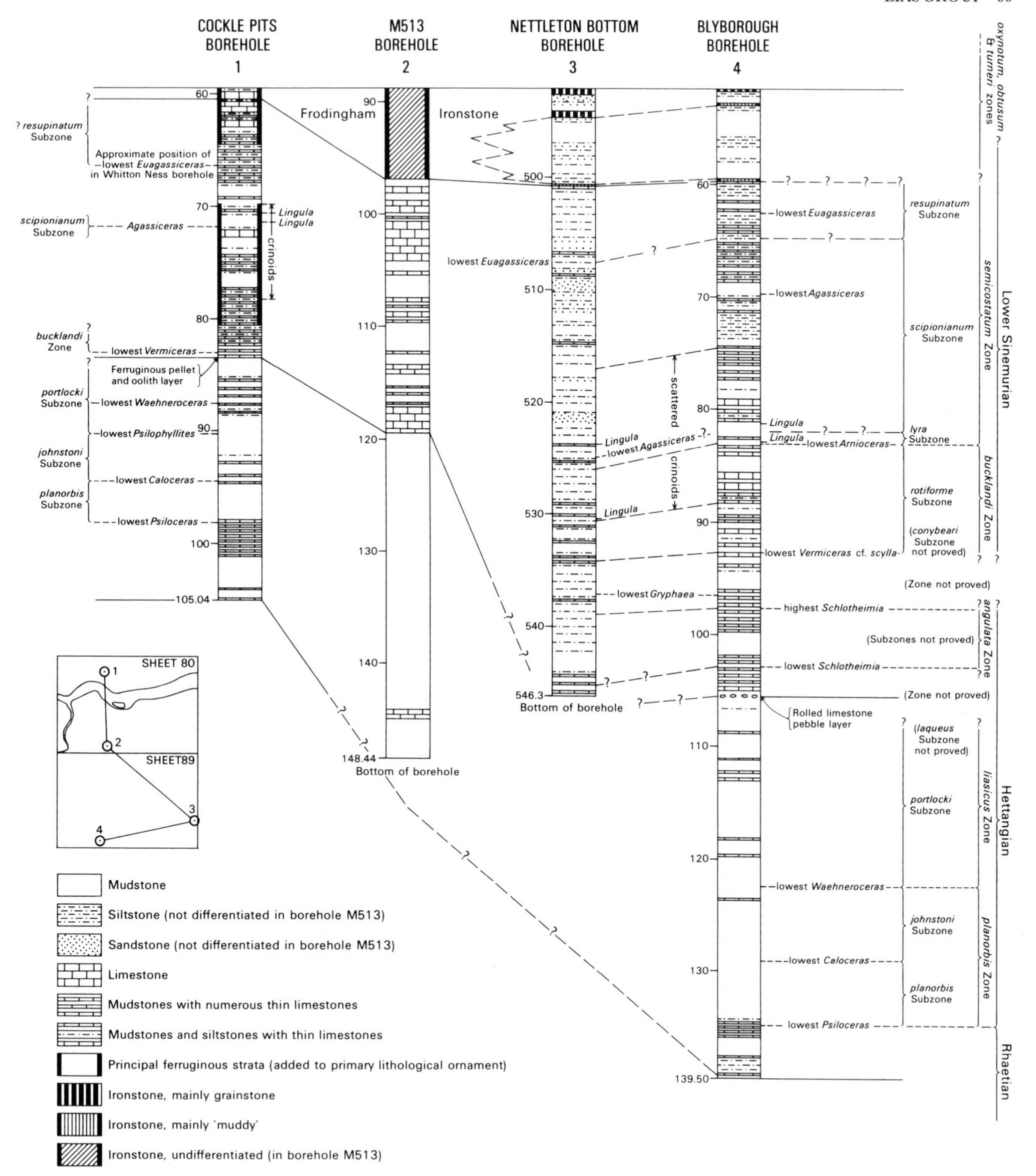

Figure 15 Borehole sections of the Scunthorpe Mudstones Formation. Depths are in metres

synaeresis movements of the original wet sediment. Pyrite crystals are common throughout, and at higher levels many of the ammonite casts are pyritised. A few goethite ooliths just above the middle of these mudstones in Cockle Pits Borehole, and coarser silty beds at a corresponding level in Blyborough Borehole, may imply a temporary lowering of sea level. The lowest *Caloceras* marks the base of the *johnstoni* Subzone, and the index species is present at higher levels. The joint ranges of *Waehneroceras* and *Psilophyllites* are used to identify the *liasicus* Zone, but these beds may only represent the *portlocki* Subzone (Figure 15).

In the remaining metre or so at the top of this sequence, only indeterminate schlotheimiids are present, so these beds may belong either to a higher part of the *liasicus* Zone or to the *angulata* Zone. The fauna from these Hettangian strata also includes echinoderms, both rugged and attached bivalves, and burrowers such as *Protocardia*, which imply continuing shallow and well-oxygenated depositional conditions.

The upper unit, comprising limestone-rich mudstones, thins northwards from nearly 46 m in Blyborough Borehole to just over 22 m in Cockle Pits Borehole. At its base in these boreholes there is, respectively, a layer of rolled limestone pebbles and a thin limestone containing ferruginous pellets and ooliths. The sequence differs from the underlying strata in three main respects: the mudstones are generally more silty; siltstones and silty limestones are more numerous and thicker; and in the northern part of the district, for example in Cockle Pits and Whitton Ness boreholes (Gaunt et al., 1980), much of the sequence is ferruginous. Although sideritic mudstones and nodular layers are present, as also is pyrite, the iron is largely represented by berthierine and its oxidised derivatives goethite and hematite, in oolith form. Much of the shell debris has also been altered to goethite. These ferruginous strata, and also the coarser, more silty beds, may reflect shallowing, and even localised emergence of an adjacent area.

Schlotheimia, indicative of the *angulata* Zone, ranges through just over 5 m of strata near the base of these beds in Blyborough Borehole and is present also in the lower beds proved in Nettleton Bottom Borehole. In neither borehole can the *extranodosa* and *complanata* subzones be distinguished, but the presence of *S.* cf. *similis* at 539.83 m depth in Nettleton Bottom Borehole suggests the latter subzone. In Cockle Pits Borehole there is only 0.35 m of strata, devoid of ammonites, above the base of the limestone-rich mudstone unit and below the lowest evidence of the *bucklandi* Zone, implying that in the northern part of the district most, and probably all, of the *angulata* Zone is cut out by a nonsequence at the base of the limestone-rich mudstones. Even in the south the rolled pebbles at the base of these mudstones in Blyborough Borehole suggest a depositional break, though clearly of less magnitude.

Some 4.0 to 4.7 m of strata without ammonites occur above the highest *Schlotheimia* in Blyborough and Nettleton Bottom boreholes; the lowest occurrences of the bivalve *Gryphaea*, 1.2 and 1.5 m respectively above the base of these strata, provide a useful local biostratigraphical marker. An exposure [SE 8879 0605] in the M180 motorway cutting just south of Bottesford yielded *Vermiceras conybeari*, the only evidence from the district of the *conybeari* Subzone of the *bucklandi* Zone. In Blyborough and Nettleton Bottom boreholes the lowest occurrences of *Vermiceras* are of *V.* cf. *scylla*, a species which ranges from the upper part of the succeeding *rotiforme* Subzone into the *lyra* Subzone of the *semicostatum* Zone. It is probable, therefore, that the *conybeari* Subzone is reduced, or cut out entirely by a nonsequence in much of the district to the north of Bottesford.

The base of the *lyra* Subzone (Ivimey-Cook and Donovan, 1983), defined by the lowest *Arnioceras*, is recognisable in Blyborough Borehole, where its appearance is associated with concentrations of crinoid debris and *Lingula*; these assist correlation with the Nettleton Bottom and Cockle Pits boreholes where *Arnioceras* is absent. The base of the *scipionianum* Subzone is normally taken at the lowest *Agassiceras*; in Nettleton Bottom Borehole this is only 0.84 m above a *Gryphaea* bed marking the top of the crinoidal sequence. In Blyborough Borehole, however, the lowest *Agassiceras* found is considerably higher (Figure 15), and the presumed base of the *scipionianum* Subzone has been estimated by reference to its probable position in Nettleton Bottom Borehole. The base of the *resupinatum* Subzone of the *semicostatum* Zone is placed at the lowest occurrence of *Euagassiceras* which, in Nettleton Bottom Borehole, appears to be about 8.4 m above the *Gryphaea* marker bed. The subzone was not proved in Cockle Pits Borehole, but its base is conjectured (Figure 15) by correlation with Whitton Ness Borehole, where a solitary *Euagassiceras?* was found (Gaunt et al., 1980, p.5, fig. 4). There is no ammonite evidence for beds of *turneri* Zone age below the Frodingham Ironstone. The ostracod *Ektyphocythere betzi*, which at Lyme Regis in Dorset has a *turneri* to *obtusum* zonal range (Lord, 1978), has been found at several localities within the district close beneath the Frodingham Ironstone. However, *Euagassiceras*, signifying the *resupinatum* Subzone, has been found locally in the basal part of the ironstone.

Frodingham Ironstone Member

In northern and central parts of the district the highest part of the Scunthorpe Mudstones Formation is markedly ferruginous and has long been known as the Frodingham Ironstone, now accorded member status. A detailed account of the ironstone is provided by Fletcher and Knox (in preparation); the following is an abridged version.

The ironstone has been worked along much of its outcrop from the Humber southwards to Ashby, on the southern side of Scunthorpe, and as ore-grade rock it continues eastwards at depth apparently as far as Grimsby. It is thinner, less ferruginous and more calcitic north of the Humber, and from Ashby southwards it is thinner, markedly less ferruginous and increasingly more argillaceous. The ironstone is traceable at outcrop by a brash of reddish brown 'ginger bread or biscuit stone' (Wedd, 1920, p.76), and it locally forms a minor topographical feature. In places, for example between the Humber and Burton upon Stather, and also at Scunthorpe, the ironstone rests on a thin limestone, but elsewhere it is separated from the limestone by up to 2 m of mudstone. This relationship may result from pre-ironstone erosion, for in the realigned Thealby Beck [from SE 903 192 to SE 906 198] the ironstone fills a channel cut 1.5 m into the underlying strata.

Because of the lack of an exposure of unweathered ironstone, the core of Roxton Wood No. 11 Borehole [TA 1651 1180], just east of the district, has been chosen as the type section of the member. The core is housed at BGS, Keyworth, and exhibits virtually all the known lithofacies and relationships. The base of the member is taken at a depth of 510.59 m in this borehole, where 0.02 m of highly ferruginous calcitic, berthierine-oolith grainstone rests with a sharp contact on greyish, poorly ferruginous, goethite-oolitic, berthierine mudstone and siltstone. The marked colour change reflects the relative amounts of iron present and is recognisable throughout the orefield. Correlation of the most northerly exposure of the base of the ironstone in the Coleby (Winterton) Mine with boreholes farther north has necessitated an upward revision of the base previously chosen farther north in Whitton Ness and Cockle Pits boreholes (Gaunt et al., 1980). There is considerable variation in overall thickness and lithology within the district, mainly from north to south, as shown in Figure 15. In Whitton Ness Borehole it is a little more than 4 m thick but reaches 8.5 m at Crosby Warren and thins to 7.8 m around Ashby. Eastwards it thickens to 9.79 m in a borehole near Thornholme [SE 970 126] and, beyond the district, to 10.24 m in Brocklesby No. 14 Borehole [TA 1429 1245].

The lithogical and sedimentary features are complex. The member comprises repeated alternations of grainstone-dominated ironstone and muddy ironstone including packstone, wackestone and mudstone. A generalised tripartite subdivision of the member is recognisable, however, consisting of a lower calcitic grainstone-dominated unit with some reddening, a middle extensively sideritised 'muddy' unit and an upper mixed grainstone and 'muddy' unit with further reddening. North of the Humber in Cockle Pits Borehole the entire ironstone is thin, largely grainstone-dominated and more calcitic, comprising 1.25 m of berthierinitic bioclastic and oolitic limestone and siltstone on 2.21 m of similar, appreciably ferruginous strata originally included in the ironstone.

South of Santon [SE 930 110] increasing amounts of 'muddy' ironstone have led to greater compaction of the ironstone and a consequent thinning towards the southern margin of the orefield. South of Brat Hill [SE 917 080], the middle 'muddy' unit expands considerably, but loses its ferruginous nature over a short distance and eventually passes into mudstones and siltstones beyond Aspen Farm [SE 907 064]. Although the thin vestigial lower and upper grainstone units remain ferruginous in this direction and are recognisable in boreholes, only the upper is traceable at outcrop; it is this highest part of the Frodingham Ironstone that has been mapped south of Bottesford.

The grainstone-dominated layers generally have sharp bases and grade upwards from coarse-grained calcitic bioclast-rich rock into finer-grained berthierine oolith-rich rock (Types D and A respectively of Davies and Dixie, 1951). Most of the grainstones are succeeded by 'muddy' ironstone layers, the intervening contact being in places sharp but more commonly gradational, with bioturbation blurring the contact. This general impression of fining-upward cycles is enhanced by an upward decrease of ooliths in many of the 'muddy' ironstones, which pass upwards from packstone through wackestone into weakly oolitic or nonoolitic mudstone (Types C and B respectively of Davies and Dixie, 1951). Many cycles are more complex. Some grainstones contain nonoolitic 'muddy' layers, and others include repeated alternations of shelly and oolitic layers. In addition, alternations of wackestone and packstone, and even thin grainstone layers, occur in some of the 'muddy' ironstones. Some cycles are incomplete, generally because the 'muddy' ironstone component is abnormally thin or absent due to intradepositional erosion. As a result, some strata, especially in the lower part of the ironstone, consist largely of superimposed fining-upward grainstones. Major bedding planes can be traced for several hundred metres in opencast faces but individual beds cannot be traced between adjacent boreholes. These variations are probably the result of repeated localised small-scale reworking, for which there is considerable evidence, particularly in the grainstones, producing the lenticular arrangement of different ironstone types noted by Davies and Dixie (1951, pp.94–95).

The iron silicate berthierine (28 per cent Fe) is a major component of the Frodingham Ironstone, and is mainly of kaolinitic type. It is the principal constituent of the 'muddy' ironstone, and in the grainstones occurs in ooliths, oncoliths, shell impregnations and cements. An increase in goethite relative to berthierine in many oolitic envelopes is associated with a decrease in the definition of the oolitic structure, an indication that goethite has replaced berthierine rather than the reverse; the co-existence of both minerals at the sediment surface is demonstrated by their alternations around goethite oolith nuclei. A berthierine precursor to the goethite is also indicated in many ooliths by a high silica content. In unweathered ironstone crystalline form goethite (50 per cent Fe) is mainly present in ooliths, oncoliths and shell impregnations, and is only a rare constituent of the 'mud' fraction. Where it occurs in the 'muddy' ironstones, it is present as thin layers at the top of berthierine mudstone units or around the margins of large-diameter dwelling-burrows in the same material, both circumstances implying syndepositional oxidation of the berthierine. All the siderite (about 40 per cent Fe) is diagenetic, occurring as an early fringing cement in grainstones, and as a replacement of berthierine in 'muddy' ironstone. Both hematite (70 per cent Fe) and pyrite (46 per cent Fe) are minor constituents that result from diagenesis, the former occurring as an alteration product of berthierine in ooliths, cements and 'muddy' ironstones, the latter as small crystals within 'muddy' layers. Calcite, present as shell debris and diagenetic cement, becomes an increasingly common constituent north of the orefield.

Ooliths, oncoliths and shell debris are the principal components of the grainstones. The ooliths have an average maximum diameter of 0.3 mm. They are of concentric type showing an oblate ellipsoidal shape due to preferential equatorial accretion (Knox, 1970). Most of the nuclei are fragments of shells (commonly altered to goethite), pre-existing ooliths, or quartz grains. The oncoliths are up to 5 mm in diameter and most are roughly spherical or oblate ellipsoidal; they generally show complex internal structures, including microstromatolites, and have suffered phases of dissolution and recementation. Bivalves and echinoderms provide most of the shell debris; many fragments are bored, the holes being accentuated by goethite and, more rarely, berthierine fillings.

The chemical composition of the Frodingham Ironstone varies widely, particularly in the proportions of iron, silica and lime (Table 5). The highest concentrations of iron (and also of manganese) are found in the central part of the orefield, between the Coleby and Trent opencast mines where, due to the dominance of oolitic packstones and grainstones, up to 26 per cent Fe is encountered. Even the richest individual beds seldom exceed 35 per cent iron. Silica, insolubles and alumina increase towards (and beyond) the margins of the orefield, a trend that can probably be related to the higher proportions of berthierine 'mud' in these areas and also to the increased amount of silt-grade quartz in the south. The greater proportion of lime towards the north is due to increase of shell debris and associated calcite cements in that direction. Sulphur, an impurity, is slightly more common in the central part of the orefield. Its highest concentrations are in pyrite-rich lenses (common as 'snap-bands' in the south-western part of the orefield), which were often separated from the processed ore. Proportions of arsenic and phosphorus, the other principal impurities, remain fairly constant across the orefield.

Ammonites indicative of several subzones have been found within the Frodingham Ironstone. The *resupinatum* Subzone seems to be the oldest, and is represented by rare occurrences of *Euagassiceras* from the basal beds in the northern opencast mines. The specimens may, however, have been reworked from earlier strata. Other species from the *semicostatum* Zone include *Coroniceras crossii* and *C. scunthorpensis* from the old Coleby Mine near West Halton, and also *C. alcinoe* (erroneously referred to the older *C. gmuendense* in Wedd, 1920). A good example of reworking within the ironstone sequence is provided by the find of *C. alcinoe* 1.83 m above the base of the ironstone in Thealby Mine, cited by Hallam (1963); this specimen derives from a level just below the middle of the member, and above strata that have yielded *turneri* Zone forms. The presence of the *turneri* Zone is proved by specimens of *Caenisites* at several localities and from strata in Blyborough and Nettleton Bottom boreholes; a specimen of *Microderoceras birchi* indicates the *birchi* Subzone (Hallam, 1963). *Arnioceras*, including *A. bodleyi* and *A. semicostatum*, continues up from the *semicostatum* Zone. The most prolific faunas come from the upper part of the *obtusum* Zone; however, the earliest part of the zone, the *obtusum* Subzone, is poorly proved. Large specimens of *Aegasteroceras* are present, but this genus continues up through the higher subzones. Most of the large specimens are of *Asteroceras* close to *A. stellare* and *A. suevicum*, but they occur with *Aegasteroceras, Eparietites, Epophioceras* and rare *Xipheroceras* (for example at Yarborough). *Promicroceras planicosta* from Yarborough and *Promicroceras* sp. occur both in the ironstone and, in Nettleton Bottom Borehole, in the coeval silty Scunthorpe Mudstones; *Promicroceras* continues above the ironstone into strata referable to the *oxynotum* Zone in Worlaby E and Elsham Wolds boreholes. The highest subzone of the *obtusum* Zone is proved by large *Eparietites denotatus* (its nominal taxon), *Epophioceras landriotii* and *Aegasteroceras sagittarium*. These ammonites occur commonly in the top 2 m of ironstone, which appears to be a condensed and reworked sequence. In the southern part of the orefield this sequence has not yielded *oxynotum* Zone ammonites, in contrast to farther north where, in both Roxby and Coleby mines, specimens of *Oxynoticeras simpsoni* occur in the top metre of ironstone. The available ammonites thus suggest that the top of the ironstone is slightly diachronous, being referable to the *simpsoni* Subzone in the north and to the *denotatus* Subzone in the south. However, further work is needed on this topic.

The ironstone is rich in other fossils. Thick-shelled and robust bivalves, especially *Gryphaea* and *Cardinia*, and large lignitic wood fragments are common locally. The fossils in the 'muddy' ironstones (apart from the berthierine-siderite mudstones, which are barren) are closely comparable to those in other Lias mudstones, with a spectrum of nektonic, epifaunal and infaunal types. Most of these fossils occur also in the grainstones, but there they are commonly fragmented and bored, the result of winnowing and reworking. Hallam (1963) has reviewed the depositional environment.

Fletcher and Knox (in preparation) consider that berthierine was the primary iron mineral. The precipitation of this mineral apparently requires conditions of low oxygenation and low salinity, yet all the petrological and fossil evidence indicates that the Frodingham Ironstone formed in well-oxygenated fully marine conditions. The implication is that the berthierine formed elsewhere and then, with some of its derivatives, was transported into the area. The berthierine muds and ooliths are believed to have formed in quiet brackish lagoons protected from the open sea by some sort of barrier (probably shell banks) and receiving dissolved and particulate iron from an adjacent low-lying well-vegetated lateritic landmass. The common occurrences of fragmented and reformed ooliths, together with shell debris as oolith nuclei and as larger fragments impregnated with berthierine, point to phases of instability when the protective barrier broke down. In severe and prolonged phases of instability, possibly resulting from sea-level changes, some of the lagoonal products—muds, ooliths and shell debris—appear to have been swept into the open sea and then redeposited under normal marine conditions. The grainstone-dominated sequences clearly result from high-energy shoal conditions in which deposition was too rapid for significant infaunal activity (periodic emergence is indicated by sporadic hematitisation), whereas the 'muddy' ironstones were formed by the settling of suspended sediment during lower-energy phases when bioturbation was active.

Coleby Mudstones Formation

The Coleby Mudstones Formation (Figure 16), although predominantly grey mudstone, contains siltstones in its lower part and two 'ironstone' members—the Pecten Ironstone and the Marlstone Rock. Numerous clay-ironstone concretions, some calcareous, are also present but, in contrast to the underlying formation, there are very few limestones. In the absence of a permanent exposure, the base of the Coleby Mudstones Formation is taken at a depth of 491.84 m in Nettleton Bottom Borehole, where calcareous mudstones rest on a hard bed equivalent to the highest Frodingham Ironstone. This slightly disconformable junction is readily identifiable in temporary exposures and borehole cores, and is easily mappable. The formation thins northwards across the district (Figure 14b and c) from about 100 to less than 50 m, due partly to sedimentary thinning and the development of nonsequences, and partly to erosion prior to deposition of the Redbourne Group.

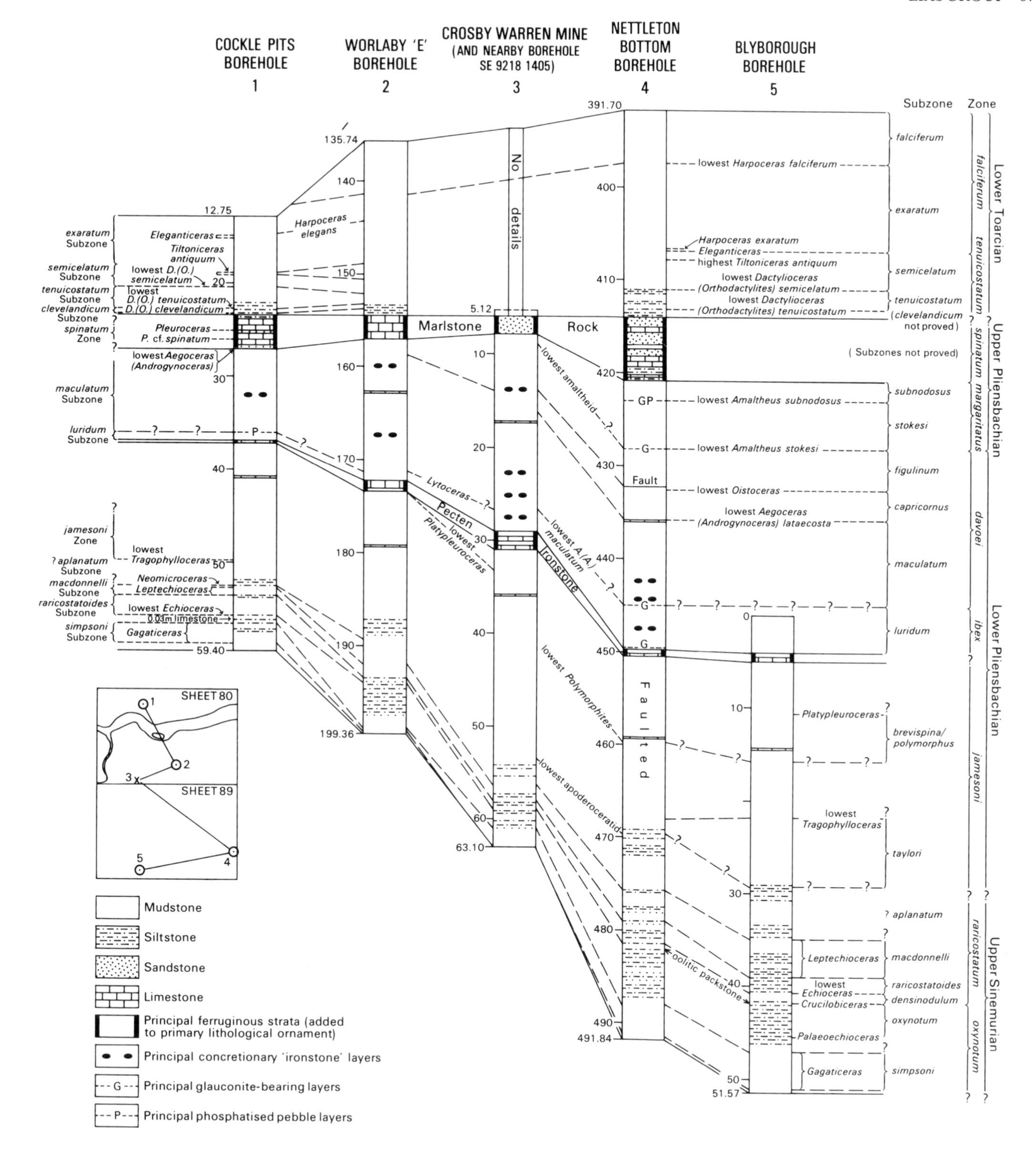

Figure 16 Sections, mainly from boreholes, of the Coleby Mudstones Formation. Depths are in metres

Beds below the Pecten Ironstone Member

During the survey the best exposure of these beds was in Crosby Warren Opencast Ironstone Mine [SE 920 127] but this is now closed and the face has been obscured. The beds thin progressively northwards from 46.57 m in Blyborough Borehole, through 32.1 m at Crosby Warren Mine and 24.4 m at Roxby Mine [SE 910 170] (see also Sellwood, 1972, fig. 9), to 22.32 m in Cockle Pits Borehole. Two informal lithostratigraphical subdivisions are recognisable, a lower one rich in siltstones, and an upper characterised by smooth-textured mudstones.

The siltstone-rich mudstones are best described from Nettleton Bottom Borehole. The lowest 4.79 m are laminated calcareous mudstones with shelly layers in the middle. Pyrite is abundant at the base, and is associated with small decomposed siderite concretions at Crosby Warren. Reworked goethite and kaolinite ooliths are also present in the lower beds, being more numerous (and generally flattened) near the base. The *simpsoni* Subzone is indicated by the range of *Gagaticeras* (see Getty, 1973), and includes part of the ranges of *Oxynoticeras* and *Promicroceras*. It continues up almost to the top of these mudstones in several well-logged sequences, and into the overlying siltstones in Cockle Pits Borehole. At Roxby Mine, however, Sellwood (1972) found *Crucilobiceras* and *Hemimicroceras*, indicative of the *raricostatum* Zone, less than 1 m above the Frodingham Ironstone (which here contains *simpsoni* Subzone ammonites); this illustrates the suggestion (see above) that the junction between the Frodingham Ironstone and the Coleby Mudstones may be diachronous.

The overlying strata, 4.95 m thick in Nettleton Bottom Borehole, are interbedded siltstones and mudstones with a few thin sandstones; bed contacts are indistinct and gradational. The siltstones are greatly disturbed by bioturbation and the mudstone–siltstone junctions are invariably masked by *Chondrites*. Shelly and coarse-grained lag beds are randomly distributed. In several borehole cores the basal layers of some of the hard coarse-grained beds contain sand-grade quartz grains and, more rarely, tiny quartz pebbles, suggesting that some fining-upward cycles are present. Sellwood (1972) interpreted these siltstones at Roxby Mine in terms of coarsening-upward cycles, but these are not evident in the Nettleton Bottom core. The presence of several layers of well-preserved goethite ooliths may indicate contemporaneous emergence in an adjacent region, and sandy layers in the middle of the sequence may reflect contemporaneous depositional breaks. Except for evidence of the *simpsoni* Subzone in the lowest siltstones in Cockle Pits Borehole, no clear subzonal indicators have been found in the sequence. Both *Oxynoticeras oxynotum* and *Palaeoechioceras* occur in the siltstones between 42 and 46.37 m in Blyborough Borehole, indicating that the *oxynotum* Subzone is over 3 m thick. The subzone is also probably present in Worlaby E Borehole where *Oxynoticeras* and *Promicroceras* occur, but is thin or absent in the north of the area.

The succeeding strata, 9.94 m thick in Nettleton Bottom Borehole, are characterised by numerous thin bioturbated siltstones in the lower part and weakly laminated mudstones in the upper part. At the base is a thin hard calcitic goethite-oolith packstone which, because it contains the lowest *Crucilobiceras* in Blyborough Borehole, is regarded as the base of the *densinodulum* Subzone, and may equate with a 0.03 m-thick limestone in Cockle Pits Borehole. This oolith packstone-limestone layer may represent a nonsequence, which is more marked in the north and which increasingly cuts out much or all the *oxynotum* and *densinodulum* subzones in that direction. A hiatus at this level is widespread, and may be related to a eustatic fall of sea level (Hallam, 1981, fig. 3). The base of the *raricostatoides* Subzone, indicated by the lowest *Echioceras*, is recognisable across the district, as also is the *macdonnelli* Subzone, which approximates to the range of *Leptechioceras* but includes also the range of *Neomicroceras*, proved at higher levels in Cockle Pits and Worlaby E boreholes. Sellwood (1972) has suggested that the upper limits of subzones are marked by the tops of the coarsest beds, but borehole evidence shows that some subzonal boundaries occur within coarser-grained beds and others within mudstones.

In Nettleton Bottom Borehole the highest 2.95 m of this siltstone-rich mudstone sequence comprises two hard siltstones separated by a thin laminated mudstone. Goethite ooliths, quartz grains and belemnites occur in the lower siltstone, and the upper one contains numerous *Gryphaea* and rhynchonellids. The equivalent beds in Blyborough Borehole are similar, but farther north the two hard siltstones are thin or absent. The presence of fresh goethite ooliths in the south may indicate a slight nonsequence in the north at this level, possibly cutting out the highest siltstones. The *aplanatum* Subzone is represented in these siltstones by the highest echioceratids (*Paltechioceras* and *Hemimicroceras*), but is poorly defined. An apoderoceratid 0.4 m below the top of the siltstone in Nettleton Bottom Borehole implies that the top of the siltstones in the south lies just within the *taylori* Subzone.

The smooth-textured mudstones forming the upper part of the beds below the Pecten Ironstone are 24 m thick in Blyborough Borehole, but are faulted in Nettleton Bottom Borehole. They consist largely of blocky mudstones with more silty beds, and are characterised by clay-ironstone layers and an abundance of *Pinna* in life position and as exhumed fragments.

The lowest beds, 3.4 m thick in Blyborough Borehole, are mainly silty and distinguished by many closely spaced, crushed shell-debris layers in which rhynchonellids and *Gryphaea* are major components; some of this debris is pyritised. Numerous layers of ferruginous concretions characterise the rest of these beds. The concretions are horizontally ovoid and contain calcite, siderite and ankerite. In borehole cores most are small buff nodules with indistinct edges but, at and near the surface, weathering and groundwater effects lead to development of the typical large clay-ironstone bullion-dogger concretions with hard outer shells (commonly embedded with pyrites cubes) encasing soft decalcified yellow-brown clayey kernels. Their origin is diagenetic (Hemingway, 1974, pp.167, 179–181), and many appear to have formed along finely bioturbated (commonly *Chondrites*-rich) silty beds. Pyrite is also present in the mudstones in finely speckled form, mostly associated with ubiquitous thin meandering burrows and trails and, at many levels in the smoothest mudstones, as framboidal clots up to 10 mm in diameter. *Pinna* is abundant in the mudstones, and crushed shell layers also occur, especially near the top. At Crosby Warren Mine, the latter form separation planes above the main layers of nodules and, in addition to *Pinna*, contain numerous *Pseudopecten*. A thin, brown-weathering, hard, tabular, silty, sideritic concretionary limestone lies 4.85 m below the Pecten Ironstone at this mine, and appears to be present across the whole opencast area; because of its well-developed joints it can be traced at outcrop by the brick-shaped blocks it yields.

In the southern part of the district, goethite ooliths are common in mudstones immediately beneath the Pecten Ironstone, and in Nettleton Bottom Borehole oolith-filled burrows penetrate down from the ironstone. The ooliths suggest a depositional break beneath the ironstone. With only sparse faunal evidence, however, the extent of the non-sequence cannot be assessed in detail, but it increases northwards because the ironstone slightly transgresses the underlying mudstones in this direction.

The smooth-textured mudstones fall entirely within the *jamesoni* Zone but, as in other areas, the overlapping ranges of *Polymorphites* and *Platypleuroceras* render differentiation of the *polymorphus* and *brevispina* subzones difficult (Ivimey-Cook, 1978, p.82; Gaunt et al., 1980, p.19). Finds of apoderoceratids (including *Apoderoceras*) and *Tragophylloceras* below the lowest *Polymorphites* in localities as far north as Worlaby E Borehole, suggest that the *taylori* Subzone is well represented. The base of the latter in Nettleton Bottom Borehole lies within the siltstone-rich mudstones, as noted above, and the subzone may be 10 m thick. At this locality the lowest *Polymorphites* coincides with a prominent concretionary layer that underlies the thin well-jointed limestone noted above (Figure 16). Within the district the lowest known occurrence of *Platypleuroceras* is taken to mark the approximate boundary between the *polymorphus* and *brevispina* subzones. In Blyborough Borehole this occurrence lies 5.6 m below the Pecten Ironstone, and is probably traceable throughout the district at the top of the main layer of large concretions. *Platypleuroceras* is also known from within the Pecten Ironstone, suggesting that the *brevispina* Subzone is represented in both the ironstone and the mudstone below. However Sellwood (1972, text fig. 9) reported ?*Jamesonites* from c.4.5 m below, and *Uptonia* from immediately below the Pecten Ironstone at Roxby Mine and concluded that the *jamesoni* Subzone was present below the Ironstone.

Pecten Ironstone Member

First named by Cross (1875, p.120) because of the abundance of *Pseudopecten equivalvis*, the Pecten Ironstone is oolitic and consists of grainstones and 'muddy' wackestones comparable to the Frodingham Ironstone, and with almost the same geographical extent as the latter. Weathering produces many of the same features, notably 'ginger-bread' enrichment following decalcification. At surface the 'muddy' wackestones are reduced to a yellow-brown 'mush', and the grainstones appear either as red-brown boxstones or as hard calcareous tufa-coated biosparite slabs. There is no consistent internal stratigraphy, much of the lateral lithological variation being lenticular. In the north, for example in Cockle Pits Borehole, the ironstone is less than 0.3 m thick, lime-rich and oolith-poor. It expands southwards to its maximum thickness of 2.2 m between Crosby Warren and Great Limber. The richest ore is in the west, probably a pointer to the original area of iron precipitation. Farther south the ironstone deteriorates as ore and thins appreciably; in Blyborough Borehole it is more silty, less oolitic and only 1 m thick.

In cores, much of the ironstone is a finely bioclastic, oolitic, siderite wackestone, mottled pale steely-grey to olive-green and khaki-brown. Most of the ooliths are of goethite (many are flattened), but berthierine ooliths are also present. Redeposited clasts and phosphatised pebbles of ironstone are common, and in the north, particularly in Winteringham Haven [SE 9353 2303] and Cockle Pits boreholes, pyrite has partly replaced the siderite cement. Oolith grainstone lenses are more calcitic, partly due to an abundance of comminuted shell debris and corroded belemnites. All the sedimentary features, including abundant cross-bedding and scour structures, testify to deposition (with some reworking) in turbulent marine conditions, followed by bioturbation. As with the Frodingham Ironstone, the Pecten Ironstone may have been deposited as a submerged longshore bar. It passes southwards beyond the district into thin calcareous siltstones. The ironstone has not been exploited as ore in recent times because of its high sulphur content. An average analysis of 12 samples (Whitehead et al., 1952, p.70) yielded 15.7 per cent iron, 26.4 per cent lime, 1.743 per cent sulphur and 10.95 per cent insolubles.

The precise age of the Pecten Ironstone is uncertain because of the scarcity of ammonites and the probability of reworking. Up to four subzones may be represented since the underlying beds, beneath a suspected nonsequence, are referable to the *polymorphus-brevispina* subzones, and the faunas of overlying strata indicate the highest (*luridum*) Subzone of the *ibex* Zone. Several specimens of *Platypleuroceras*, including *P. bituberculatum*, have been found in the ironstone, suggesting that the *brevispina* Subzone is represented. Specimens attributed to *Uptonia* and *Liparoceras* cf. *bronni* are recorded and, although neither is subzonally diagnostic, the former may imply the *jamesoni* Subzone and the latter suggests the *ibex* Zone. Dense accumulations of *Pseudopecten equivalvis*, many in current-stable positions, characterise the ironstone; alteration of nacreous aragonite to powdery calcite has rendered the shells fragile.

At the time of the survey the Pecten Ironstone was exposed in Roxby Mine [SE 912 170], behind the disused Dragonby Mine office [SE 903 146], throughout the Crosby Warren Mine and in the eastern wall of the former Wing Mine [SE 934 090 to 934 085]. Ploughed-up debris occurs on both sides of the railway at Santon Mine [SE 930 122] and in fields east of Yarborough Mine [SE 9345 1010 to SE 9340 0900]. A Romano-British site [SE 9040 1385] south of Dragonby stands on the ironstone outcrop, and there is evidence of smelting here. Two other old (but undated) smelting sites lie close to a small exposure [SE 9275 0347] south-west of Middle Manton. Exposures [SE 927 057] near Twigmoor Hall suggest that locally the ironstone is represented by no more than a horizon of nodules. Farther south the ironstone is virtually unexposed, although a weathered section was formerly visible in the railway cutting south-west of Kirton in Lindsey Station (Ussher, 1890, p.47).

Beds between the Pecten Ironstone and the Marlstone Rock Members

The finely graded, blocky, micaceous, variably silty mudstones occurring between the Pecten Bed and the Marlstone Rock may be up to 40 m thick in the south (the sequence in Nettleton Bottom Borehole is faulted), but they thin northward to less than 10 m in Cockle Pits Borehole (Figure 16). This is the most marked northerly attenuation within the Lias of the district, the correlative features suggesting that much of it is due to an unconformity beneath the Marlstone Rock. Numerous layers of clay-ironstone nodules are present; some of the larger nodules are septarian, with green drusy calcite, and the contained ammonites have iridescent shells and green calcite in their chambers, like their contemporaries in the Green Ammonite Beds of Dorset (Lang, 1936). The more prominent concretionary layers, mainly in the lower part of the sequence, assist in correlation, as also does a thin limestone in the middle of the sequence that passes southwards into a shelly layer. Exposures in the Frodingham Ironstone opencast mines are near-vertical, unstable and wet. A 'clay' pit [SE 9325 0045] near Cleatham (Howarth and Rawson, 1965), which contains the uppermost 12 m of these beds (including those referable to the *stokesi* Subzone), is the only good exposure. Nettleton Bottom Borehole, despite faulting, provides the most complete and detailed sequence, and a number of thin layers containing glauconite grains have been identified in it. They provide informal subdivisions for descriptive purposes and probably represent quasi-hardgrounds; some appear to equate with subzonal bases.

The glauconite grains in the basal layer of these beds are associated with degraded flattened berthierine ooliths, small pyritised gastropods and other pyritised masses. The succeeding 4.7 m in Nettleton Bottom Borehole comprise three fining-upward cycles which, at their most complete, pass up from silty mudstone through smooth blocky mudstone into weakly laminated mudstone. They contain discrete burrows (some pyritised) and more general bioturbation at several levels. These beds and their correlatives elsewhere contain *Beaniceras*. Although this genus can range higher, for example at Crosby Warren Mine where *B. luridum* has been found 0.6 m above an *Aegoceras (Androgynoceras) maculatum*, generally it can be ascribed to the *luridum* Subzone in the absence of any ammonites referable to the *davoei* Zone.

Resting on these beds in Nettleton Bottom Borehole is a thin silty layer containing abundant glauconite grains and pyritic masses. From the relationship of this layer to three prominent concretionary ironstone layers (Figure 16), its approximate stratigraphical level at Crosby Warren can be deduced, and there it yielded *Aegoceras (Androgynoceras) maculatum*, indicative of the *maculatum* Subzone. A single *Lytoceras* in a corresponding low position in Worlaby E Borehole may also lie at this approximate subzonal level and, although Cockle Pits Borehole gave no comparable diagnostic fauna, a phosphatised pebble layer (and a berthierine-oolith layer in Winteringham Haven Borehole) probably represent the same quasi-hardground as the glauconitic layer in Nettleton Bottom Borehole.

The succeeding 16.7 m in this last borehole contain at least seven cycles, most of which pass up from blocky to weakly laminated mudstones. Numerous concretionary ironstone layers, some septarian, are present and there is abundant pyritisation, commonly of burrows within or immediately below the coarsest beds. A thin pyritised shelly limestone in the middle of the sequence (which can be correlated with a thin shelly limestone at Crosby Warren, a thin ferruginous limestone in Worlaby E Borehole, and a yellow and green burrowed layer elsewhere) underlies the only significant development (about 3 m) of very silty mudstone at the base of any of these cycles.

The *maculatum* Subzone probably continues up to this shelly limestone and its correlatives farther north, but specimens of *Aegoceras (Androgynoceras) lataecosta* within the layer in Nettleton Bottom Borehole and just over a metre above it at Crosby Warren signify the *capricornus* Subzone, whilst, higher up in several sections, the lowest *Oistoceras* indicates the *figulinum* Subzone. In Nettleton Bottom Borehole the uppermost 3 m of these mudstones are intensely burrowed and contain several pyritic layers, and at Crosby Warren there is evidence of quasi-hardgrounds at a comparable level.

Another prominent glauconitic layer rests on these beds in Nettleton Bottom Borehole and contains the subzonal index *Amaltheus stokesi*, marking the base of the *margaritatus* Zone. It probably equates with a closely spaced sequence of quasi-hardgrounds at Crosby Warren, which lies about a metre below the Marlstone Rock and contains the lowest *Amaltheus* in that locality. In Nettleton Bottom Borehole the succeeding 5.2 m of bioturbated silty mudstones contain numerous pyritised glauconitic layers, particularly near the top, where much burrowing is evident. These beds are overlain by 1.7 m of mudstone devoid of glauconitic quasi-hardgrounds and containing *A. subnodosus*, the only known representative of the *subnodosus* Subzone in the district; also present here are numerous *Balanocrinus* columnals. At the top, pyritised burrows extend down from the Marlstone Rock. There is no evidence of the *gibbosus* Subzone of the *margaritatus* Zone in the district.

Marlstone Rock Member

The term 'Marlstone' originated in Dorset (Smith, 1817) for a 'Middle Lias' succession of 'sands' capped by a hard sandy limestone. The latter was referred to by Hull (1857) as a 'rock-bed' and subsequently both 'Marlstone' and 'Marlstone Rock-bed' were applied as names to the same unit by various authors. In the district the equivalent hard sequence includes a variety of lithologically contrasting beds, so the suffix 'bed' is dropped and it has been given member status. In southern Lincolnshire (Whitehead et al., 1952, p.97) the Marlstone Rock consists of calcareous sandstone overlain by berthierine oolite (the latter having formerly been worked for iron). This bipartite nature does not continue northwards into the district. Here, although informally referred to as ironstone, the Marlstone Rock is essentially a fine-grained calcitic sandstone with only a low iron content in the form of berthierine, siderite and goethite muds and ooliths. There is considerable lateral variation and a sharply defined unconformable base (Figure 16); the latter characteristic also exists in Dorset and at the base of the coeval Main Seam of the Cleveland Ironstone Formation in Yorkshire. The member is thickest in the east, being 7.62 m in Limber Borehole [TA 1239 0928] and rather less thick in the north, totalling 3.78 m in Cockle Pits Borehole. In the west it thins to 1.9 m between Roxby and Crosby Warren, and remains less than 2 m thick as far south as the Manton area. There is a slight increase to 2.44 m in the 'clay' pit near Cleatham, but farther south it is less than 1 m in a temporary exposure [SK 9361 9522] at Grayingham.

In Nettleton Bottom Borehole the Marlstone Rock, 7 m thick, contains 40 distinct beds. The lowest 2.47 m resemble parts of the Frodingham Ironstone, with thin bioturbated berthierinitic/sideritic, berthierine-oolith packstones and grainstones predominating. Interbedded with these are olive bioclastic/berthierine-oolitic, silty, siderite wackestones and thin calcitic sandstones containing wackestone intraclasts and phosphatised mudstone pebbles and ooliths. In contrast to the older ironstones in the Lias, numerous grains of glauconite are present, some of which may have been derived. The succeeding 2.94 m are mainly bioturbated sideritic/calcitic, fine-grained, silty sandstones, with less glauconite. Serpulids and 'nests' of *Tetrarhynchia tetrahedra* and *Lobothyris punctata* are conspicuous, as they are in the lower part of the Marlstone Rock of north-eastern Leicestershire (Hallam, 1962; Howarth, 1980). The highest 1.59 m comprise bioturbated berthierine-oolitic siderite siltstones and weakly cross-bedded sandy sparry limestones, with numerous olive glauconite grains near the top. The contact with the overlying silty mudstones appears to be gradational.

A similar and only slightly thinner sequence (5.16 m) is present as far north as Elsham Wolds Borehole [TA 0653 1333], but a short distance farther north-east, in Worlaby E Borehole, the Marlstone Rock is only 2.37 m thick. This may be due partly to absence of some of the higher beds and partly to a condensed conglomeratic basal layer. This basal layer characterises the western exposures; at Roxby and Crosby Warren mines, for example, there is a shelly basal conglomerate containing hard, flat, phosphatised mudstone pebbles up to 0.17 m long. The sequence in Cockle Pits Borehole has no basal conglomerate and resembles that in Elsham Wolds Borehole.

Surface oxidation and decalcification, producing 'ginger-bread' enrichment, have been greatly assisted by the presence of open vertical joints. Joint faces, fissures and bedding planes have become coated with tufa and/or iron-pan minerals, a process that has led ultimately to the development of boxstone structures around cores of less weathered rock. In some deeper excavations and temporary exposures the boxstone cores remain as unaltered hard bluish green rock with sparry poikilitic textures. Although fresh Marlstone Rock is of poor ore quality (the iron percentage ranges between 4.9 and 15.6, with as much as 36.3 per cent lime), it is sufficiently iron-enriched in its 'ginger-bread' state at outcrop to be siliceous ore. Unfortunately, since it does not form wide dip slopes, the area of 'ginger-bread' development is small and mining is impracticable. In historical times shallow pits have been dug in a few places, but only for smelting trials or small-scale production.

There is no evidence that any subzone of the *margaritatus* Zone is present in the Marlstone Rock, even in the south. Several occurrences of *Pleuroceras* cf. *spinatum*, for example in Nettleton Bottom and Cockle Pits boreholes, and in a railway cutting [SE 9041 3204] just north of the district (Cole, 1886, pp.24–25), indicate that most of the Marlstone Rock is referable to the *spinatum* Zone. The *apyrenum* Subzone has not been proved, but *P. cf. hawskerense* has been recorded by Howarth and Rawson (1965, p.262) near Cleatham, and indicates the *hawskerense* Subzone.

A few ammonites referable to the *tenuicostatum* Zone are known from the upper part of the Marlstone Rock or from just above it in this area. Their distribution indicates that its top is diachronous, possibly due to shoals being locally scoured so that parts remained as highs whilst reworked ferruginous sediments, mud and ammonites were deposited between the shoals. At Roxby Mine a specimen of *Protogrammoceras paltum* occurs in mudstones above the Marlstone Rock; this indicates a change in conditions in the earliest part of the *tenuicostatum* Zone (Howarth, 1980, p.645). At 31.62 m in Risby Warren Borehole [SE 9175 1330] (c.4 km away) *Dactylioceras* (*Orthodactylites*) cf. *clevelandicum* indicates the *clevelandicum* Subzone in calcareous siltstone. However, *Dactylioceras* sp. is found near the top of the Marlstone Rock at Cleatham (Penny and Rawson, 1969, p.197), *D.* (*O.*) *tenuicostatum* of the later *tenuicostatum* Subzone occurs on the top of the Marlstone Rock in Elsham Wolds Borehole and *D.* (*O.*) cf. *clevelandicum* and *D.* (*O.*) *tenuicostatum* were recorded near or at the top in Scotney Farm No. 2 Borehole [SE 9525 1664] between 36.7 and 37.2 m depth.

The fauna of the Marlstone Rock largely comprises brachiopods, bivalves and crinoids. The brachiopods include *Gibbirhynchia* sp., *Tetrarhynchia tetrahedra* and *Lobothyris punctata*, typical of the 'Midland Province' (Ager, 1956), and also the southern form *L. edwardsi*, recorded from near Cleatham and Ellerker [SE 992 297] (Ussher, 1890, pp.53–54).

The fossils clearly indicate a marine environment of normal salinity, the sandy lithology implying an appreciable detrital input, and the lenticular bedding and conglomeratic facies suggesting high-energy depositional conditions. None of these circumstances is conducive to berthierine precipitation, which must have taken place elsewhere. As with the Frodingham Ironstone (p.34) the berthierine is believed to have formed in lagoons fringing lateritic landmasses. Judging from the amount of iron in this part of the Lias throughout England, these landmasses must have been widespread. They probably lay to the west of the district because of the westward increase in the proportion of the conglomeratic facies. Constant reworking was probably responsible for the lack of internal stratigraphical consistency, and may also have produced the diachronous top postulated above on ammonite evidence. The Marlstone Rock does not appear to represent a regressive facies as envisaged by Howarth (1980, p.655), but to be a coarse basal transgressive facies.

Beds above the Marlstone Rock Member

The predominantly argillaceous rocks that succeed the Marlstone Rock range in thickness within the district from 28 m in a borehole [TA 0377 0639] east of Brigg to 10.57 m in Cockle Pits Borehole in the north (Figure 14c). Most of this northerly thinning is due to differential erosion prior to deposition of the Redbourne Group. In the thicker southerly sections an informal tripartite lithostratigraphical subdivision is possible into (in ascending order) silty mudstones, bituminous paper shales and laminated mudstones. Nettleton Bottom Borehole provides the best section for descriptive purposes.

The silty mudstones, 5.99 m thick in Nettleton Bottom Borehole, contain wispy-bedded siltstones in the lower half and a few thin bituminous paper shales in the upper part.

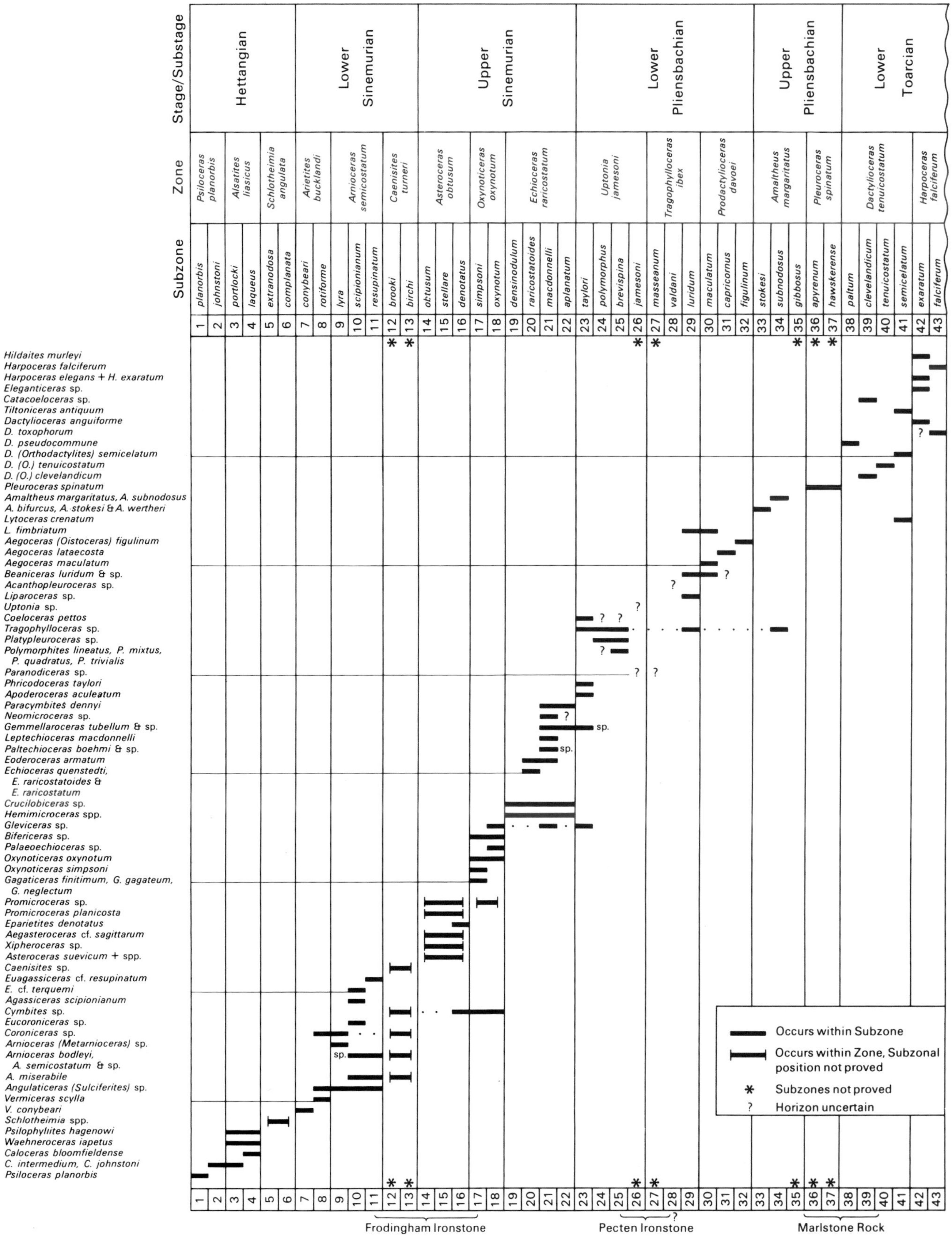

Stage/Substage: Hettangian; Lower Sinemurian; Upper Sinemurian; Lower Pliensbachian; Upper Pliensbachian; Lower Toarcian

Zone: *Psiloceras planorbis*; *Alsatites liasicus*; *Schlotheimia angulata*; *Arietites bucklandi*; *Arnioceras semicostatum*; *Caenisites turneri*; *Asteroceras obtusum*; *Oxynoticeras oxynotum*; *Echioceras raricostatum*; *Uptonia jamesoni*; *Tragophylloceras ibex*; *Prodactylioceras davoei*; *Amaltheus margaritatus*; *Pleuroceras spinatum*; *Dactylioceras tenuicostatum*; *Harpoceras falciferum*

Subzone: 1 *planorbis*; 2 *johnstoni*; 3 *portlocki*; 4 *laqueus*; 5 *extranodosa*; 6 *complanata*; 7 *conybeari*; 8 *rotiforme*; 9 *lyra*; 10 *scipionianum*; 11 *resupinatum*; 12 *brooki*; 13 *birchi*; 14 *obtusum*; 15 *stellare*; 16 *denotatus*; 17 *simpsoni*; 18 *oxynotum*; 19 *densinodulum*; 20 *raricostatoides*; 21 *macdonnelli*; 22 *aplanatum*; 23 *taylori*; 24 *polymorphus*; 25 *brevispina*; 26 *jamesoni*; 27 *masseanum*; 28 *valdani*; 29 *luridum*; 30 *maculatum*; 31 *capricornus*; 32 *figulinum*; 33 *stokesi*; 34 *subnodosus*; 35 *gibbosus*; 36 *apyrenum*; 37 *hawskerense*; 38 *paltum*; 39 *clevelandicum*; 40 *tenuicostatum*; 41 *semicelatum*; 42 *exaratum*; 43 *falciferum*

Hildaites murleyi
Harpoceras falciferum
Harpoceras elegans + *H. exaratum*
Eleganticeras sp.
Catacoeloceras sp.
Tiltoniceras antiquum
Dactylioceras anguiforme
D. toxophorum
D. pseudocommune
D. (Orthodactylites) semicelatum
D. (O.) tenuicostatum
D. (O.) clevelandicum
Pleuroceras spinatum
Amaltheus margaritatus, *A. subnodosus*
A. bifurcus, *A. stokesi* & *A. wertheri*
Lytoceras crenatum
L. fimbriatum
Aegoceras (Oistoceras) figulinum
Aegoceras lataecosta
Aegoceras maculatum
Beaniceras luridum & sp.
Acanthopleuroceras sp.
Liparoceras sp.
Uptonia sp.
Coeloceras pettos
Tragophylloceras sp.
Platypleuroceras sp.
Polymorphites lineatus, *P. mixtus*, *P. quadratus*, *P. trivialis*
Paranodiceras sp.
Phricodoceras taylori
Apoderoceras aculeatum
Paracymbites dennyi
Neomicroceras sp.
Gemmellaroceras tubellum & sp.
Leptechioceras macdonnelli
Paltechioceras boehmi & sp.
Eoderoceras armatum
Echioceras quenstedti, *E. raricostatoides* & *E. raricostatum*
Crucilobiceras sp.
Hemimicroceras spp.
Gleviceras sp.
Bifericeras sp.
Palaeoechioceras sp.
Oxynoticeras oxynotum
Oxynoticeras simpsoni
Gagaticeras finitimum, *G. gagateum*, *G. neglectum*
Promicroceras sp.
Promicroceras planicosta
Eparietites denotatus
Aegasteroceras cf. *sagittarum*
Xipheroceras sp.
Asteroceras suevicum + spp.
Caenisites sp.
Euagassiceras cf. *resupinatum*
E. cf. *terquemi*
Agassiceras scipionianum
Cymbites sp.
Eucoroniceras sp.
Coroniceras sp.
Arnioceras (Metarnioceras) sp.
Arnioceras bodleyi, *A. semicostatum* & sp.
A. miserabile
Angulaticeras (Sulciferites) sp.
Vermiceras scylla
V. conybeari
Schlotheimia spp.
Psilophyllites hagenowi
Waehneroceras iapetus
Caloceras bloomfieldense
C. intermedium, *C. johnstoni*
Psiloceras planorbis

Key:
- ▬ Occurs within Subzone
- ⊢⊣ Occurs within Zone, Subzonal position not proved
- * Subzones not proved
- ? Horizon uncertain

Frodingham Ironstone (subzones 12–17); Pecten Ironstone (subzones 25–28?); Marlstone Rock (subzones 35–37)

Figure 17 Principal ammonite occurrences in the Lower Jurassic (not all taxa cited in text have been included)

The basal siltstone, 0.79 m thick, is notably hard and calcitic. Elsewhere, for example in Scotney Farm No. 2 Borehole, layers of phosphatised pebbles occur in the basal bed. Several layers of sideritic/calcitic nodules occur in the succeeding siltstones, some with burrows filled with sand containing glauconite; scattered glauconite grains also occur in a few other layers and may indicate depositional pauses. Pyrite is abundant as small concretions, drusy nodules, cubes and discontinuous rims around fossils, being particularly associated in the weathered zone with thin ironpans, well-preserved but fragile dactylioceratids, and lignitic wood fragments. The upper half of the silty mudstones sequence is generally more micaceous, finer grained and less pyritic than the lower strata. Thin dark brown bituminous paper shales occur at four levels, suggesting a passage into the strata above. These beds retain their fissile character even when weathered. Calcareous concretions occur at outcrop but have not been proved in boreholes.

The subzonal position of the basal strata may be variable. The subzonal index *Protogrammoceras paltum* was found in the lowest overlying silty mudstones at Roxby Mine (Howarth, 1980), and *Dactylioceras (D.) pseudocommune*, which occurs in the *paltum* Subzone in north-eastern Yorkshire, is present in the mudstones above the Marlstone Rock in Worlaby E Borehole. At a number of localities, however, the succeeding *clevelandicum* Subzone is established near the base of these lower, siltstone-bearing beds by its index ammonite, though this was not found in Nettleton Bottom Borehole. The succeeding *tenuicostatum* and *semicelatum* subzones are both well represented by their index species. The former is also characterised by numerous long stout belemnites. The latter includes in its higher part the range of *Tiltoniceras* (including *T. antiquum*), with locally abundant *Pseudomytiloides* sp. and *Steinmannia* sp. (Figure 17).

The overlying bituminous paper shales thin northwards from 10.6 m in Nettleton Bottom Borehole to about 7 m along the Humber. In most places farther north they are partly or entirely cut out by erosion beneath the Redbourne Group. In Cockle Pits Borehole, for example, less than 5 m are preserved, and in South Cave Borehole [SE 9366 3230], just north of the district, the presence of *Tiltoniceras?* only 0.04 m below the base of the Redbourne Group indicates the highest part of the *tenuicostatum* Zone and implies (by comparison with Nettleton Bottom Borehole, see below) that the highest Lias present in the north is either in the topmost silty mudstones or in the basal bituminous paper shales. These paper shales, of similar lithology to the Jet Rock of north-eastern Yorkshire (Howarth, 1962; Hemingway, 1974), are finely laminated, with numerous fining-upward cycles that terminate in 0.1 to 0.3 mm shale laminae. The coarser laminae are mainly of angular detrital quartz. Mica, finely comminuted wood, pyrite hoppers, shell and phosphatic fish debris are all fairly common. Larger pieces of lignitic wood are present also, and in the north, for example in Cockle Pits Borehole, fragments of pyrite-veined jet occur in the lower beds. Where weathered, much of the sequence has a brownish hue and a bituminous odour.

At outcrop and in shallow boreholes (but apparently not farther east in deeper boreholes) layers of calcareous concretions occur at three more silty levels in the middle to upper part of the sequence. Some of the concretions are similar to the 'Whale Stones' in the Jet Rock of north-eastern Yorkshire (Howarth, 1962, p.385), referred to as 'extraordinary calcareous concretions' by Hallam (1962). They were formed by diagenetic segregation prior to compaction, producing bituminous calcitic siltstone nodules that preserve the original bedding. They were subsequently fractured, disorientated, and subjected to several stages of viscous cementation (Hallam, 1962). Some of the best examples were seen on the upper slopes [SE 9178 1774] of Roxby Mine during the survey.

In Nettleton Bottom Borehole the highest *Tiltoniceras* was found 0.15 m above the base of the bituminous paper shales, confirming the continuation of the *semicelatum* Subzone up to this level. In the same borehole *Eleganticeras* occurs 1.15 m higher and *Harpoceras exaratum* a further 0.39 m above, both proving the *exaratum* Subzone of the *falciferum* Zone. The base of this subzone in Worlaby E and Cockle Pits boreholes is, in the absence of the index fossil, less accurately known, but evidence from Nettleton Bottom, Brough No. 1 and Risby Warren boreholes suggests that it lies about a metre above the highest *Tiltoniceras*. The presence of *Harpoceras elegans* at higher levels in these boreholes indicates the upper part of the *exaratum* Subzone. Almost all of the bituminous paper shales are referable to this subzone, but the lowest *Harpoceras falciferum* at Nettleton Bottom places the base of the *falciferum* Subzone in the very highest laminae of the bituminous paper shales. The comparable level is proved in Worlaby E and Risby Warren boreholes, but in Cockle Pits and Brough No. 1 boreholes it is absent, having been cut out northwards by the unconformity at the base of the Redbourne Group.

The succeeding, laminated mudstones comprise medium grey, calcareous, silty mudstones, which are slightly bituminous, but much less so than the underlying paper shales. In Nettleton Bottom Borehole a thin bed of silty cone-in-cone limestone is present 2.9 m above the base, and layers of small calcitic siltstone nodules occur about 2.5 m higher. At several levels there are thin shell coquinae, many of which are pyritised. Along the outcrop, for example in Roxby Mine and at the south-western portals of Kirton in Lindsey railway tunnel [SE 970 001], the calcitic fissile nature of the laminated mudstones is clear even after many years of exposure. The mudstones weather to pale purplish grey flakes with a coating of sulphurous yellow material. They are progressively cut out northwards by the unconformity at the base of the Redbourne Group, but localised thickness variations occur, probably due to gentle folds being truncated by the unconformity. For example, they are 5.7 m thick in Nettleton Bottom Borehole and only slightly thinner in Worlaby E Borehole, but are thicker in Elsham Wolds (about 8 m) and Gokewell Strip boreholes [SE 9460 0957] (nearly 8 m), reaching a maximum within the district of 12 m in a borehole [TA 0377 0639] east of Brigg. Ammonites (see below) suggest that similar structural variations north of the Humber have preserved thin sequences of these laminated mudstones beneath the unconformity in a few places.

The subzonal index *H. falciferum* occurs throughout the laminated mudstones in Nettleton Bottom and Worlaby E boreholes and at a number of other localities. Whether the thicker sequences proved elsewhere, for example east of Brigg and in Elsham Wolds Borehole, extend upwards into the *bifrons* Zone is not known, for no faunas were collected. The *bifrons* Zone is believed to be present farther south near Lincoln (Ussher et al., 1888, pp.34–35; Trueman, 1918, p.106). Although in most places north of the Humber the *falciferum* Subzone is cut out, in at least two localities it is apparently preserved. From Melton Bottoms Borehole [approximate site SE 9752 2743] '*H. falciferum*' was reported in at least 1.53 m of mudstone (J Pringle, *in litt.*), and from a borehole [SE 9278 2911] near Ellerker (just north-west of Cockle Pits, where no *falciferum* subzonal ammonites were found) Dakyns et al. (1886, p.17) record '*Ammonites serpentinus*', probably indicating the *falciferum* Subzone. The possibility exists that even higher Lias strata are locally preserved just north of the district in view of the record of 'immense numbers of '*A.*' *communis* and '*A.*' *bifrons* from between Sancton and Haughton Hall (Tate and Blake, 1876, p.185). These ammonites (= *Dactylioceras (D.) commune* and *Hildoceras bifrons*) suggest the preservation of the *commune* and possibly even the *fibulatum* subzones of the *bifrons* Zone.

REDBOURNE GROUP

The Redbourne Group, comprising limestones, sandstones and argillaceous strata, thins northwards across the district towards the Market Weighton Structure from some 70 m to about 24 m, a thinning that is accompanied by lithological changes and by overstepping. For this reason the established stratigraphical subdivisions of the south-eastern Midlands are not appropriate to the district. A modified lithostratigraphical classification has been employed and, together with its chronostratigraphical correlations, is shown diagrammatically in Figure 18.

The macrofaunas collected from several horizons were extensive but appear to be of limited biostratigraphical value, so are not quoted in detail.

Northampton Sand Formation

The Northampton Sand (Figure 19) is a ferruginous sandstone of Aalenian age, present at outcrop south of New Cliff Farm [SE 9131 1874], near Winterton. The formation rests unconformably on the Lias, transgressing on to progressively older levels of the *falciferum* Zone northwards, and contains pebbles of Lias rocks, some of which have a phosphatic coating, at or near its base. Its top is locally cut into by channels filled with the overlying strata; in places these channels cut through the formation into the underlying Lias, for example along the Scunthorpe Bypass [SE 945 065]. In some southern areas and at depth to the south-east, the Northampton Sand is over 10 m thick. There is a gradual but irregular thinning to the north-east, largely the result of penecontemporaneous erosion attributable to activity on the Market Weighton Structure.

The Northampton Sand consists largely of hard, pale greenish, sideritic sandstone (commonly referred to as 'limestone' by drillers) which becomes yellow or red-brown goethitic sandstone on weathering. Nettleton Bottom Borehole provides the best unweathered sequence (Figure 19), which is 10.13 m thick and comprises numerous beds of bioturbated pale olive to yellowish grey, berthierinitic/sideritic,

FORMATION	MEMBER		PRESUMED AMMONITE ZONE	STAGE
Cornbrash	Upper Cornbrash		Macrocephalus	Callovian
	Lower Cornbrash		*discus*	Bathonian
Glentham	Blisworth Clay			
	Snitterby Limestone			
	Priestland Clay		*?aspidoides* *hodsoni*	
	Thorncroft Sands (with Nettleton Bottom Limestone (NBL))	NBL	*humphriesianum* *sauzei*	? ? ?
Lincolnshire Limestone	Cave Oolite	Hibaldstow Limestones	*laeviuscula*	Bajocian
		Kirton Cementstones (with Scawby Limestone (SyL)) SyL		
	Santon Oolite		*discites*	
	Raventhorpe Beds (with Ellerker Limestone (ElL) and Cleatham Limestone (CmL))	ElL CmL		
Grantham	(undivided)			Aalenian
Northampton Sand	(undivided)		*?murchisonae*	

Figure 18 Lithostratigraphical classification of the Redbourne Group, with chronostratigraphical correlation

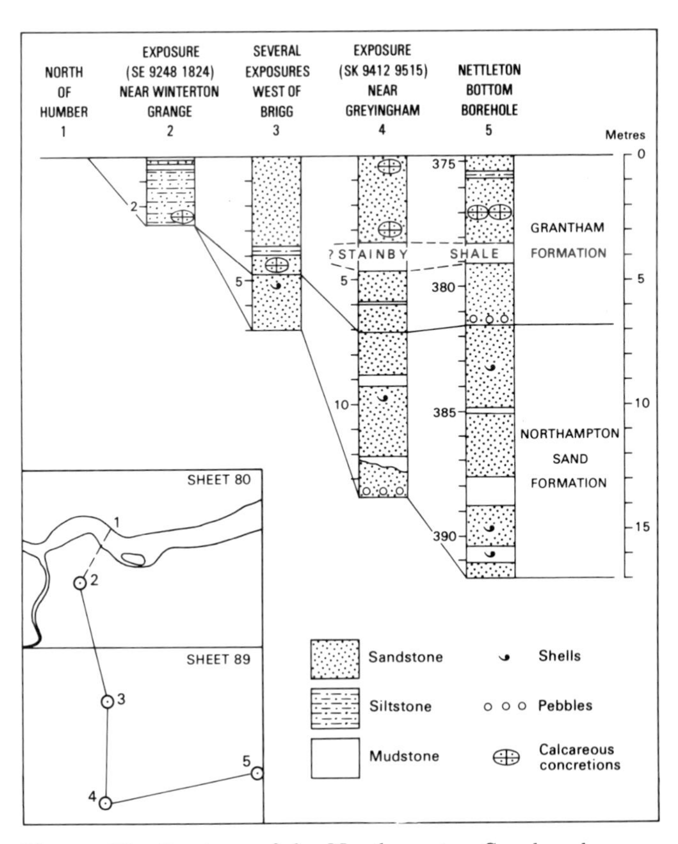

Figure 19 Sections of the Northampton Sand and Grantham formations

medium-grained, subangular-quartz sandstone containing degraded berthierine ooliths. At some levels white, powdery, kaolinite ooliths (with hard shiny outer coatings), blackish carbonaceous debris, finely divided mica, pyrite and shell lags are conspicuous; rare glauconite and small pebbles of sideritic sandstone also occur. Interbedded with the sandstone in the south are two beds of brownish grey, partly laminated mudstone. Numerous impersistent thin layers, laminae, chips, galls, pebbles and ripple-surface drapes of dark mudstone occur throughout. Bedding is generally planar-ripple to cross-laminated with no signs of slumping. In Worlaby E Borehole, 0.28 m of silty sandstone containing ooliths and shells, described by Richardson (1979, p.5) as Grantham Formation, is probably at least partly Northampton Sand.

Oxidation of the berthierinitic and sideritic rocks by meteoric waters has considerably changed the original sediments. At outcrop and at shallow depth, oxidation and decalcification have produced a friable iron-enriched 'ginger-bread' rock with boxstones, iron pans and tufa-coated joints. Alteration to goethite of the iron minerals is extensive and oxides have been redeposited on the surface of shells and in rock partings. Subsequent shell decalcification has left oxide moulds in which even the most delicate surface ornament is preserved. Some parts of the rock are completely leached of iron to form whitish orthoquartzite patches rimmed by a hard brown iron pan.

The lithology of the Northampton Sand in the district suggests deposition in shallow tidal water, marginal to land farther west. Desiccation cracks in the associated mudstones suggest shallow fringing lagoons. The environment may have ranged from freshwater to marine, and the berthierine ooliths and cements indicate that ephemeral enclosed iron-precipitating lagoons existed at times. Lagoonal beach and barrier backshore, foreshore and middle shoreface environments all seem to be represented, and the sequence is reminiscent of the Whitby facies of the Dogger Formation of north-eastern Yorkshire (Hemingway, 1974). Iron-rich oolith grainstones, packstones and wackestones, typical of the deeper-water transition zone (Hollingworth and Taylor, 1951), are not present north of Lincoln.

Apart from plant debris in the mudstones, most of the fossils are marine and generally occur as shell lag beds in the sandstones. Except for trace fossils such as *Diplocraterion*, the assemblages are detrital accumulations from more 'muddy' environments. The only ammonites are *Ludwigia* and *Dactylioceras*, the latter derived from the Lias. There is a rich fauna of bivalves with some bryozoa, brachiopods, gastropods, crinoids and echinoids.

In the southern Midlands the Northampton Sand Ironstone falls mainly within the *scissum* Subzone of the *opalinum Zone* (Hollingworth and Taylor, 1951, p.14), although the occurrence of *Leioceras opalinum* at Harlaxton [SK 895 311] near Grantham (Kent, 1975, p.314) indicates the presence of the preceding *opalinum* Subzone. In the present district *Ludwigia* has been found in the Northampton Sand near Scunthorpe suggesting the presence of the *murchisonae* Zone and implying contemporaneity of deposition with the Dogger Formation of north-eastern Yorkshire (Hemingway, 1974, pp.186 and 188) and with the Variable Beds of the Northampton Sand in Northamptonshire (Cope et al., 1980b, p.14).

The outcrops are commonly marked by ochreous sandy soil with 'ginger-bread' fragments containing empty shell moulds. Cambering, soil creep, hill wash and Head obscure them locally, but lines of springs, locally breaking through the Quaternary deposits, mark the base of the formation in some places. Exposures are rare; most are in small bluffs or faces above intermittent springs as at Manton [SE 9347 0270], or in stream-beds as at Beaulah Wood [SE 9580 0609] and Castlethorpe [SE 9804 0776].

Grantham Formation

Originally known as the White Sands (Sharp, 1870; Horton, 1977) and later called the Lower Estuarine Series (Jukes-Browne, 1885a), the Grantham Formation (Kent, 1975) consists largely of sandstones, siltstones and mudstones (Figure 19) and is of Aalenian age. It rests with slight unconformity on the Northampton Sand and, where this is absent, on the Lias; there is a concurrent northerly overstepping of successive beds in the formation, which thins in that direction. Differential erosion prior to the deposition of the Lincolnshire Limestone has further reduced its thickness so that it wedges out north of Winteringham and also to the north-north-east of Brigg. In the southern part of the district it is almost 8 m thick (7.11 m in a borehole at Grayingham [SK 9412 9515] and 6.81 m in Nettleton Bottom Borehole), but elsewhere south of the Humber it averages only 4 m. In some areas it is particularly thin or even absent, as near Castlethorpe [SE 9862 0737]; these areas may have been relatively upstanding during either deposition or subsequent erosion.

The base of the formation is normally marked by a thin layer of carbonaceous mudstone, in places no more than a veneer, or by a conglomerate of decalcified sandstone pebbles (some of which are completely pyritised) and hard, pale brown, partially oxidised, sideritic siltstone fragments. Above this layer the formation is characterised by pale grey, friable and locally unconsolidated quartzose sandstones and smooth-textured micaceous siltstones. In the thickest sections south of Scawby 1.0 m of medium grey, laminated, micaceous mudstone with silty laminae and one or two layers of greyish brown sideritic siltstone nodules are conspicuous about 2.4 m above the base. These strata have not yielded any marine shells, but they are presumed to equate with the Stainby Shale known farther south (Swinnerton and Kent, 1976, p.14). At Grayingham similar strata are 1.1 m thick, and in Nettleton Bottom Borehole 0.87 m thick; they have not been proved north of Scawby. Sandstones below the presumed Stainby Shale are commonly iron-stained but, apart from pyritic burrow-fills and a few layers containing glauconite, the iron has migrated from the underlying Northampton Sand. The coarser-grained beds are finely cross-laminated with isolated large subangular grains of quartz, and the finer grained are commonly ripple to flaser-bedded with numerous anastomosing 'muddy' laminae.

Above the presumed Stainby Shale the strata are generally bioturbated and almost white. Analysis of a random sample showed 83 per cent quartz and 12 per cent feldspar. Large-scale channelling, with associated syndepositional faulting, cross-bedding, mudstone-draped surfaces and rip-up flakes have all been recognised; the depositional environment was clearly an agitated one. The high carbonaceous content emphasises bedding in the finer-grained strata and thin plant-bearing clays probably mark the tops of channel sets, although subsequent scouring has left only truncated carbonised rootlets which extend down in places to 2 m. A major feature of the upper part of the formation is the presence of scattered calcite-cemented masses ranging from spheres 0.40 m in diameter to large blocks 5 m long by 1 m deep. They are similar to the 'pot lids' of Thompson (1921, p.224), and may be concretions or remnants of an overall calcite cement. North of Scawby 'pot lids' rest locally upon the eroded Northampton Sand or, as near Winterton Grange [SE 9144 1955], on the Lias.

The formation contains abundant terrestrial plant material, notably fusainised pinnules of the fern *Phlebopteris woodwardii* and pteridophyte and gymnosperm spores, implying proximity to densely vegetated land. *Botryococcus* algae suggest a freshwater imput, and the acritarchs *Caddasphaera halosa* and *Micrhystridium* spp. indicate marine influences, an implication supported by the presence of *Cardinia* in the highest sandstones. Although the assemblages do not provide precise dating, the formation was deposited before the fall in sea level which, at the close of Aalenian times, led to its erosion prior to deposition of the Lincolnshire Limestone.

A shallow-water depositional regime is indicated, with meandering channels that were temporarily abandoned and converted to vegetated marshes. Occasional marine incursions flooded the marshes and may have been responsible for the overall calcite cementation. Subsequent decalcification was accompanied by the deposition of iron salts along bedding surfaces.

Lincolnshire Limestone Formation

The Lincolnshire Limestone extends from near Market Weighton to Northamptonshire and eastwards at depth to the edge of the East Midlands Shelf. It is entirely of early Bajocian age. The succession within the district is transitional between the partly deltaic sequences of north-eastern Yorkshire (Hemingway, 1974) and the limestone-dominated sequences in the east Midlands (Ashton, 1980). Four members can be recognised south of the Humber, separated from each other by minor unconformities. In ascending order they are the Raventhorpe Beds, Santon Oolite, Kirton Cementstones and Hibaldstow Limestones. North of the Humber the Santon Oolite is absent, and both the upper part of the Kirton Cementstones (from the Scawby Limestone upwards) and the Hibaldstow Limestones pass laterally into the Cave Oolite Member (Figure 20).

The formation was deposited in fairly shallow marine water not far from land, and minor influxes of sand and silt suggest a marginal zone of transient muddy lagoons and oolite shoals. The effects of the Market Weighton Structure are reflected by northward depositional thinning and by erosional interludes. Slight earth movements and subsequent channelling have given rise to a haphazard pattern of thickness variation. In general, however, the greatest thicknesses are in two areas. One is indicated by the westerly bowed outcrop of Hibaldstow Limestones in the Kirton in Lindsey–Redbourne area, which apparently continues eastwards to Nettleton Bottom; the other lies around Haverholme [SE 950 125]. Thicknesses of over 26 m have been recorded locally in both these areas. The thinnest sequence recorded is 10.44 m in Brough No. 1 Borehole, where much of the Cave Oolite is absent (p.51).

Raventhorpe Beds Member

The Raventhorpe Beds Member consists of bioturbated, variably calcareous mudstones and siltstones containing upward-coarsening cycles, and also two thin limestones, the Cleatham Limestone and the Ellerker Limestone, which were formerly equated and named the Hydraulic Limestone. The Raventhorpe Beds Member of this account encompasses a greater stratal range than the Raventhorpe Member of Ashton (1975, fig. 3, in unit C), which was confined to the beds above the Cleatham Limestone.

The beds below the Cleatham Limestone consist mainly of mudstones and form a single upward-coarsening cycle. They are generally less than 2 m thick and their narrow outcrop is included with that of the Cleatham Limestone on the 1:50 000 geological sheets. Only four permanent exposures are known. On the southern side [SE 9469 0100] of the eastern Kirton Tunnel cutting these strata are less than 0.3 m thick. They comprise pale grey, calcareous, slightly micaceous mudstones, laminated or finely flaser-bedded with paler siltstones. Several whitish shell-grit layers are packed with thin-shelled bivalves and echinoderm debris, and the contact with the underlying Grantham Formation is generally sharp. There is, however, some burrowing across the junction, and thin neptunean dykes also occur. In other exposures, notably those examined in trenches near Grayingham [SK 942 952] (0.87 m thick) and along the Scunthorpe Bypass [SE 958 065] (up to 1.6 m thick), thin cavernous nodular iron-pans with associated calcareous tufa coatings and manganese staining lie immediately above the base and at higher levels. There are also numerous thin hard calcite-cemented shelly lenses of micaceous siltstone and fine-grained quartzose sandstone containing scattered irregular calcareous peloids, small carbonised wood laths and tiny orange specks (?after pyrite). The fossils, probably of *discites* Zone age, comprise a diverse microflora and fauna of dinoflagellate cysts, spores, foraminifera, brachiopods, gastropods, bivalves, ostracods, crinoids, echinoids and fish.

In Nettleton Bottom Borehole a mudstone 0.8 m thick is present at the base of the Raventhorpe Beds. It is separated from the Cleatham Limestone by 2.06 m of partly cross-bedded, shelly, bioturbated siltstones and quartzose sandstones that have not been recognised elsewhere.

The Cleatham Limestone is generally less than 1.5 m thick and thins northwards. It contains an abundance of fine shell debris and irregular ooliths. Foraminifera, tiny stunted or juvenile bivalves and gastropods, ostracods (referable to the early to middle parts of the *discites* Zone; Bate, 1980, p.8) crinoid columnals, echinoid spines and crushed shell fragments are tightly packed in silty, calcitic, micritic or sparry matrices. Irregular patches of mudstone are present also. In those layers that have the greatest shell concentrations (generally in the upper part of the limestone), thin-skinned calcite ooliths encapsulate complete shells, many of which are in part altered to aggregated pyrite. Sporadic peloids form a minor component. A calcite mudstone with a few shells is widespread at about the middle of the limestone; locally this is represented by softer mudstone separating two

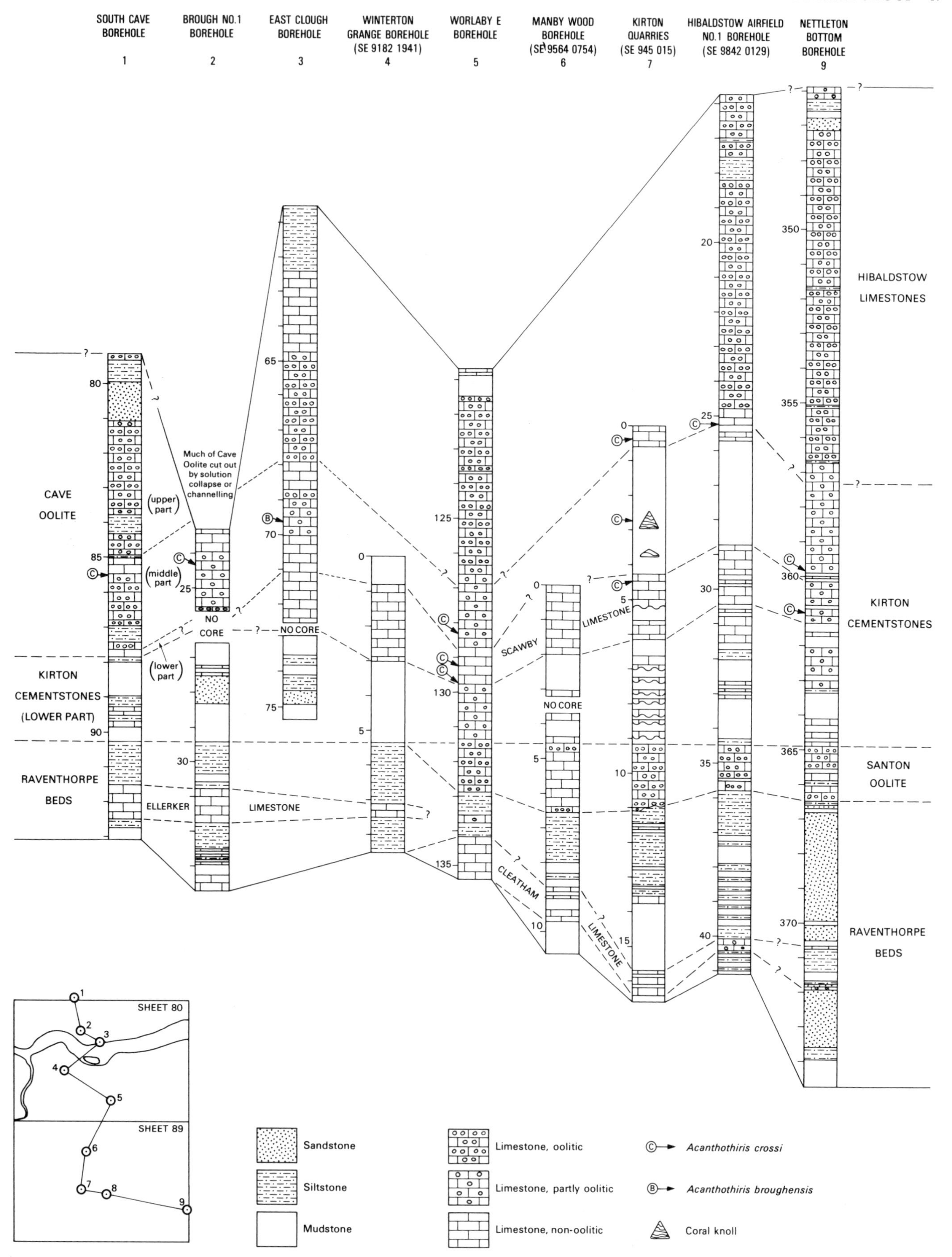

Figure 20 Sections of the Lincolnshire Limestone Formation. Depths are in metres

hard limestone layers. Thin interbedded silty layers contain much fine-grained quartz sand, and are associated with small carbon laths and with orange goethitic patches after pyrite and siderite. The tendency for the pyrite and siderite in the limestone to oxidise to orange-brown goethite is so common as to give its basal layers the appearance of an ironstone; similar ferruginous boxstones are present as far south as Low Santon. In Nettleton Bottom Borehole (Figure 20) much of the Cleatham Limestone consists of siltstone; the associated limestones are extremely sandy, and clasts of shelly limestone and calcite mudstone up to 0.4 m long are conspicuous.

Good surface exposures occur at the eastern portals of the Kirton Tunnel cutting (1.27 m thick), in a stream at Castlethorpe [SE 9804 0776] (0.67 m thick), and in quarries at Manton (1.45 m thick) and Gokewell Strip [SE 9415 0963]. The sequence at Manton has been described by Ashton (1975, fig. 3, units A and B), although differently interpreted. South of Kirton in Lindsey Airfield, the limestone is extensively exposed on the dip slope of the west-facing scarp capped by the Northampton Sand. At surface it is traceable northwards as a low brash-covered feature along the Lincoln Edge, and also around incised valleys to the east, almost as far as Winterton Grange [SE 913 189]. It is absent in the Coleby Mine section, close to Grange Farm.

The beds above the Cleatham Limestone (including the Ellerker Limestone) are divisible into a lower, mainly mudstone sequence that weathers to a topographical slack and an upper siltstone/sandstone sequence forming the lower part of the scarp capped by the Santon Oolite or, where the latter is absent, below the steeper slopes of the Kirton Cementstones.

The mudstone, 1–2 m thick, commonly contains scattered thin-skinned ooliths and tiny oysters in its lower part. Typically there is a bioturbated layer of coarse flaser-bedded siltstone with small shells at the base; some of the coarser patches are cemented by calcite. The mudstone is slightly micaceous and finely flaser-bedded, and contains numerous shelly patches. Chondritic burrowing is prominent and marked by pale calcite micrite infills. The best sections occur in the eastern cutting of Kirton Tunnel and in the Kirton lime (Figure 20) and Manton stone quarries, where hard flattish concretions and beds of calcite mudstone are conspicuous. Northwards from Winterton Grange one such porcellanous bed thickens to about 1 m. It was considered by Fox-Strangways (in Ussher, 1890, p.63) to be the northern continuation of the Hydraulic Limestone of Lincolnshire (renamed the Cleatham Limestone), but is now distinguished as the Ellerker Limestone (Gaunt et al., 1980, p.8, fig. 5). Unlike similar but thinner limestones, it is traceable north of the Humber as a subsidiary feature on the scarp below the Cave Oolite, and its type section is a waterfall [SE 9244 2967] in Ellerker Beck at Ellerker. It is correlated with the Hydraulic Limestone of the Howardian and Hambleton Hills, north of Market Weighton (Gaunt et al., 1980, p.29, fig. 14), now renamed the Gormire Limestone (Powell and Rathbone, 1983, p.372, fig. 2).

Fossils from the silty layer beneath the Ellerker Limestone are larger than in the underlying beds; they include disarticulated valves of *Gervillella, Parallelodon* and *Pseudotrapezium*. Burrowing forms such as myaceids, *Pholadomya, Protocardia* and crustaceans are present, along with encrusting serpulids, coral fragments, echinoderm and vertebrate debris which suggest transport from less 'muddy' environments.

The micaceous siltstone/sandstone sequence above the mudstone is generally less than 3 m thick and comprises up to nine upward-coarsening cycles. Thin dark grey mudstones, some no more than a veneer, at the bases of these cycles are succeeded sharply by coarse siltstone or fine-grained angular to subangular quartz sandstone. The upper boundaries of the mudstones show considerable burrowing, and speckled pyrite is associated with beautifully preserved trace fossils. Some pockets of comminuted shell grit have acted as centres of calcite cementation. The lowest cycles tend to be flaggy-bedded. Bioturbation is general, but the tops of some cycles are planar cross-laminated with carbonaceous micaceous laminae and are cemented by calcite. The highest beds show influxes of calcite mud, accompanied by fragmented shells, small solitary corals and micrite clasts containing poorly formed ooliths. The sequence appears to have been differentially eroded before deposition of the overlying Santon Oolite, and the topmost beds are penetrated in places by neptunean dykes. A feature of the northern succession, where the Santon Oolite is not preserved, is its marked iron staining, with some intervals weathering out as prominent goethite bands. Most of the shells in this siltstone/sandstone sequence are derived, but serpulids, '*Corbula*' and *Neomiodon* are indigenous. The trace fossils include *Chondrites*, *Thalassinoides* and *Zoophycus*. Lignitic fragments, some of large size, also occur, and the microflora comprises species of the dinoflagellate cysts *Nannoceratopsis* and *Sentusidinium*.

Santon Oolite Member

The Santon Oolite forms a small but prominent scarp at the top of the Raventhorpe Beds in the area south of Winterton, and several streams issue from its base. Its brash generally consists of yellowish coarse oolite with irregular, commonly flattish ooliths. Three oolitic limestone/calcite mudstone cycles can be recognised in places, mainly in the south of the district. Near its northern limit the Santon Oolite is predominantly coarsely oolitic, while to the south-east it thins and has a greater sand content. The limestone beds are notable for their large drusy cavities and irregular pink patches. The basal contact appears to be gradational, but this is the result of reworking and the member rests disconformably on an eroded surface of the Raventhorpe Beds, small clasts of which are incorporated in the basal limestones, particularly in the central part of the district. Thicknesses of over 3 m are recorded in boreholes near the southern boundary of the district; farther north the Santon Oolite thins to less than 2 m in the central area, probably being represented in Worlaby E Borehole (Figure 20) by much or all of the 1.37 m of thick argillaceous shelly limestone which Richardson (1979, p.6) considered to be a condensed deposit, and it gradually wedges out between Winterton and Winteringham. An easterly thinning is also evident in several parts of the district and it is locally absent near Appleby.

Good exposures exist in the Manton (Ashton, 1975, fig. 3, unit D) and main Kirton quarries, and a typical section is provided by Hibaldstow Airfield No. 1 Borehole (Figure 20).

The fossils include compound and solitary corals, a diverse fauna of gastropods and bivalves, some crinoids and rare echinoids. Wood fragments, some of considerable size, also occur, and south of the district at Greetwell, near Lincoln, the ammonite *Hyperlioceras*, indicative of the *discites* Zone, is known from the Santon Oolite (Kent and Baker, 1938, pp.168–170).

Kirton Cementstones Member

The Kirton Cementstones comprise up to 9.5 m of calcareous mudstones and thin limestones which, at outcrop south of the Humber, form a tripartite sequence of (in ascending order) interbedded mudstones and thin limestones, the Scawby Limestone, and mudstones containing coral mounds and locally a few thin limestones (formerly differentiated as the Kirton Cement Shale (Richardson, 1940; Kent, 1941). North of the Humber only the lowest of these subdivisions is recognisable, the Scawby Limestone and overlying strata having passed laterally into the lower and middle parts of the Cave Oolite (p.51). At depth in eastern parts of the district south of the Humber there is a similar passage of much of the higher part of the Kirton Cementstones into limestones, for example in Worlaby E and Nettleton Bottom boreholes (Figure 20).

The Kirton Cementstones are best exposed in the Manton–Cleatham area, particularly in the extensive Kirton quarries [SE 945 015] (Plate 1). Fortuitously, this is the area where the member is thickest. Lateral changes occur in the included limestones (except those near the base) and a wide range of lithologies, including thin siltstones and sandstones, can be present at any stratigraphical level. Consequently the interpretation of borehole records is difficult, and it is not alway possible to identify the Scawby Limestone in them, one of the reasons why the name Kirton Cementstones has been extended upwards from previous usage.

The beds below the Scawby Limestone are equivalent to the Leadenham Member of central Lincolnshire (Ashton, 1980) and also incorporate channels of that author's Cathedral Beds facies. The basal beds are calcareous silty mudstones which rest on an uneven surface of the underlying strata and in places contain calcite ooliths and small pebbles of Santon Oolite type. Thin, hard, shelly, oolitic limestones are present in some localities; one such bed can be traced from north of Roxby almost to Winteringham and, being

Plate 1 Lincolnshire Limestone in Kirton Quarries. The Santon Oolite at the base is overlain by the lower part of the Kirton Cementstones, comprising grey interbedded mudstones and limestones below and the Scawby Limestone above. A 14376

appreciably pisolitic in places, is superficially similar to the Santon Oolite. The main part of these beds consists of cyclic sequences of finely flaser-bedded, dark grey, calcareous, silty mudstones and pale grey, bioturbated, calcite mudstones. Diagenesis of the latter has produced up to fourteen highly irregular (0.1–0.6 m) layers of nodular porcellanous calcilutite, traditionally called 'cementstone'. Large dark-rimmed ooliths of irregular shape are a major component of the more laminated mudstones. They are less numerous and unevenly scattered in the calcilutites, in which they produce spotty textures. Quartz silt, shell debris and thin lenses of shelly, sparry limestone are common in the calcite mudstones. The micrite matrix is generally recrystallised to microspar, and tiny spar-filled shrinkage cracks within and around the margins of up to three generations of chondritic burrows are conspicuous. Finely disseminated pyrite occurs in both mudstones and micrites, and pyrite marks some of the burrow and sediment interfaces, but most commonly occurs as aggregate crystals in shell fragments. Much of the fauna comprises comminuted gastropods, ostreids and crinoids probably swept in from adjacent areas. Foraminifera and ostracods are also common, and there is a rich in-situ fauna of gastropods and bivalves. Dinoflagellate cysts (*Nannoceratopsis*) also occur.

The largest of the Kirton quarries [SE 943 015] exposes the southern flank of a slightly sinuous north-westerly-trending channel cut into interbedded mudstones and limestones and filled with coarse bioclastic limestones, siltstones and mudstones. The channel-fill is typical of the Cathedral Beds described by Ashton (1980, p.211) from the Lincoln–Metheringham area. On the northern side of the quarry a layer of shelly calcite mudstone nodules just below the transgressive base of the Scawby Limestone passes laterally into a wedge of hard thin-bedded shelly limestone that cuts down to within 0.27 m of the Santon Oolite. The individual beds of shelly limestone are separated by thin lenticular mudstones and grade into planar beds of calcilutite towards the middle of the channel. The fauna in the channel-fill is largely comminuted and represents a clear-water, low-diversity, lime-mud, coral/brachiopod/echinoderm community, contrasting with the environment of tidal, 'muddy', lagoonal conditions indicated by the enclosing strata. Boreholes to the east record the presence of siltstones, sandstones and oolites which may represent other channel fills.

North of the Humber the lower part of the Kirton Cementstones consists of up to 3.1 m of shelly calcareous silty mudstones with, in the middle, a variable sequence of calcareous siltstones, fine-grained sandstones and sandy limestones, seen in some of the Humber boreholes (Gaunt et al., 1980, p.5, fig. 5), Eastfield Quarry [SE 915 323] (Sylvester-Bradley, 1947; Bate, 1967) and a road cutting [SE 9188 3039] south of South Cave. The fauna includes *Ceratomya*, '*Corbula*', *Neomiodon, Pholadomya* and *Chondrites*.

The Scawby Limestone forms a prominent topographical feature at outcrop south of the Humber, with the village of Scawby on its dip-slope. At the top of the main working face in quarries in the Manton–Cleatham area a dark grey, brown-weathering, well-laminated, slightly bituminous oolitic–oncolitic mudstone, about 0.6 m thick, is succeeded by six closely spaced limestone layers. This mudstone is included in the Scawby Limestone because of its close cyclical relationship to the overlying limestones. Mudstones also occur between the limestone layers, that above the third lowest layer being the most persistent.

The range of lithologies in the Scawby Limestone differs little from those in the limestones in the underlying strata except that, in places, the lowest and highest layers are shelly pelletal limestones. Shelly calcite mudstones predominate; they contain scattered dark bluish grey elongate ooliths and some aggregated pyrite. In the Kirton quarries the Scawby Limestone is 2 to 3 m thick, being thinnest over the channel in the underlying strata. It remains at least 2 m thick as far south as Lincoln, and as far north as Appleby, but from there to the Humber it is 0.7 to 1.3 m thick, being 0.99 m in Worlaby E Borehole (Figure 20). Because of inadequate descriptions, it is difficult to identify in many borehole records. North of the Humber the lateral equivalent of the Scawby Limestone is recognisable as the lowest part of the Cave Oolite.

The fauna is dominated by bivalves, among which the frilly-profiled *Trigonia hemisphaerica* is common. Large solitary corals are prominent, and a few small *Acanthothiris crossi* also occur, the earliest known. At a locality [SK 9685 9348] near Atterby an ammonite of *Witchellia* type has been collected, suggesting the *ovalis* Subzone of the *laeviuscula* Zone.

The beds above the Scawby Limestone, distinguishable only south of the Humber, consist of up to 4 m of mudstones containing coralline knolls of patch-reef type and one or more mappable impersistent thin limestones. They form a slack below the scarp of the Hibaldstow Limestones and are well exposed in the southernmost of the Kirton quarries, where they are about 3 m thick. The mudstones are dark to medium grey, finely flaser-bedded or laminated, calcareous and silty, and contain thin shelly calcite-mudstone lenses. Many of these lenses were produced by micritic cementation around colonies of corals or large bivalves, and these occur, apparently randomly, at various stratigraphical levels. There is a tendency for the lenses to be stacked directly above each other, and to become progressively smaller and more pod-like upwards (the 'crog-balls' of Richardson, 1940, p.215), overlapping one another with little or no intervening mudstone. In this way small conical knolls up to 2 m in diameter and 1.5 m in height have been built up. Where exhumed along the outcrop these knolls stand proud as calcite sparry patches, and in some places, for example, near Broughton Crossroads [SE 9638 0665] they have been mistaken for burial mounds.

From the locality [SE 929 200] containing the most northerly knolls a thin pale to medium grey fissile limestone with a poikilitic texture and containing small amounts of detrital quartz, scattered small ooliths, abundant small shell fragments and a few burrowing structures is traceable almost to the Humber. Boreholes in this area suggest that much of the associated mudstone becomes increasingly calcareous and passes into limestones northwards. North of the Humber the beds above the Scawby Limestone continue as the middle part of the Cave Oolite (p.51), and they pass eastwards into limestones possibly corresponding to the Scottlethorpe Beds of central Lincolnshire (Ashton, 1980, fig. 7).

The mudstones above the Scawby Limestone yield foraminifera, rare corals, *Acanthothiris crossi*, some bivalves and a diverse fauna of ostracods. The limestones and coralline knolls contain a diverse bivalve fauna with compound corals, brachiopods, gastropods, the cirripede *Eolepas* and cidarids.

Hibaldstow Limestones Member

Between Lincoln and the Humber the Hibaldstow Limestones rest with slight disconformity on the underlying rocks. Along the outcrop the junction is marked by a hard basal pelletal calcite-mudstone which rests upon the mudstones of the Kirton Cementstones, but at depth to the east (where limestones are more common in the Kirton Cementstones) the junction is less clearly defined. The Hibaldstow Limestones are equivalent to the Upper Lincolnshire Limestone of older classifications and are northern correlatives of the Ancaster Beds and Great Ponton Beds of southern Lincolnshire (Ashton, 1980).

The Hibaldstow Limestones mainly comprise shelly oolitic limestones that accumulated in the barrier bar and tidal inlets of an outer-shelf carbonate shoal. Their thickness is extremely variable because of intra-Jurassic erosion. Where fully preserved they are up to 13.2 m thick, but thinning to the north and north-east is apparent, 6 m being present in Worlaby E Borehole (Figure 20). Erosion, channelling and solution prior to and during deposition of the overlying Thorncroft Sands have considerably reduced their thickness, and in places they are absent.

Two main rock types occur as fragments in soil on the Hibaldstow Limestones outcrop; a fine-grained nonoolitic limestone containing spar-filled shells (among which those of the small spiny brachiopod *Acanthothiris crossi* are distinctive) and a friable oolite. The nonoolitic limestone lies at the base of the sequence and its outcrop is therefore confined to the western edge of the Hibaldstow Limestones scarp and to small outliers farther west. It is only about 1 m thick. The abundance of spiny rhynchonellids led to its being formerly called the Crossi Bed and, despite scattered occurrences of these rhynchonellids down to and within the Scawby Limestone, this remains a useful informal term. The nonoolitic limestone is well exposed in the southern Kirton quarry and in a railway cutting near Station Farm [SE 9740 0390]. Elsewhere, most exposures of the Hibaldstow Limestones are of the oolite, though all are low in the sequence, for Quaternary deposits obscure the higher beds. In the past, several quarries exposed these higher beds (Ussher, 1890), but during the survey they could be seen only in a quarry [SE 9726 0075] to the west of the Hibaldstow – Redbourne road and in a nearby stream [SE 9774 0110].

The oolites comprise well-rounded, small to medium-sized, concentrically laminated, calcite ooliths with subrounded shell fragments, large deformed oncoliths and tiny micritic pellets of possible faecal origin, closely packed in a calcite spar matrix. Angular echinoderm and flat bivalve fragments, many of which are enveloped by a thin calcite skin, are also common; complete and articulated shells, other than of foraminifera and ostracods, are rare. Some upward-fining cycles are present. Angular quartz silt is present at many levels and in places the tops of the cycles are composed of sparitic siltstones which, weathered-out, form marly shell-sand partings in quarries along the outcrop. Rare low-angled cross-lamination is accompanied by much carbonaceous material which causes dark bituminous streaks in the quartz-rich matrices. Irregularly shaped pebbles of oolite with calcite rims of possible algal origin are common in the basal layers of some of the limestones, the tops of which have sporadic vertical silty burrow-like patches. In the Hibaldstow – Nettleton area a very distinctive oncolitic layer, which serves as a local marker, is present above beds that contain bryozoa and high-spired gastropods. Farther south the layer may correlate with upward-coarsening cycles in the Sleaford Member of southern Lincolnshire above the Great Ponton Gastropod Beds (Ashton, 1980, p.217); it has not been noted to the north. The oncoliths are large (3 to 25 mm), and so closely packed as to give the rock a conglomeratic appearance; some are finely burrowed and most show signs of plastic deformation. Some incorporate both ooliths and pellets, and an algal origin is suggested by the presence of tiny globules of spar resembling bird's-eye texture.

The macrofauna of the Hibaldstow Limestones is dominated by bivalves with some nerinellid gastropods. The rich ostracod fauna was noted by Bate (1967, p.124, beds 13 – 21) and suggests an age no younger than the *laeviuscula* Zone for the topmost beds. Foraminifera are also present, together with palynomorphs, and indicate a nearshore marine environment. Ammonites implying the *ovalis* Subzone of the *laeviuscula* Zone are known from correlatives of the 'Crossi Bed' farther south, and others indicative of the higher, *laeviuscula* Subzone have been recorded from correlatives of the upper part of the Hibaldstow Limestones (Ashton, 1980, p.221).

Cave Oolite Member

From the Humber northwards almost to Market Weighton the upper part of the Lincolnshire Limestone has long been called the Cave Oolite (Phillips, 1835). Up to at least 12 m thick, it forms a prominent west-facing scarp. Most past and present exposures, including numerous old quarries, suggest a sequence of massive oolites, but the Humber boreholes (Gaunt et al., 1980, pp.8 – 10, 29 – 30, figs. 5 and 14) have enabled appreciable variations to be identified and relationships with the equivalent strata south of the Humber and in north-eastern Yorkshire to be elucidated. The Cave Oolite is laterally equivalent to the Scawby Limestone, the upper part of the Kirton Cementstones and the Hibaldstow Limestones south of the Humber. For descriptive and correlative purposes it can be divided into lower, middle and upper parts.

The lower part of the Cave Oolite thins northwards from 1.37 to 0.25 m between East Clough and South Cave boreholes (Figure 20) and consists mainly of hard, pale bluish grey, fine-grained and locally thin-bedded oolith-poor limestone, silty near its base. Where weathered, as in a road cutting [SE 9188 3039] south of South Cave, it is orange-brown, apparently slightly ferruginous, partly nodular and bioturbated. In Eastfield Quarry [SE 915 323], just north of the district, it was called the 'Bottom Blue' by the quarrymen and was described by De Boer et al. (1958, pp.164 – 165, fig. 2, item B) as 0.48 m of hard, blue, compact, sparsely oolitic limestone; in the record by Bate (1967, pp.124 – 125, fig. 2, section 3) of this quarry it is probably represented by items 6 and 7—ferruginous limestone full of shells 0.28 m, succeeded by rubbly pellety limestone with bivalves (seen to) 0.25 m. The contained macrofauna, mainly molluscs, apparently increases in abundance northwards and, despite bioturbation, is largely epifaunal. This thin limestone, which rests on calcareous silty mudstones closely comparable to the lower part of the Kirton Cementstones, is considered to be laterally equivalent to the Scawby Limestone south of the Humber.

The middle part of the Cave Oolite is 3.39 and 2.56 m thick in East Clough and South Cave boreholes respectively. It changes northwards from dark grey, thin-bedded, fine-grained, silty and silt-laminated limestones with some loosely compacted, moderately sorted ooliths in the middle 0.5 m of the sequence, to closely compacted well-sorted oolite containing thin calcareous siltstones near the base and at the top. In Eastfield Quarry these strata are cross-bedded oolites containing some pisoliths, scattered shell debris and a few wood fragments, being mainly item C of de Boer et al. (1958, fig. 2) and items 8 to 11 of Bate (1967, pp.124–125). Most of the recognisable macrofauna comes from the oolith-poor strata, and although an epifauna that includes *Gervillella* and *Parallelodon* is dominant, some infaunal forms, including deep-burrowing myaceids, are present. *Acanthothiris broughensis* was found in the lower part of the sequence in East Clough Borehole and, on the evidence of the primary geological survey field map and papers by Sheppard (1901) and Drake and Sheppard (1909), the type locality of this species is probably a former small quarry [c. SE 9357 2764] in the grounds of Ruby Lodge, near Brough. *A. crossi* is recorded from slightly higher strata elsewhere in the area. The specimen of *Hyperlioceras* (*H. ruddidiscites* according to Senior and Earland-Bennett, 1974) from Eastfield Quarry, suggestive of the *discites* Zone, appears to have come from the middle part of the Cave Oolite, which the available evidence suggests is laterally equivalent to those upper beds of the Kirton Cementstones lying above the Scawby Limestone south of the Humber. The zonal implication of the *Hyperlioceras*, if it was derived from the middle part of the Cave Oolite, therefore appears to conflict with the finding in the Scawby Limestone of *Witchellia* sp. in the southern part of the district and of *Sonninia* (*Fissilobiceras*) sp. farther south (Ashton, 1980, p.212), both of which indicate the younger *laeviuscula* Zone.

The upper part of the Cave Oolite, considered to be the lateral equivalent of the Hibaldstow Limestones south of the Humber, is at least 7.5 m thick locally, apparently with a slight northward depositional thinning, but in a few places it is reduced or, as in Brough No. 1 Borehole, cut out entirely (Figure 20) by subsequent channelling or solution (p.54). Except locally near the top, it consists mainly of closely compacted, moderately to well-sorted oolites and contains shell debris, especially in 'marly shell-sand' partings. In Eastfield Quarry these oolites form the highest 4 m of limestone and exhibit low-angle cross-bedding dipping to the north-east. In a few more southerly boreholes, notably East Clough, thin, shelly, fissile, fine-grained, silty, poorly or nonoolitic limestones occur at the top of the sequence. In South Cave Borehole 0.55 m of slightly shelly and sparsely oolitic calcareous siltstone is present in the lower part of the sequence, and thin calcareous siltstones and mudstones occur near and at its top in several of the Humber boreholes. These more argillaceous strata contain a distinctive macrofauna of nerineids, *Pseudomelania* and bivalves, and also numerous ostracods (Bate in Gaunt et al., 1980, appendix I, p.34).

Resting on the massive oolites at South Cave are (in ascending order) sandstone 1.11m; laminated, micaceous, slightly calcareous siltstone with scattered ooliths and fragmented shells 0.59 m; oolitic limestone 0.20 m. The sandstone and siltstone are reminiscent of the local Thorncroft Sands and these uppermost beds either pass northwards by interdigitation into the Thorncroft Sands (see below) or should be considered part of that member. In view of the evidence of erosion and channelling elsewhere at this stratigraphical level it is not inconceivable that in the latter case the 0.20 m of oolitic limestone and the shells and ooliths in the siltstone could represent debris derived from the top of the Cave Oolite elsewhere.

Glentham Formation

In the southern part of the district the strata comprising the Glentham Formation are similar to those occurring farther south in Lincolnshire, but variations to the north have necessitated changes in nomenclature (Figure 21). The lower, more arenaceous, part of the old Upper Estuarine 'Series' is differentiated as the Thorncroft Sands, and the upper argillaceous part and succeeding argillaceous equivalents of the lower part of the Great Oolite (or Blisworth) Limestone are called the Priestland Clay. The higher part of the Great Oolite Limestone becomes the Snitterby Limestone, and the highest member retains the name Blisworth Clay. North of Brigg, where the Snitterby Limestone cannot be recognised, the Priestland Clay and the Blisworth Clay are not distinguishable. Much of the formation is nonmarine, having been deposited in deltaic, fluvial, lacustrine and marsh environments, and most, if not all of it is considered to be the southerly equivalent of the Gristhorpe Member of the Cloughton Formation together with the succeeding Scalby Formation of north-eastern Yorkshire (Gaunt et al., 1980, p.30, fig. 14).

Thorncroft Sands Member

The Thorncroft Sands consist largely of white, brown, grey and locally black friable sandstones that are extremely variable in thickness—over 18 m in places but absent in others. This variation is attributable to several factors, including an overall southward depositional thinning, localised occupation of deep solution hollows and channels in the underlying Lincolnshire Limestone, lateral passage into Priestland Clay in some areas, and pre-Callovian erosion in the most northerly part of the district.

The best evidence of Thorncroft Sands occupying solution hollows in the underlying Lincolnshire Limestone is in Eastfield Quarry [SE 9158 3241], where the base of the sands cuts down through 1.8 m of Cave Oolite to produce a channel-like profile 8 m wide. However, the thin bedding, laminations and small-scale cross bedded layers within the Thorncroft Sands are parallel to the junction with the Cave Oolite, showing that they are not part of a channel fill but, having originally been deposited on a level surface, have subsided into a hollow at the top of the Cave Oolite, presumably due to solution of the limestone by acidic groundwater. Three large masses of Thorncroft Sands within the quarry, with their bases down to 3 m below the general top of the Cave Oolite, probably also occupy solution

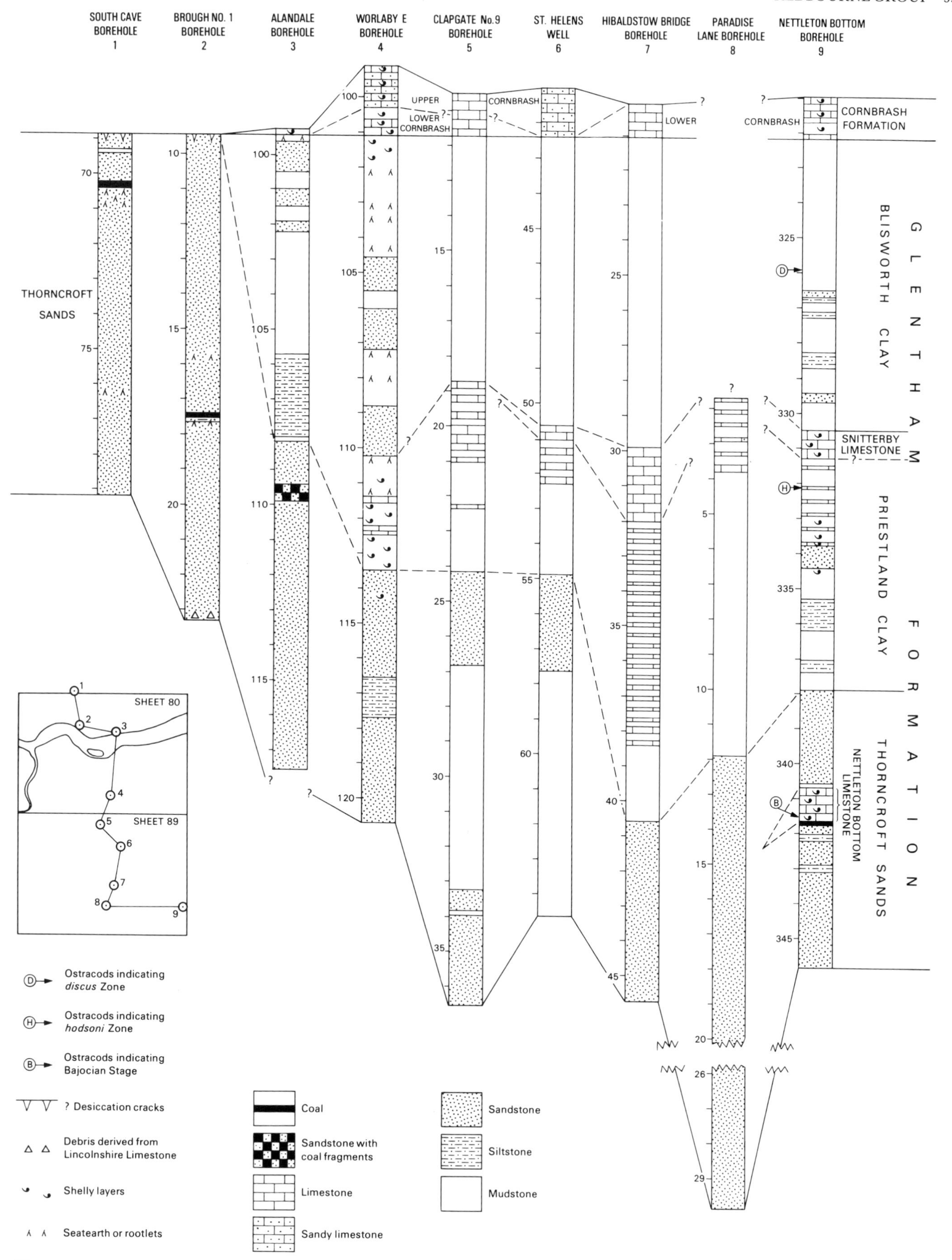

Figure 21 Sections of the Glentham and Cornbrash formations. Depths are in metres

hollows. Boreholes such as Brough No. 1, where thick Thorncroft Sands rest on the middle part of the Cave Oolite, also suggest solution hollows.

In some localities farther south, however, the Thorncroft Sands clearly occupy channels incised into Lincolnshire Limestone. The most obvious example is in an area south-east of Risby Warren [SE 945 130] where they rest on Kirton Cementstones in a channel aligned north-west to south-east. Abnormally thick Thorncroft Sands, locally in excess of 18 m, were proved in Paradise Lane [SK 9892 9878], Clapgate No. 9 [SE 9778 1159] (Figure 21) and Appleby Station [SE 9514 1283] boreholes.

From Welton Waters 'A' Borehole [SE 9544 2476] (Gaunt et al., 1980, p.10, fig. 6), on the northern side of the Humber, to Scabcroft [SE 9724 1810] there is an elongate area where Thorncroft Sands are thin or absent; on the evidence of boreholes this is apparently due to lateral passage into the Priestland Clay. This general area may have been the site of a long-lasting lake, lagoon or interdistributary floodplain. A similar situation is apparent farther south in Clapgate No. 9 Borehole and in St Helen's Well [TA 0138 0772] (Figure 21).

The base of the Thorncroft Sands is locally marked by an irregular thin layer of blue-black carbonaceous silty mudstone that has been likened to a lake-bottom deposit (Aslin, 1968), but may be a solution residue. At outcrop this mudstone softens to a sticky 'sooty' clay, and an iron pan (the Ironstone Junction Bed of Judd, 1875; Thompson, 1930) normally marks its base; it is particularly prominent south of Snitterby where the Hibaldstow Limestones had previously been removed. The succeeding 1 to 2 m of strata commonly contain debris derived from the Lincolnshire Limestone, including limestone pebbles, shell fragments, pisoliths and ooliths, as seen in Eastfield Quarry and in some of the Humber boreholes; Richardson (1979, p.7) also recorded quartz pebbles and thin coarse-grained sandstone layers in the basal beds in Worlaby E Borehole.

The remainder of the Thorncroft Sands sequence consists of two main lithologies. The more extensive comprises grey to yellow, locally bioturbated, massive, parallel or cross-bedded, fine-grained, patchily iron-cemented but commonly friable quartzose sandstones, with plant debris, rootlets and thin vitrainous lignitic coals in places. Small ridges between Snitterby and Waddington may owe their origin to impersistent hard beds of cemented sandstone. The other lithology consists of thinly bedded and laminated, locally sphaerosideritic and pyritic sandstones, siltstones and mudstones with seatearths and thin coals. There are scattered sideritic concretions, and plant debris is common. Although largely quartzose, the sandstones contain some feldspar and heavy minerals such as garnet and zircon, with lesser amounts of staurolite, chloritoid, apatite, rutile and tourmaline.

Much of the bedding in the Thorncroft Sands is planar-laminated, but scour-and-fill trough cross-bedding and ripple-bedding are present, particularly in the upper part, and from these structures and others, such as burrows and rootlets near the top, Brasier and Brasier (1978) regard the member as having been deposited in a marshy fluvial environment. This interpretation is compatible with the available fossils and also with the plant debris, coal layers and lenses, coal debris-rich sandstones, and seatearths. Fossils collected from the basal part of the member in Nettleton Bottom Borehole include foraminifera, gastropods and marine bivalves; dinoflagellate cysts from 0.6 m above the base are referred to the *sauzei* Zone by Riding (1987).

In Nettleton Bottom Borehole 1.04 m of silty, flaser-bedded limestone, here named the Nettleton Bottom Limestone, were encountered approximately in the middle of the Thorncroft Sands. Burrows truncated by this limestone may indicate a depositional break. The limestone is divisible into five layers, possibly representing storm or wave-induced cycles. It is abundantly shelly, containing gastropods and marine bivalves. Leiospheres, acritarchs and dinoflagellate cysts were recorded by Bradshaw and Penney (1982, p.121). Ostracods from the lower part of the limestone (Bradshaw and Bate, 1982) indicate the *polita* ostracod Biozone and are comparable with those from the Scarborough Formation of north-eastern Yorkshire. This suggests that the Nettleton Bottom Limestone is at least partly referable to the *humphresianum* Zone and is of late Lower Bajocian age, a conclusion supported by dinoflagellate cysts (Riding, 1987).

In Worlaby E Borehole '*Corbula*', *Cuspidaria ibbetsoni* and '*Neomiodon*' were found 0.86 m below the top of the Thorncroft Sands, suggesting brackish or freshwater deposition. The Thorncroft Sands exposed in a ditch near Redbourne Church [SK 9780 9895] yielded well-preserved miospores, and *Botryococcus*. Exposures in Redbourne Park [SK 9781 9936] and on Hibaldstow Airfield [SE 9806 0048] yielded spores as well as marine dinoflagellate cysts, including *Dichadogonyaulax* sp. and *Nannoceratopsis gracilis*, an association implying a probable late Bajocian age (Fenton and Fisher, 1978). No dinoflagellate cysts were found in the upper part of the Thorncroft Sands (above the Nettleton Bottom Limestone) in Nettleton Bottom Borehole (Riding, 1987). These upper strata are considered by Bradshaw (1978) to be of Upper Bajocian age. How much, if any, of the Thorncroft Sands extend up into the Bathonian Stage is at present unknown.

Priestland Clay Member

South of Brigg up to 11 m of mainly argillaceous strata, here termed the Priestland Clay, are present between the Thorncroft Sands and the Snitterby Limestone. The best exposure [SK 9839 9762], near Priestland Covert, reveals 6 m of yellow-green mudstones and blue-green siltstones. Complete sequences are recorded only in a trench [SK 9833 9427] near Snitterby and in boreholes, notably Nettleton Bottom (Figure 21).

The Priestland Clay can be divided into lower and upper parts for descriptive purposes. The sharp basal contact is level, except where a few blackish clay-filled channels cut down into the Thorncroft Sands. The lower part is cyclic. Many of the cycles begin with marine strata and pass up through brackish to freshwater (probably lacustrine) strata. Their tops show evidence of emergent or near-emergent conditions. Erosion at the base of each cycle and rapid lateral variations in lithology preclude precise correlations between sections. A complete cycle consists (in ascending order) of marine limestone, brown and greenish berthierinitic, illitic and kaolinitic mudstones, grey and purple-grey shaly mudstones, and pale grey siltstones. Carbonaceous debris, rootlets, local pseudobreccias and sideritic or calcitic concretions in the upper part of the cycles point to the presence of soils and imply the establishment of coastal swamps. The strata strongly resemble the Rhythmic Sequence (Aslin, 1968) farther south, although *Lingula kestevenensis* has not been found in the district. A locality [SK 980 940] near Thorncroft Farm yielded numerous marine bivalves, some of which are also tolerant of brackish conditions. The accompanying ostracods include *Fastigatocythere juglandica*, *Micropneumatocythere quadrata* and *Schuleridea trigonalis*. They are indicative of the *hodsoni* Zone in the late Bathonian. This age is confirmed by an ostracod assemblage from Nettleton Bottom Borehole that comprises *Lophocythere propinqua*, *L. transversiplicata* and *Progonocythere triquetra* with *Glyptocythere guembeliana* and *M. quadrata*.

The upper part of the Priestland Clay consists predominantly of fossiliferous marine mudstones, with some barren, probably less-marine beds. Its base is drawn where shells become consistently abundant, commonly coincident with the presence of patches of quartz sand in the mudstones. Crushed, coarse-ribbed rhynchonellids, including *Kallirhynchia sharpi*, are particularly common close above this base, suggesting a possible correlation with the Passage Beds of Northamptonshire. However, only in Nettleton Bottom Borehole, where the two parts of the member are separated by a quasi-hardground, is there a distinct and sharp junction. The fossiliferous mudstones are generally bluish grey and the barren mudstones more greenish. Thin siltstones occur within the mudstones, and calcitisation of these has produced hard patches and cone-in-cone lenses which lead to some of them being referred to as limestones in certain borehole records. The carbonate is generally dissolved at surface, although some discontinuous thin bands have been mapped between Waddingham and Pyewipe [SK 981 985]. Crushed fossils abound but only the larger oysters are well preserved. Ostracods from Nettleton Bottom Borehole are indicative of the highest part of the *hodsoni* Zone. A small coral bioherm lying just above the base of these beds near Snitterby [SK 9805 9396] yielded ostracods implying a position close to the *hodsoni*–*aspidoides* zonal boundary. The bioherm contains a wide variety of marine dinoflagellate cysts associated with miospores (mainly bisaccate pollen grains), and the alga *Botryococcus*.

Snitterby Limestone Member

From Brigg southwards a thin, shelly, flaggy, sandy limestone with a few shaly layers separates the Priestland Clay from the Blisworth Clay. The limestone forms a minor scarp and a long dip-slope from Hibaldstow southwards, and is here called the Snitterby Limestone. It is equivalent to the highest part of the Great Oolite (or Blisworth) Limestone of the Midlands. No clear evidence of the limestone has been found either north of Brigg or, except in Nettleton Bottom Borehole, at depth to the east, and in these directions it apparently passes into calcareous argillaceous strata in the highest part of the Priestland Clay. Exposures of the limestone are fewer and poorer than during the original geological survey (Ussher, 1890, pp.82–84). Incomplete sections less than 2 m thick were seen during the survey in several deep drains, notably near East Field [SK 9875 9714], on the roadside [SK 9775 9705] north of Waddingham and at localities around Hibaldstow Airfield [SE 9860 0050; 9875 0180], Cherry Farm [SE 9900 0290] and Island Carr [SE 9945 0650].

The limestone is abundantly fossiliferous, with *Liostrea hebridica* and *Modiolus imbricatus* being predominant. Crushed rhynchonellids and terebratulids are patchily distributed. Other fossils include compound corals (e.g. *Isastraea* and *Trigerastraea*), a variety of marine and some more brackish water bivalves, and crinoid debris which indicate a dominantly marine environment. In Nettleton Bottom Borehole (331.47 m) the foraminifera *Ammodiscus* and the ostracod *Glyptocythere* were accompanied by large numbers of dinoflagellate cysts.

Blisworth Clay Member

The Blisworth Clay is a dominantly argillaceous sequence separating the Snitterby Limestone from the Cornbrash. It is highly variable, due mainly to differences in channel depths at the base and also to pre-Cornbrash erosion. From about Hibaldstow Airfield northwards, but not to the south, there is increasing evidence of erosion prior to the deposition of the Blisworth Clay, possibly due to a marine regression in late *hodsoni*–early *aspidoides* zonal times.

The Blisworth Clay is poorly exposed, having been seen chiefly in ditch sections noted on the 1:10 000 maps. Only one complete section [a gas pipeline trench at SK 9957 9305] has been recorded along the outcrop where it is 6.5 m thick. This, with cores from Worlaby E and Nettleton Bottom boreholes, has been used to interpret other records.

Boreholes indicate that the basal beds contain channels filled with dark grey highly carbonaceous mudstones and/or quartz sands. In a trench near Thorncroft, lacustrine mudstones, in which thin discontinuous silty sideritic lenses are present, lie at the base of the member. At the same locality a channel filled with sandstone is incised through several mudstone beds, and similar sandstone-filled channels may be the cause of isolated masses of sandstone along the outcrop, for example near Gander Farm [SE 9920 0200]. Most of the overlying strata comprise green or brown, waxy, illitic mudstones showing widespread evidence of chemical and textural diagenetic change. They contain numerous listric surfaces and abundant carbonaceous debris. This part of the member is similar to the cyclic mudstones in the lower part of the Priestland Clay, with thin marine layers and laterally impersistent beds of laminated pale siltstone, dark sooty mudstone and pure quartz sand. It suggests deposition in freshwater swamps and brackish 'muddy' creeks. In some sections there is an upward change to darker grey shelly mudstones, in which crushed rhynchonellids and small oysters are conspicuous. The junction with the overlying Cornbrash appears to be gradational in some localities, but some of the boreholes show a much sharper boundary, with the Cornbrash resting directly upon unfossiliferous green listric mudstones. This sharp contact and the absence below it of the uppermost shelly mudstones seen elsewhere may indicate some pre-Cornbrash erosion.

The fossil assemblages in the Blisworth Clay reflect marine incursions into a brackish and freshwater environment. This is clearly shown by the presence of marine dinoflagellates cysts, foraminifera, bivalves and ostracods on the one hand and brackish-water bivalves and brackish and freshwater ostracods (especially at the top of the sequence) on the other. The foraminifera and ostracods suggest a *discus* Zone age.

GLENTHAM FORMATION (UNDIVIDED)

In the absence of the Snitterby Limestone north of Brigg, the Priestland Clay and Blisworth Clay members cannot be satisfactorily distinguished, and their northern equivalents are shown on the geological maps as Glentham Formation undivided. They comprise a mainly argillaceous sequence that thins generally northwards, being about 12.4 m in Worlaby E Borehole but expanding locally to at least 15 m in the area between Scabcroft [SE 9724 1810] and Welton Waters 'A' Borehole, where the sequence extends down to include argillaceous strata probably equivalent to the Thorncroft Sands (p.54). It thins more markedly north of the Humber and cannot be proved north of Brough. In a few boreholes there is evidence that more calcareous, and locally more arenaceous strata occur impersistently in the middle of the sequence, and it seems probable that these are lateral equivalents of the upper part of the Priestland Clay, or the Snitterby Limestone, or both.

The lower part of the sequence, underlying the more calcareous strata where the latter are detectable, consists mainly of grey to black, partly silty or sandy mudstones. In Welton Waters 'A' Borehole, where these mudstones rest directly on Cave Oolite and enclose small fragments of oolite and derived shells at the base, they are grey to mauve grey, partly silty and contain thin cross-bedded lenses, laminae and burrow fills of white fine-grained sandstone and also, locally, abundant carbonaceous plant debris. Thin cross-bedded siltstones, and to a lesser extent sandstones, are interbedded with the mudstones in places, more commonly in northern localities. Fossil evidence is sparse. Some of the mudstones in a borehole [SE 9716 1716] west of Saxby All Saints are described as 'shelly', and palynomorphs, including dinoflagellate cysts, occurring in siltstones just above the Thorncroft Sands in Welton Crossing Borehole [SE 9552 2613] suggest brackish conditions and a Bathonian age. These strata vary considerably in thickness, from 0 to 9 m. They were seen in a few small exposures east of The Follies [SE 9586 1281], which revealed greenish and black carbonaceous mudstones and some pyritic sandstones, lithologically not unlike the lower part of the Priestland Clay farther south.

The middle, more calcareous part of the sequence contains calcareous mudstones and thin argillaceous limestones, locally sandy; it is abundantly fossiliferous and described in some borehole records as 'shell marls'. In Worlaby E Borehole, Richardson (1979, p.7, fig. 5) recorded 3.25 m of 'Calcareous Beds', largely comprising greenish poorly laminated calcareous mudstones with various bivalves, some of which suggest a brackish environment. They rest here directly on Thorncroft Sands. Near the top of these calcareous strata there are several thin nodular 'ironstone' layers, some 'seatearth-like' rootlet layers and increasing numbers of quartz grains. At outcrop, green 'oyster-rich' calcareous sandstone, possibly partly decalcified, is exposed in a dyke [SE 9666 1112] near Clapgate Farm and, on the evidence of fragments in the soil, a thin flaggy shell-bearing limestone forms a minor west-facing scarp feature [SE 954 161] near Mickleholme. The northern extent of these calcareous strata is uncertain. North of the Humber, in Welton Crossing Borehole, 0.11 m of slightly fissile mudstone about 3.5 m above the Thorncroft Sands contains dinoflagellate cysts, spores, pollen, serpulid tubes, *'Corbula' hebridica*, *Meleagrinella?*, mytilid and other bivalve fragments, and ostracods suggestive of marine deposition and Bathonian age (R J Davey and R H Bate, in Gaunt et al., 1980, p.12). This mudstone may, on the evidence of the shelly fauna, represent the calcareous strata occurring farther south.

The upper part of the sequence, although largely argillaceous, contains thick sandstones in places, especially towards the east and north. Included are the 'Upper Sandy Beds' of Richardson (1979, pp.7–8, fig. 5)—thickly interbedded, poorly cemented, fine-grained sandstones and slightly carbonaceous grey mudstones with rootlet layers suggestive of seatearths. The succeeding beds in Worlaby E Borehole are largely green mudstones, with grey and brown patches in the lower part, but becoming mainly dark grey upwards. They contain plant debris, rootlets, sparse fish debris and, in the upper part, scattered bivalves, mainly *Neomiodon* sp. and *Entolium* sp. The thin shelly limestone originally included in these beds by Richardson (*ibid*) is now recognised as being Lower Cornbrash. To the north, the interbedded sandstones occur mainly in the upper part of the sequence (Gaunt et al., 1980, pp.10–12, fig. 6). In Low Farm Borehole dinoflagellate cysts, spores, pollen and fish scales, which according to Dr R J Davey are suggestive of brackish conditions and Bathonian age, were found in the mudstones. Rootlet layers suggestive of seatearths were recorded from these strata in some of the other Humber boreholes, particularly at the top of the sequence where the presence also of shell-filled burrows and possible desiccation cracks suggest a long-lived land surface. The only exposures of note are in deep drains around Thornholme Priory [SE 9650 1238], which reveal bluish green and mauve, locally calcite-cemented, mudstones, 'sooty' grey mudstones containing concretions, and some pyritic fine-grained sandstones. Some of the concretions are calcite-siderite, pseudoseptarian and discoidal; they contain fish debris and wood fragments, and exhibit onion-skin weathering. There is a gradual northern thinning of these upper strata from over 9 m in Worlaby E Borehole to 4–5 m at the Humber, with further attenuation towards Brough, beyond which no argillaceous strata attributable to the Glentham Formation have been proved.

Cornbrash Formation

The shallow-water marine limestones of the Cornbrash are generally less than 2 m thick and rest on the slightly eroded top of the Blisworth Clay. The formation comprises two members, the Lower Cornbrash of Bathonian age and, unconformably above, the Upper Cornbrash of Callovian age. The two members are lithologically distinct and characterised by different fossil assemblages. The Lower Cornbrash consists mainly of shelly calcite mudstones and limestones, and the more widespread Upper Cornbrash is commonly sandy and conglomeratic and locally includes calcitic sandstones. Erosion affecting both members produces significant differences between individual localities; in some places only the Lower Cornbrash is present, in others only the Upper Cornbrash. Both members are present, although not completely, at most localities south of Redbourne Hayes [SE 9937 0005], but the Upper Cornbrash is absent over the Askern–Spital Structure in the extreme south.

At outcrop the Cornbrash forms a low scarp and, despite its limited thickness, a long dip-slope. As far north as Gander Farm [SE 9940 0233] the soils derived from the Cornbrash are charactistically red-brown and contain abundant fossil debris. Drains and irrigation ponds occasionally provide good fossiliferous exposures, for example near Woofham Farm [SK 9948 9856], at River Head [SE 9948 0075] and near Gander Farm [SE 9942 0231]. Exposures around Thornholme [SE 9652 1253] resemble dry stone walls due to irregular splitting of the tufa-coated flaggy sandstones. The most northerly outcrops [SE 958 164, SE 958 168] lie just east of Mickleholme, but borehole records (Gaunt et al., 1980, p.12, fig. 7) show that the Cornbrash continues to the northern side of the Humber, and derived Cornbrash debris is present locally, at least as far north as South Cave, in neptunean dykes at the top of the Glentham Formation and also in the basal part of the Kellaway Beds.

The fullest sequences of the Lower Cornbrash occur in the south near Atterby; exposures are practically confined to the area south of Hibaldstow. The best section seen was in a trench [TF 0001 9277] in the extreme south near Beck Farm, Atterby, which was cut through the Lower Cornbrash and showed sharp disconformable contacts with the underlying Blisworth Clay and overlying Kellaways Beds. Five beds are present as follows: argillaceous limestone with shell debris (decalcified to clay near top) 0.45 m, on oolitic limestone 0.30 m, on silty mudstone 0.10 m, on silty mudstone with rhynchonellids 0.10 m, on shelly sandy limestone with rhynchonellids and oysters 0.40 . Fossils collected from these beds along the outcrop as far north as Sand Hills [SE 9935 0230] show that both of the Lower Cornbrash brachiopod zones, the *Cererithyris intermedia* and *Obovothyris obovata* zones of Douglas and Arkell, 1928) indicative of the Bathonian *discus* Zone, are present.

Good exposures of the Upper Cornbrash occur in drains on Appleby Carrs [SE 9706 1248], and also near Thornholme [at SE 9654 1223], where 0.7 m of bioturbated sandstone overlies 0.41 m of Lower Cornbrash. The best section of the Upper Cornbrash, however, is in Worlaby E Borehole, where it totals 1.25 m and overlies Lower Cornbrash which was assigned to the Blisworth Clay by Richardson (1979, pp.7–8). Here, as commonly elsewhere, the basal part of the Upper Cornbrash contains a conglomerate of Lower Cornbrash pebbles. The youngest strata containing a Cornbrash fauna are preserved near Haverholme (Ussher, 1890, p.89).

Three of the Humber boreholes proved Upper Cornbrash of a less sandy facies (Gaunt et al., 1980, p.12, fig. 7). In Low Farm (South Ferriby) Borehole on the southern edge of the estuary the member comprises 0.70 m of greenish grey bioclastic limestone containing unbroken shells concentrated in distinct lenticular layers. On the northern edge, East Clough Borehole proved a similar bioclastic limestone only 0.14 m thick but underlain by 0.41 m of dark grey calcareous mudstone which also contains an Upper Cornbrash shelly fauna concentrated in layers. Farther east along the northern edge of the estuary the Upper Cornbrash is represented by only 0.18 m of similar shell-layered dark grey calcareous mudstone in Alandale Borehole.

The Cornbrash is well known for its extensive brachiopod and bivalve faunas, which indicate the widespread development of marine conditions. The Bathonian ammonite zonal index *Clydoniceras discus* has not been found north of Sudbroke in central Lincolnshire, but the Lower Cornbrash brachiopod zone faunas characterised by *Cererithyris intermedia* and *Obovothyris obovata* are recorded in the district together with *Kallirhynchia yaxleyensis* and abundant bivalves. There are also corals (*Chomatoseris, Isastraea*), gastropods and echinoids. The Upper Cornbrash yields the brachiopods *Obovothyris stiltonensis*, *Ornithella subcalloviensis* and *Rhynchonelloidea inflata*, some gastropods, many bivalves, *Macrocephalites (Dolikephalites) typicus* and belemnites. The *siddingtonensis* brachiopod Zone, the lowest in the Callovian, is, however, known only around Redbourne.

ANCHOLME CLAY GROUP

In the eastern Midlands the strata above the Cornbrash Formation and below the Spilsby Sandstone Formation are predominantly argillaceous and are up to 300 m thick. They have traditionally been divided, in ascending order, into the Kellaways Beds, Oxford Clay, Ampthill Clay and Kimmeridge Clay. In the Fenland, Gallois and Cox (1977) differentiated the more silty West Walton Beds between the Oxford Clay and the Ampthill Clay and, using a combination of faunal and lithological characters, these authors (Gallois and Cox, 1976, 1977; Cox and Gallois, 1979, emend. 1981) established standard bed-numbered sequences for the West Walton Beds and for the higher parts of the succession. These standard sequences have been identified in some cored boreholes within the district, notably Nettleton Bottom Borehole (Wilkinson, 1983; Penn et al., 1986; Thomas and Cox, 1988).

In order to put the entire succession on a formal lithostratigraphical basis it has been designated the Ancholme Clay Group, and within it three mainly arenaceous formations are distinguishable (Figure 22), namely the wide-ranging Kellaways Beds Formation at the base and, at successively higher levels, the Brantingham Formation and the Elsham Sandstone Formation, which are restricted to the northern and central parts of the district respectively. The argillaceous parts of the succession, although much thicker, contain lithological variations of insufficient magnitude to warrant any formal lithostratigraphical subdivision, and they remain undifferentiated in this respect.

Biostratigraphical zonation of the Ancholme Clay Group is based on ammonites and the zones are used chronostratigraphically to define stages. Three stages are recognised, in ascending order the Callovian (which extends downwards to include the Upper Cornbrash in the highest part of the Redbourne Group), the Oxfordian and the Kimmeridgian (Figure 22). The base of the Upper Jurassic, formerly equated in Britain with the base of the Callovian, is now placed at the base of the Oxfordian (Cope et al., 1980b, pp.1–2).

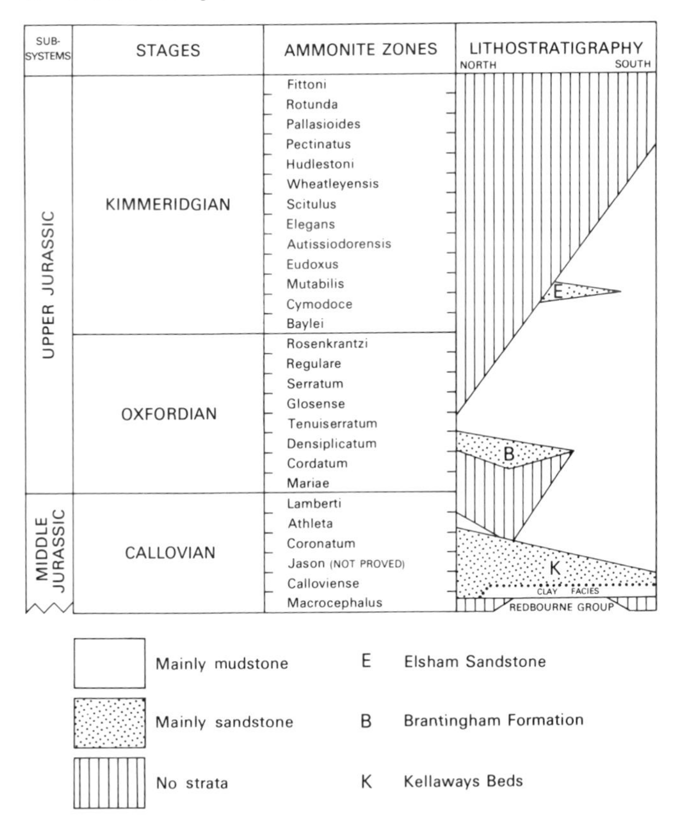

Figure 22 Chronostratigraphical correlation of the Ancholme Clay Group

Figure 23 Occurrences of selected ammonite taxa in the Ancholme Clay Group of the district. Detailed biostratigraphy of the Middle Oxfordian to Upper Kimmeridgian from Gallois and Cox (1976, 1977) and Cox and Gallois (1979, emend. 1981)

The mudstones comprising most of the Ancholme Clay Group vary from pale to dark grey, from very silty to non-silty, from strongly calcareous to noncalcareous, and from homogeneous and massive with a blocky fracture to laminated and fissile. The fossil content varies in diversity, abundance and preservation. Although different combinations of lithological and fossil characteristics occur, four basic types of mudstones are recognisable.

(i) Pale to medium grey, calcareous, silty mudstone with blocky fracture; phosphatic nodules present locally; shelly fauna scattered to locally abundant, in places occurring as concentrations of fragments, some of which are worn, bored and encrusted; burrows and bioturbation common; plant debris, mainly lignitic wood fragments, generally present but sparse; few of the fossils pyritised.

(ii) Pale to medium grey, calcareous, non-silty (or only slightly silty) mudstone with blocky fracture; shelly fauna diverse but generally scattered and not abundant, with some shells in burrowing position; trails and burrows present but not common; some fossils pyritised.

(iii) Medium to dark grey, noncalcareous (or only slightly calcareous) mudstone; silt, if present, confined to weakly developed laminae that produce slight fissility; shelly fauna diverse and fairly abundant, commonly concentrated in distinct layers and including iridescent ammonites; serpulids, trails and burrows generally abundant; fossils commonly pyritised.

(iv) Dark grey, brown and (rarely) black, noncalcareous, non-silty, fissile mudstone; shelly fauna limited in diversity and abundance, and consisting mainly of nektonic taxa and lingulids; fish debris and trace fossils locally abundant; some fossils pyritised. Some of these mudstones, together with a few of type iii above, are bituminous and have an oily smell.

Gallois and Cox (1976; 1977) and Cox and Gallois (1979) have described comparable types of mudstone in the Upper Jurassic of Fenland.

Siltstones are uncommon in the group except in the Brantingham Formation and in a thin but wide-ranging sequence not far above it. Sandstones are virtually limited to the three named formations. There are numerous borehole references to thin limestones some at least of which are concretions, and difficulties of correlation suggest that most are impersistent. The lithologies of these limestones are influenced by the lithology of the host rock. In the Kellaways Beds and the Brantingham formations the limestones are almost all sandy or coarsely silty, being in effect strongly calcite-cemented sandstones and siltstones. Elsewhere, limestones are largely confined to the beds between the Brantingham and Elsham Sandstone formations. A few are associated with coarsely silty strata and some have formed along concentrated shelly layers, but the majority appear to be layers of finely crystalline concretions, the 'cementstones' of some authors. Some concretions have shelly nuclei and some contain greenish calcite-filled septarian cracks. Most occur within pale to medium grey silty and/or calcareous mudstones (types i and ii above), but smaller and more ferruginous carbonate concretions occur in the medium to dark grey mudstones (type iii above). An abundant and diverse marine macrofauna is present, in which the ammonites are the most significant for biostratigraphical purposes (Figure 23).

The lithology and fossil content indicate warm fully marine depositional conditions in a region that, except for a few transient and mainly localised episodes, received 'muddy' sediment only. In the absence of substantial contemporaneous tectonic activity, this marked change from the conditions in which the underlying Redbourne Group was deposited implies a major eustatic sea-level rise early in the Callovian, for which there is evidence elsewhere in the world. Even so, wood fragments, saurian remains, phosphatic nodules and other features suggest that land was not too distant, and that it was well vegetated and possessed a mature low relief.

Despite its thickness and the width of country across its outcrop, the Ancholme Clay Group is rarely exposed within the district. It weathers to a clay, commonly with diagenetic selenite crystals, and forms low ground largely covered by Quaternary deposits. Except near the Humber, where it is extensively worked for cement-making, it has no present economic value. As a consequence, evidence from scattered

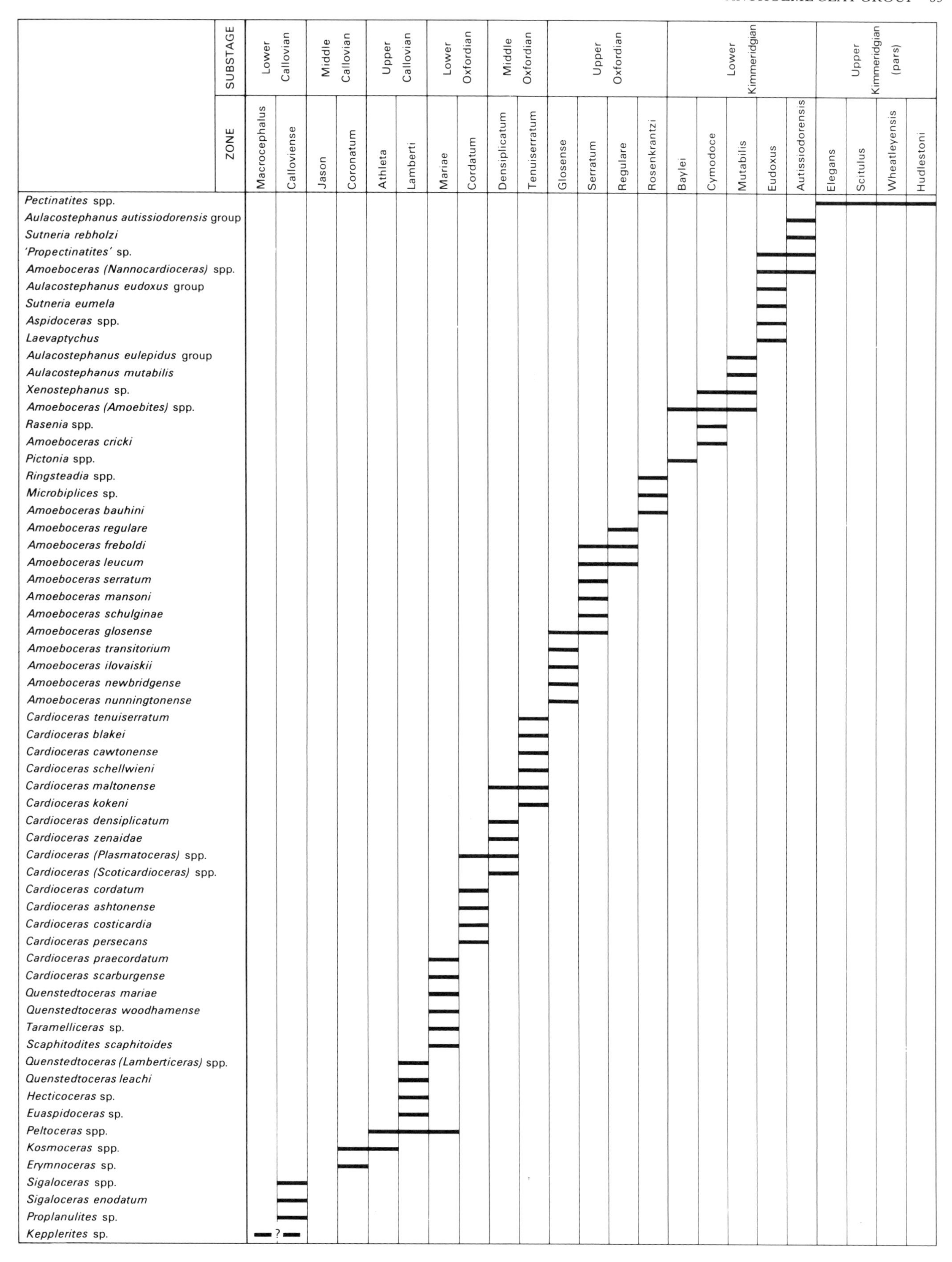
SUBSTAGE
Lower Callovian
Middle Callovian
Upper Callovian
Lower Oxfordian
Middle Oxfordian
Upper Oxfordian
Lower Kimmeridgian
Upper Kimmeridgian (pars)
ZONE
Macrocephalus
Calloviense
Jason
Coronatum
Athleta
Lamberti
Mariae
Cordatum
Densiplicatum
Tenuiserratum
Glosense
Serratum
Regulare
Rosenkrantzi
Baylei
Cymodoce
Mutabilis
Eudoxus
Autissiodorensis
Elegans
Scitulus
Wheatleyensis
Hudlestoni
Pectinatites spp.
Aulacostephanus autissiodorensis group
Sutneria rebholzi
'Propectinatites' sp.
Amoeboceras (Nannocardioceras) spp.
Aulacostephanus eudoxus group
Sutneria eumela
Aspidoceras spp.
Laevaptychus
Aulacostephanus eulepidus group
Aulacostephanus mutabilis
Xenostephanus sp.
Amoeboceras (Amoebites) spp.
Rasenia spp.
Amoeboceras cricki
Pictonia spp.
Ringsteadia spp.
Microbiplices sp.
Amoeboceras bauhini
Amoeboceras regulare
Amoeboceras freboldi
Amoeboceras leucum
Amoeboceras serratum
Amoeboceras mansoni
Amoeboceras schulginae
Amoeboceras glosense
Amoeboceras transitorium
Amoeboceras ilovaiskii
Amoeboceras newbridgense
Amoeboceras nunningtonense
Cardioceras tenuiserratum
Cardioceras blakei
Cardioceras cawtonense
Cardioceras schellwieni
Cardioceras maltonense
Cardioceras kokeni
Cardioceras densiplicatum
Cardioceras zenaidae
Cardioceras (Plasmatoceras) spp.
Cardioceras (Scoticardioceras) spp.
Cardioceras cordatum
Cardioceras ashtonense
Cardioceras costicardia
Cardioceras persecans
Cardioceras praecordatum
Cardioceras scarburgense
Quenstedtoceras mariae
Quenstedtoceras woodhamense
Taramelliceras sp.
Scaphitodites scaphitoides
Quenstedtoceras (Lamberticeras) spp.
Quenstedtoceras leachi
Hecticoceras sp.
Euaspidoceras sp.
Peltoceras spp.
Kosmoceras spp.
Erymnoceras sp.
Sigaloceras spp.
Sigaloceras enodatum
Proplanulites sp.
Kepplerites sp.
?

trenches, ditches, site investigation boreholes and old degraded brick pits is crucial. Age-determinations made by Dr B M Cox on the fauna from such localities in the district, and the biostratigraphy of the Callovian, Oxfordian and Kimmeridgian sequence as represented in Nettleton Bottom Borehole are presented in Appendix 4.

Kellaways Beds Formation

The name Kellaways Beds is here used as a term of formation status to include strata traditionally referred to, in ascending order, as the Kellaways Clay, Kellaways Sand and Kellaways Rock. It was first used in Lincolnshire (Ussher et al., 1888, pp.73–76) and originally excluded the Kellaways Clay (then referred to as the Basement Clays), but this 'clay' was subsequently included by Woodward (1895, pp.5–10).

Kellaways Clay Member

The Kellaways Clay is present in much of southern and eastern England south of the Humber. It is mainly dark grey mudstone generally not more than 4 m thick except in parts of Dorset and Wiltshire where it expands to nearly 29 m. Equivalent mudstones in north-eastern Yorkshire are called 'Shales of the Cornbrash', and Wright (1977) included them in his Cornbrash Formation. The Kellaways Clay is largely referable to the Kamptus Subzone of the Macrocephalus Zone, although in Dorset and Wiltshire it ranges up into the Koenigi Subzone of the Calloviense Zone.

In the southern half of the district the Kellaways Clay is medium grey to black (locally bluish or brownish) and variably laminated, the lower part being generally darker and finer grained. Pyrite is present as small crystals and amorphous masses and also replaces some shell debris and burrow fills. In Nettleton Bottom Borehole [TF 1249 9820] there are pebbles of Cornbrash at and just above the base, possibly a result of sub-aqueous channelling. Except in Snitterby Carr Lane Borehole [TF 0028 9471] in the extreme south, which proved 3.6 m, thicknesses are generally between 0.5 and 2.5 m (Figure 24). In places thicknesses vary considerably over short distances, and subsequent channelling has locally removed the Kellaways Clay entirely. South of Redbourne, where the Kellaways Clay rests on Lower Cornbrash, there is clearly a depositional hiatus, possibly due to slight movement along the Askern–Spital Structure. Elsewhere, where Upper Cornbrash is present, there may have been little or no depositional break, but merely an increase in depth of water and distance from land. The only exposures seen are in a weathered state in ditches.

Farther north the Kellaways Clay becomes more silty and even locally sandy, particularly near its top. It also thins irregularly northwards and becomes impersistent; it is absent, for example, in Worlaby E Borehole (Richardson, 1979, p.8)

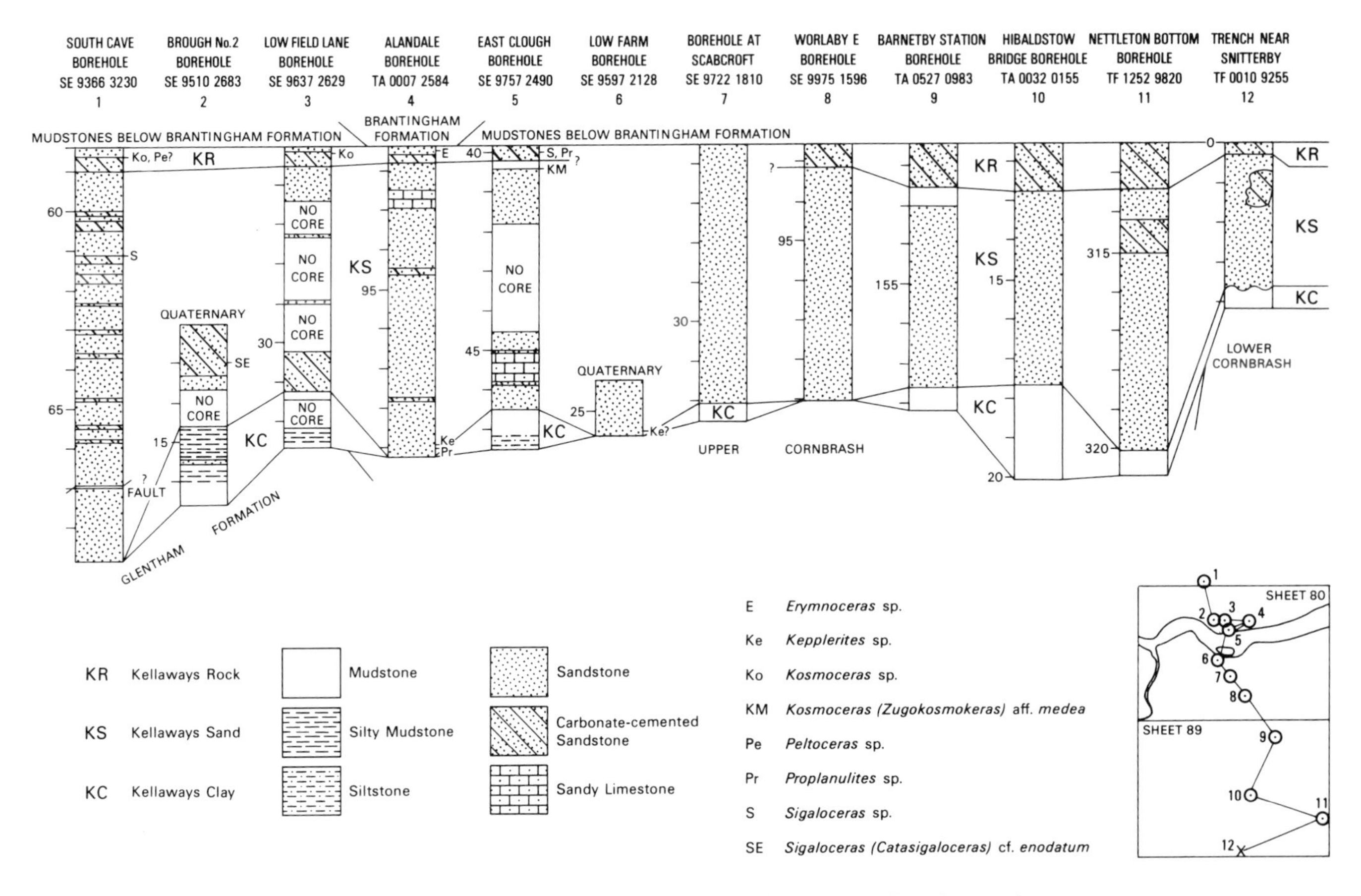

Figure 24 Sections, mainly from boreholes, of the Kellaways Beds Formation. Depths are in metres

and Low Farm (South Ferriby) Borehole (Gaunt et al., 1980, fig. 7), and has not been proved north of a line from Brough to Welton. Near its northern limit the Kellaways Clay becomes locally paler, greenish and brownish; where it rests mainly on the Glentham Formation its basal part is silty and sandy, and the presence within it of thin sandy mudstones and sandstones suggests a northward lateral passage into Kellaways Sand.

The shelly fauna is largely concentrated in thin layers and lenses, and contains bivalves (a few in burrowing positions), belemnites and rhynchonellid brachiopods. Ostracods, trails, burrows, general bioturbation including chondritic mottling, and (in the north) a few wood fragments have also been recorded. No determinable ammonites have been found within the district, but those from the lowest part of the overlying Kellaways Sand in the Humber boreholes (see below) suggest that, at least in these northern areas, the Kellaways Clay lies within the Macrocephalus Zone.

Kellaways Sand Member

Everywhere the Kellaways Beds include an arenaceous sequence. In much of southern England this sequence is called the Kellaways Rock, as it is also in north-eastern Yorkshire, where Wright (1978) includes it as the lowest member of his Osgodby Formation. In the eastern Midlands the name Kellaways Rock is largely restricted to a thin hard indurated layer at the top of the sequence, and the rest of the arenaceous sequence is called the Kellaways Sand. The Kellaways Sand and its lateral equivalents range through much of the Calloviense Zone.

Within the district the Kellaways Sand consists mainly of pale to medium grey (locally greenish), well-sorted, fine- to medium-grained silty quartzose sandstone. Its friable nature leads to poor recovery in boreholes and to rapid weathering to yellow and pale brown goethite-mottled sand at outcrop. Coarse-grained layers have been noted in its upper part in several of the Humber boreholes (where locally they are poorly sorted), but farther south they are known only in Clapgate No. 8 Borehole [SE 9712 1134]. The grain shape is predominantly subangular. Despite widespread bioturbation, some of the Kellaways Sand is recognisably thin-bedded and laminated. Ovoid, lenticular and irregular masses of hard, calcite-cemented sandstone up to 1.3 m thick seen in trenches, old sand pits and boreholes possibly represent the unleached remnants of original calcite-cemented beds. Clusters of calcareous ooliths, forming thin lenses and infillings of *Gryphaea* and ammonites, were noted in some of the Humber boreholes. In several scattered boreholes, including Barnetby Station [TA 0527 0983], St Helens Well [TA 0138 0772], Brigg Bridge Farm [SE 9845 1062] and East Clough (Gaunt et al., 1980, p.12, fig. 7), mudstone up to 0.46 m thick is present at or near the top of the Kellaways Sand. South of the Humber the thickness of the Kellaways Sand varies, apparently randomly, from 3.0 to 6.5 m, but farther north it increases considerably, being 9.85 m in South Cave Borehole (Figure 24).

The base of the Kellaways Sand was seen to be channelled into the Kellaways Clay in a trench [TF 0010 9255] near Beck Farm, and in Nettleton Bottom Borehole the contact was described as 'irregular, mudstone below churned up and interlensed'. In Worlaby E Borehole, where the Kellaways Clay is absent, Richardson (1979, p.8) noted 'pebbles of grey limestone probably derived from the underlying Cornbrash' in the basal Kellaways Sand. 'Pebbles' were also recorded from the Kellaways Sand in a borehole [SE 9645 1966] on Winterton Carrs, where the Kellaways Clay is absent, and the basal Kellaways Sand contains pebbles from the immediately underlying Thorncroft Sands in Brough No. 1 Borehole (Gaunt et al., 1980, p.12).

Most of the contained shelly fauna is in the calcite-cemented sandstones, probably due to decalcification elsewhere. Shells, especially belemnite guards, found within the unconsolidated beds are considerably corroded. The more common taxa are belemnites (including *Cylindroteuthis*) and 'oysters' (including *Gryphaea*), but other bivalves (including burrowing forms), rhynchonellid brachiopods and a few ammonites have also been found. Burrows, general bioturbation and wood fragments are fairly common. In the trench near Beck Farm, plant debris occurs in a belemnite-rich layer, and roots, thought to be intraformational, penetrate down from another layer. Sheppard (1900) recorded bones thought to be of '*Cryptocleidus*', a relative of *Plesiosaurus*, from 'sand' pits [probably SE 940 278] on Mill Hill, Elloughton. The few ammonites found have come from the Humber boreholes (Gaunt et al., 1980, p.12). Occurrences of *Kepplerites* sp. in the basal part of the Kellaways Sand in Low Farm and Alandale boreholes could indicate the Macrocephalus Zone, but *Sigaloceras* sp. at higher stratigraphical levels in Welton Waters 'B' and South Cave boreholes indicate the Calloviense Zone, and the subzonal index *S. (Catasigaloceras)* cf. *enodatum* at a comparable level in Brough No. 2 Borehole indicates the uppermost part of that zone.

The lithology and fossil content of the Kellaways Sand suggest higher energy depositional conditions and a closer proximity to land than during deposition of the Kellaways Clay. Brasier and Brasier (1978) have concluded from sedimentary and biogenic evidence that the Kellaways Sand in the South Cave area originated as a nearshore subtidal deposit. The most likely source of the arenaceous sediment in this northern area is from emergent Thorncroft Sands.

Kellaways Rock Member

In the Midlands the highest arenaceous strata in the Kellaways Beds are, where hard and indurated, differentiated as the Kellaways Rock. Ammonites show that these strata are diachronous, becoming progressively younger to the north. In the south they are referable to the Calloviense Subzone of the Calloviense Zone, the succeeding Enodatum Subzone being represented by the overlying mudstones. Near South Cave, however, *S. (C.) enodatum* is found in the arenaceous strata and occurrences in sand quarries [unspecified, but probably SE 920 329] were used by Callomon (1955; 1964) in the original definition of the Enodatum (formerly Planicerclus) Subzone.

The Kellaways Rock crops out mainly south of Hibaldstow and near South Cave, forming a low but distinct scarp. The strong calcite cement accounts for its being described as limestone in some borehole records. Where fresh it is pale grey, locally bluish or greenish, and weathers to brown goethitic hues. The sand content is fine to medium, and locally coarse grained, moderately to poorly sorted, and subangular to rounded; in Worlaby E Borehole rounded grains are particularly noticeable (Richardson, 1979, p.8). There are vestigial traces of silty lenses and laminae, but most of the Kellaways Rock has been severely disrupted by burrows, commonly filled or lined with clay, and by general bioturbation. The thickness of the Kellaways Rock varies from 0.3 m to just over 1.3 m, apparently randomly and in places over short distances.

The fossils are largely belemnites and 'oysters' (mainly *Gryphaea*), but other bivalves, a few ammonites (*Sigaloceras* and *Proplanulites* in East Clough Borehole, indicating the Calloviense Zone), rhynchonellid brachiopods and wood fragments are also present. These fossils increase in abundance towards the top, where they form a concentrated layer in which many are broken, worn, bored, encrusted and pyritised. In South Cave, Low Field Lane and Alandale boreholes this fossil layer is succeeded by up to 0.24 m of generally less well-cemented, fine-grained sandstone containing dark grey silty lenses and laminae and a few fossils; burrowing has also introduced masses of mudstone. On lithological grounds this sandstone has been included in the Kellaways Rock (Gaunt et al., 1980). However, occurrences of *Kosmoceras (Kosmoceras)* sp. and *Peltoceras* sp. in South Cave Borehole, *Kosmoceras* sp. in Low Field Lane Borehole, and *Erymnoceras* sp. in Alandale Borehole indicate that it is much younger in age (Coronatum–Athleta zones) and that it correlates with the Acutistriatum Band–Comptoni Bed near the top of the Lower Oxford Clay in areas farther south (Callomon, 1968). In Nettleton Bottom Borehole, Coronatum Zone mudstones intervene between this bed and the underlying Kellaways Rock.

Except for the carbonate cement and the concentrated fossil layer, the Kellaways Rock is lithologically and faunally closely comparable to the Kellaways Sand, and a similar peritidal origin is presumed. Because the concentrated fossil layer contains the more robust and durable elements from a variety of habitats, and because the fossils are well degraded, it may be a *remanié* layer accumulated by winnowing and sifting of an originally thicker arenaceous sequence. Much, if not all, of this process was probably due to tidal effects, but possible emergence is suggested by the rounded grains (although there is no other evidence of an arid hinterland) and by the 'mud cracks' noted by Brasier and Brasier (1978, p.13). The fossil layer was well burrowed prior to emplacement of the calcite cement, and any loose sand still present was reworked into the overlying thin sandstone which represents the initial sediment formed by the transgressive episode responsible for the succeeding mudstones.

Strata between the Kellaways Beds and the Brantingham Formation

The strata between the Kellaways Rock and the Brantingham Formation are identifiable only in the northern half of the district, where the Brantingham Formation is present. They are mainly medium to pale grey (locally brownish, greenish and bluish), non-silty or only slightly silty, variably calcareous mudstones with a blocky fracture. In the extreme north up to 1.99 m of mudstone have been proved. In some places along the northern edge of the Humber they are absent, for example in Alandale and Capper Pass C [SE 9712 2556] boreholes, where the Brantingham Formation rests directly on the Kellaways Beds Formation. The mudstones are present farther south, however, and thicken markedly beyond the definable limits of the Brantingham Formation. In Nettleton Bottom Borehole (Appendix 4) there are a few interbedded layers up to 0.25 m thick of dark grey and brownish grey laminated fissile mudstone of 'paper shale' type which are typical of the Lower Oxford Clay to the south of the district. A few thin beds in the upper part of the mudstones in Worlaby E Borehole are sufficiently calcareous to be considered argillaceous limestones.

The fairly diverse and locally pyritised fauna in the mudstones is partly concentrated in lenticular layers, but is otherwise well dispersed. It includes numerous bivalves, scattered gastropods and ammonites, a few serpulids, belemnites and crustaceans. Wood and other plant debris, commonly pyritised, is present, mainly in the north, and fish debris is more noticeable (probably because of the relative sparsity of other fauna) in the darker, more laminated mudstones. Trails and burrows, many pyritised, are ubiquitous.

Approximately the lowest 1.3 m of mudstone in Nettleton Bottom Borehole are referable to the Coronatum Zone (Figure 25), but most of the thick sequence above and the thinner sequences north of the Humber lie within the Athleta Zone. Both East Clough and South Cave boreholes yielded *Kosmoceras (Spinikosmokeras)* sp., and *K. (Lobokosmokeras)* cf. *duncani* was found in the former (Gaunt et al., 1980, fig. 7). The higher mudstones in Worlaby E Borehole and farther south extend up into the Lamberti and Mariae zones, and so span the Callovian–Oxfordian boundary (Appendix 4).

Brantingham Formation

In the Humber area strata which Gray (1955) described as 'Corallian Limestone' have subsequently been shown in the Humber boreholes to include sandstones, siltstones and limestones. These rocks are of local occurrence only and do not correspond to the Corallian rocks of north-eastern Yorkshire, despite some lithological similarities. The Quaternary cover is so extensive that their only surface manifestation is a slight topographical feature near Brantingham, from where their name is derived (Gaunt et al., 1980, p.14, fig. 8). The formation is thickest and coarsest along an approximate line between Brough and Hessle. The greatest thickness (15 m) was proved in Alandale Borehole where it is most arenaceous (Figure 25). The formation here rests on the lithologically similar Kellaways Beds; the base of the Brantingham Formation is defined as the base of the thin sandstone (resting on mudstone) at 56.43 m in South Cave Borehole. The top of the formation in Alandale and East Clough boreholes has been taken slightly lower than that illustrated by Gaunt et al. (1980, fig. 8). Southwards from the Humber the formation becomes progressively more difficult to delineate and loses its identity near Worlaby.

Sandstones are largely limited to the lower part of the formation near the area of maximum thickness. They are pale to medium grey, massive, calcareous, fine-grained, silty and appreciably bioturbated. In places, notably south of Melton, the calcite cement is sufficient to produce thin sandy and silty limestones which, because they cannot be correlated between boreholes, may be lenticular or concretionary. Small

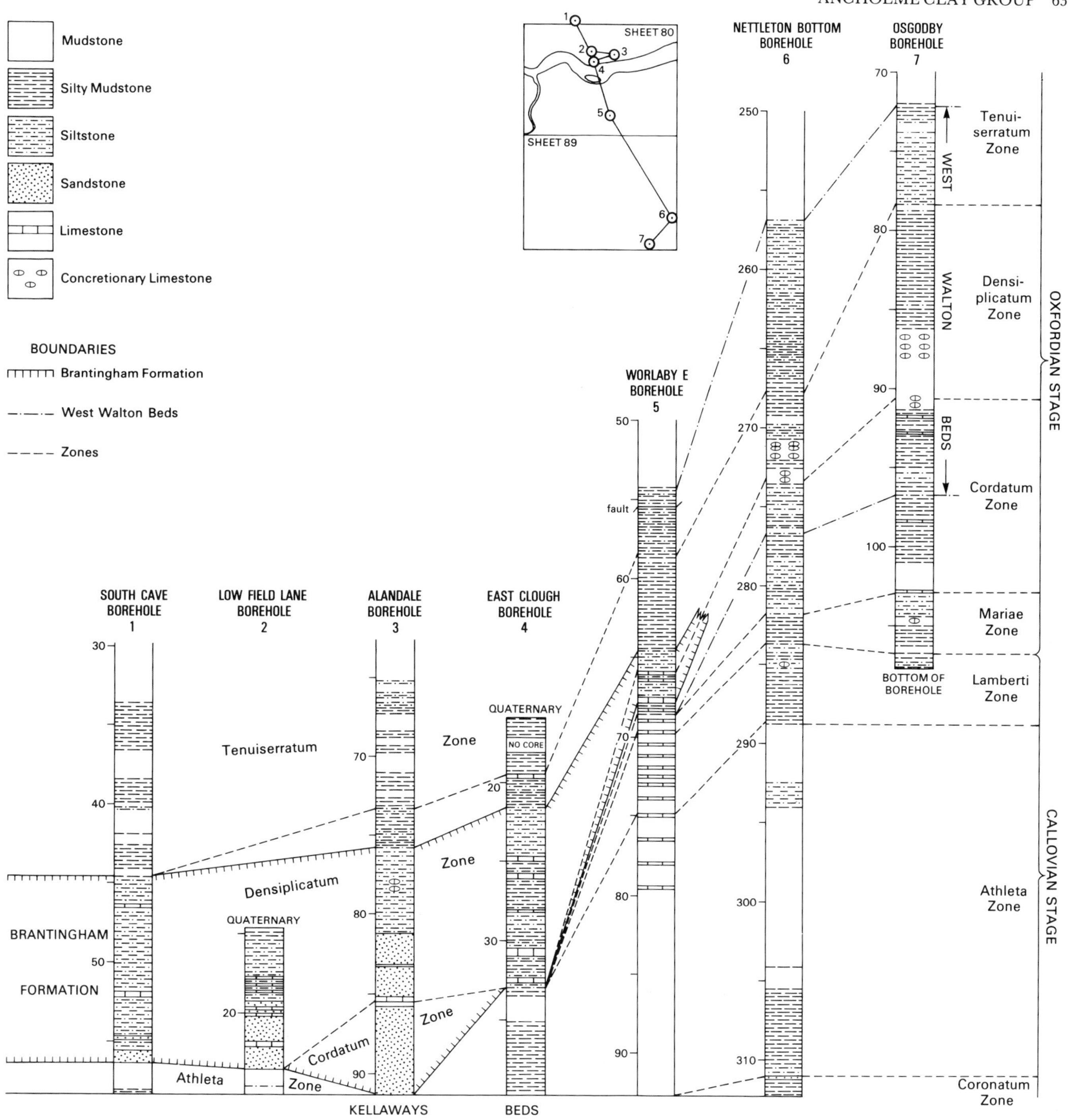

Figure 25 Borehole sections of the lower part of the Ancholme Clay Group above the Kellaways Beds Formation. Depths are in metres

phosphatic pebbles were found in the basal sandstone in Alandale Borehole. The upper part of the formation is composed mostly of siltstones, as also is most of the formation away from the area of maximum thickness. These, too, are pale to medium grey, commonly calcareous, and contain thin apparently lenticular or concretionary, silty and sandy limestones. A few thin medium to dark grey calcareous mudstones are interbedded with the siltstones.

A diverse, but not particularly abundant, shelly macrofauna is scattered through the formation. It includes epifaunal and infaunal bivalves (some of the latter in burrowing positions), ammonites and belemnites. *Rhaxella* spicules were used by Gray (1955, p.29) in his comparison of these rocks with the Corallian rocks (notably with the Lower Calcareous Grit) of north-eastern Yorkshire. Chondritic and other burrowing structures are widespread, particularly in the sand-

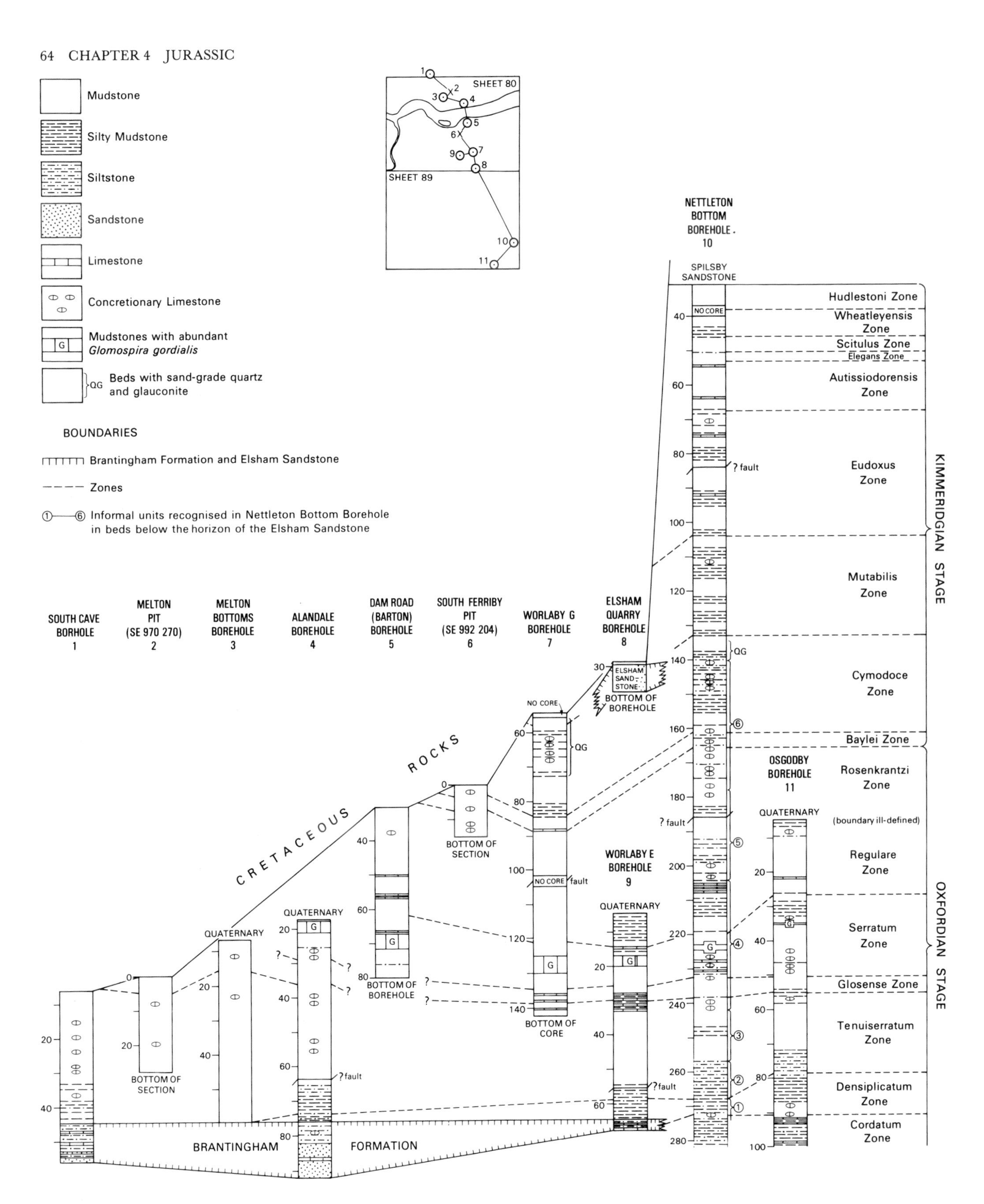

Figure 26 Sections, mainly from boreholes, of the upper part of the Ancholme Clay Group. Depths are in metres

stones, and wood and other plant debris also occur. Some of the shells, burrow fills and plant debris are pyritised. *Cardioceras (Subvertebriceras?)* aff. *costellum*, *C. (C.) ashtonense* and *C. (C.) persecans* in the sandstone in Alandale Borehole indicate the Cordatum Zone, which has also been proved in the lower part of the formation in Worlaby E Borehole. The higher beds in these and other boreholes, and apparently the entire formation in South Cave Borehole, are referable to the Densiplicatum Zone. *C. (Subvertebriceras) densiplicatum* occurs in the middle of the formation in several boreholes, and in East Clough Borehole *C. (S.)* cf. *zenaidae* in the basal beds and *C. (Maltoniceras)* cf. *maltonense* in the upper beds imply the Vertebrale and Maltonense subzones respectively of the Densiplicatum Zone. *C. (Plasmatoceras)* spp. and *C. (Scoticardioceras)* spp. have also been found in the formation.

In the Humber area, where its basal strata are sandy and locally pebbly, the Brantingham Formation oversteps almost on to the Kellaways Beds and cuts out the Lamberti and Mariae (and in places the Cordatum) zones (Figure 25). The formation thins and becomes less arenaceous to both north and south. In the latter direction it is reduced to a thin siltstone and limestone sequence, apparently without a basal hiatus, in Worlaby E Borehole, and it cannot be differentiated with certainty farther south.

The thickest and coarsest development is coincident with the area of maximum previous erosion, which is broadly aligned with the Brough fault complex (p.106). It is probable that the relatively coarse sediment which characterises the Brantingham Formation resulted from exposure and erosion of the Kellaways Beds as a result of movement along the faults of this complex in terminal Callovian and/or earliest Oxfordian times.

Certain of the Corallian rocks of north-eastern Yorkshire (Hemingway, 1974), particularly the various 'calcareous grits' and the North Grimston Cementstones, are lithologically very similar to the Brantingham Formation. The West Walton Beds of the southern parts of the east Midlands (Gallois and Cox, 1977) are less similar lithologically, for they largely comprise an interbedded mudstone/siltstone sequence, but their stratigraphical limits can be identified in the more southerly boreholes in the district (Appendix 4) and they are included in Figure 25 for comparison.

Strata between the Brantingham Formation and the Elsham Sandstone formations

The mudstones above the Brantingham Formation in the north, and above its approximately equivalent stratigraphical level in the south, are interrupted by only a few thin siltstone and limestone layers, and most of the latter are concretionary (Figure 26). They range zonally from the Middle Oxfordian to the Lower Kimmeridgian. The sequence in Nettleton Bottom Borehole is summarised in Appendix 4, Figures 46 and 47. No formal lithostratigraphical subdivision is warranted, but for descriptive purposes the succession has been split into six informal units, largely on the incidence of silty strata and hard calcareous layers. There are only minor variations in depositional thickness across the district; the present marked northward thinning of the succession is due to later erosion. The lithological and faunal characteristics of the mudstones imply that, after cessation of the localised effects that produced the Brantingham Formation in the north, quiescent open-sea depositional conditions, which had persisted farther south, spread northwards again across the entire district.

The lowest of the six informal units largely comprises pale to medium grey, calcareous, silty mudstones with a blocky fracture. Some thin silt-laminated, fissile beds are present, and thin siltstones in the lower part reflect a passage by alternation up from the Brantingham Formation. Montmorillonite has been reported in the lower part of these mudstones near Brantingham by Mr K S Siddiqui. A few buff calcareous concretions, some septarian, have been noted, and pyrite is widespread as small crystals and replacing fossils. The fauna consists largely of layered concentrations of bivalves (although a few are in burrowing positions). The ammonites *Cardioceras* and *Perisphinctes*, fish debris and serpulids (both mainly near the top of the unit), numerous trails and burrows (some of the latter being chondritic), and a few wood fragments are also present. In the south, where the Brantingham Formation is absent, no assessment of thickness variations of this unit can be made, but from Worlaby E Borehole northwards to South Cave Borehole there is a slight irregular thinning.

The succeeding informal unit consists mainly of pale grey, hard, massive, calcareous siltstones, locally with thin silty mudstones. In most boreholes there appears to be an upward passage into the siltstones, but in Nettleton Bottom Borehole small ?phosphate pebbles may reflect a pause in deposition. Sparse pyrite is present. The fossils include a few wood fragments, gastropods, bivalves, species of *Cardioceras* and *Perisphinctes*, burrows and general bioturbation. The siltstones thin northwards from 5.85 m in Nettleton Bottom Borehole to only 0.22 m in South Cave Borehole; they form the highest part of the West Walton Beds (Tenuiserratum Zone) of Gallois and Cox (1977) (Figures 25 and 26; Appendix 4).

Resting on the siltstones, in places sharply but elsewhere with a gradational passage, are mudstones with, north of the Humber, a few sideritic concretions. These mudstones vary from pale grey, hard, blocky and calcareous with a diverse fauna containing species of *Cardioceras* and perisphinctids, trails and burrows, to dark grey, fissile and pyritic, with a few jet-like wood fragments and sparse fish debris. The latter type of mudstone is largely confined to the lower half of the unit and has not been found north of the Humber. In its most extreme form it consists of bituminous, oily smelling 'paper shale', three layers of which were recorded in Nettleton Bottom Borehole. The unit thickens northwards from 14.35 m in Osgodby Borehole to over 20 m north of the Humber, where its upper limit is ill defined. It is referable to the Tenuiserratum Zone.

The overlying unit contains all four types of mudstone mentioned above (p.58) but is characterised by thin beds or concretionary layers of pale grey, hard, silty, finely crystalline limestone at up to four levels including its basal and topmost strata. The siltier mudstones (type i) are more common (but atypically are weakly laminated in places) in the upper part of the unit, for example the highest 8 m of Jurassic strata in Worlaby E Borehole (Figure 26), and the dark grey fissile mudstones (type iv) are most common in the middle beds. Otherwise the various types are generally intimately interbedded, and may indicate cycles. Perisphinctids are still present amongst the ammonites, but *Cardioceras* is replaced by *Amoeboceras*. A thin sequence of mudstones in the middle of the unit, mainly of the dark grey fissile type (type iv), contains 'flood' occurrences of the foraminifer *Glomospira gordialis* which provides a useful biostratigraphical marker in Beds 19–20 of Cox and Gallois' (1979) standard Ampthill Clay sequence (Figure 26; Medd, 1976). With a few exceptions the thin limestones are only sparsely shelly, the most notable feature being an abundance of serpulids in limestones near the top of the unit. The unit as a whole is about 38 m thick and ranges from the uppermost part of the Tenuiserratum Zone to the lower part of the Regulare Zone.

The succeeding unit largely comprises medium to dark grey mudstones (corresponding to type iii and, to a lesser extent, type iv). Some of the fissile mudstones are of 'paper shale' lithology. A few small carbonate concretions, some septarian, occur near the base and top. The shelly fauna includes species of *Amoeboceras* and perisphinctids, and serpulids are common. The only complete section of the unit examined is from Nettleton Bottom Borehole, where it is about 27 m thick. It ranges from the Regulare Zone into the Rosenkrantzi Zone.

The mudstones of the highest unit, estimated to be 36 m thick in Nettleton Bottom Borehole, are mainly pale to medium grey and silty, and contain numerous layers of carbonate concretions, some septarian. The silt content increases upwards and thin siltstones and sand-grade quartz grains occur in the higher beds. During the survey these mudstones were exposed in the chalk and clay pit at South Ferriby [SE 992 204] (Smart and Wood, 1976). The lower part of these beds is particularly rich in serpulids and the bivalve *Oxytoma*; *Amoeboceras* also occurs. Ostracods are common in some layers (Ahmed, 1987). In ascending order the ammonite genera *Ringsteadia*, *Pictonia* and *Rasenia* indicate a range from the Rosenkrantzi Zone through the Baylei Zone into the Cymodoce Zone, showing that the unit spans the Oxfordian–Kimmeridgian stage boundary. Correlation between the section at South Ferriby and Nettleton Bottom borehole is shown in Appendix 4, Figure 48.

Elsham Sandstone Formation

The sandstone cropping out in the Elsham area was originally thought to be an outlier of Spilsby Sandstone (see p.67) preserved in a syncline beneath the sub-Carstone unconformity (Fox-Strangways *in* Ussher, 1890, p.108; Ingham, 1929; Wilson, 1948, fig. 18). This correlation was proved to be incorrect when the Lower Kimmeridgian ammonites *Xenostephanus* sp. (which is known to occur in both the Cymodoce and Mutabilis zones, Gallois and Cox, 1976) and *Aulacostephanoides* cf. *mutabilis* were found in the sandstone by Kent and Casey (1963), who named the rock Elsham Sandstone. During the survey the sandstone was traced from near Bonby, where it is truncated northwards beneath the sub-Carstone unconformity, southwards almost to Barnetby le Wold, where it wedges out. From Elsham southwards it passes beneath mudstones high in the Ancholme Clay Group which were originally interpreted as the (Lower Cretaceous) Tealby Clay (Ussher, 1890, pp.109–110). Elsham Quarry Borehole (Gaunt et al., 1980, p.14) proved *Aulacostephanus* sp. in the topmost part of the sandstone, and *A. (Aulacostephanites) eulepidus*, indicative of the Mutabilis Zone, in mudstone immediately above it. In a pipeline trench near Elsham, Sir Peter Kent (pers. comm.) found *Rasenia* sp. and *Pictonia* sp., indicative of the Cymodoce and Baylei zones, in about 15 m of mudstone immediately below the sandstone. The sandstone thus probably spans the Cymodoce–Mutabilis zonal boundary.

The Elsham Sandstone is generally pale grey to greenish grey, but brown iron-pan staining is apparent in weathered sections. It is medium to coarse grained, moderately to poorly sorted, and comprises subangular to (rarely) rounded grains loosely to closely compacted in an argillaceous or calcareous matrix. Small pebbles are locally present. Large-scale decalcification has rendered much of the sandstone friable and incohesive, but irregular masses of hard calcareous rock remain and preserve extensive bioturbation structures. These masses form bluffs in old quarries on the southern side of Elsham. Small red, green, dark blue and black detrital grains are scattered within the sandstone, and Ingham (1929), although assuming that he was working on Spilsby Sandstone, records the heavy minerals. He also illustrates (*ibid*, plate 1, fig. 2) a thin section showing small-scale calcite veining. Diagenetic pyrite typically occurs around shelly fossils and there are lignitic wood fragments, largely at the top of the sandstone. The shelly fauna, which includes gastropods, bivalves and ammonites, is preserved mainly in the more calcareous parts of the sandstone and, according to Kent and Casey (1963), shows no evidence of abrasion despite the coarse nature of the host rock. Scarce poorly preserved foraminifera (*Lenticulina* sp.) have also been found. These authors describe the Elsham Sandstone in a pit [TA 043 104] near Gallows Farm, Wrawby, as 'finer-grained'; this location is near the known southern limit of the rock, and has yielded *Xenostephanus* sp. and *Rasenia?*. The sandstone is about 9.2 m thick in a trench [TA 038 115] at Elsham (Kent and Dilley, 1968); this accords well with the incomplete 8.43 m proved in Elsham Quarry Borehole.

No Elsham Sandstone was found farther south, but sand-grade quartz and glauconite are present in Nettleton Bottom Borehole in about 6 m of calcareous siltstones and silty mudstones that are referable to the upper part of the Cymodoce Zone (Figure 26; Appendix 4, Figure 47). Worlaby G Borehole, to the north of Elsham, did not prove the sandstone, but Richardson (1979, p.14) suggested that it may have been represented by 0.76 m of 'sand' lost between the Ancholme Clay and the Carstone. Significantly, sand-grade quartz and glauconite occur here in the highest 17.45 m of Ancholme Clay preserved, being most abundant in the highest 4 m which span the Cymodoce–Mutabilis zonal boundary. These sandy mudstones and siltstones are considered to be lateral equivalents of the Elsham Sandstone. The sandstone is likely to have been thicker and coarser to the north-west or west, suggesting derivation from one of these directions. Whether it was transported 'by a flush of fast-flowing water' (Swinnerton and Kent, 1976, p.54) or was a migrating sublittoral sand bar is uncertain, but both possibilities imply local shallowing during early Kimmeridgian times.

STRATA ABOVE THE ELSHAM SANDSTONE FORMATION

The succession above the Elsham Sandstone and its lateral equivalent is known in detail only from Nettleton Bottom Borehole, where more than 100 m of mudstones are proved, ranging up into the Hudlestoni Zone (Figure 26; Appendix 4, Figure 47). Farther south, however, there are mudstones of the succeeding Pectinatus Zone, the youngest rocks of the Ancholme Clay Group proved in the district.

Lithological variation within the thinly interbedded mudstones apparently represents cycles of the kind described by Gallois (1973) in the coeval Kimmeridge Clay farther south. Four basic types of mudstone are present in a full cycle. The lowest is pale to medium grey-brown, calcareous and shell-laminated; it is best described as a bituminous 'paper shale', and is thought to be the northern equivalent of the 'oil shales' described by Gallois (1978) and Gallois and Cox (1976). Above is mainly dark grey mudstone with only a sparse shelly fauna, concentrated on bedding planes. This is succeeded by pale to medium grey, variably calcareous mudstone that is weakly laminated by silt and/or shell layers. The highest type is similar but more massive, with silty patches and a blocky fracture, and it contains scattered shells throughout. The two lower types occur only as thin layers (up to 0.2 m thick). A few thin argillaceous limestones are present, and some are demonstrably concretionary.

Mudstone immediately below the Spilsby Sandstone in a temporary exposure [TF 1085 9889] on Nettleton Hill has yielded *Pectinatites (Arkellites)* spp. including *P. (A.)* cf. *hudlestoni*, indicating the Hudlestoni Zone. However, a specimen from a location [TF 1152 9633] in Acre House (Claxby Ironstone) Mine (locality 55 in Figure 49), collected in 1918 by G W Lamplugh and recorded as 'bituminous shaly clay' close beneath the Spilsby Sandstone, contains fragments of *Pectinatites* including *P. (P.) eastlecottensis?*, indicating the Eastlecottensis Subzone of the Pectinatus Zone.

Spilsby Sandstone Formation

The Spilsby Sandstone (Strahan, 1886), a formation distinct from and resting unconformably on the Ancholme Clay Group, is present from Grasby southwards beyond the district, and in this direction eventually passes into the Sandringham Sands Formation of Norfolk (Casey and Gallois, 1973). From Caistor southwards within the district the formation is estimated to be generally about 8 to 11 m thick, but it may reach 18 m in places.

Two members are recognised, the Lower Spilsby Sandstone and the Upper Spilsby Sandstone. They are almost identical lithologically and each has a prominent basal 'nodule' bed containing phosphatised pebbles and derived fossils (Figure 27; Casey, 1962; 1963). The formation spans the Jurassic–Cretaceous boundary and the junction between the two members is a non-sequence indicative of a depositional break early in the Ryazanian. For convenience, and especially because the Upper Spilsby Sandstone is known at only one locality in the district, the entire formation is described in this chapter.

The Spilsby Sandstone represents the north-western part of a marginal marine quartzose sand facies deposited, following substantial erosion, between the uplifted Market Weighton Structure and the London Platform. Its fauna shows that marine connections with the southern side of the London Platform were slowly severed, whilst a northerly arm of the sea remained in continual contact with the Greenland Province of the Arctic Boreal Ocean (Krimgolts et al., 1968).

LOWER SPILSBY SANDSTONE MEMBER

Much of the Lower Spilsby Sandstone is unconsolidated, but it encloses irregular masses of hard sandstone with a calcite cement. Within the district the member crops out along the Lincolnshire Wolds scarp. Because of its high porosity it is a useful aquifer and produces copious springs which have contributed to the development of landslips and mudflows. It gradually thickens southwards to between 8 and 10 m south of Caistor, but may reach 18 m where it apparently fills incised channels.

The Basal Spilsby Nodule Bed at the base of the Lower Spilsby Sandstone, although less than 0.15 m thick, is significant because of its fossil assemblage and economic potential. It contains numerous 6 to 12 mm-diameter phosphatised pebbles, or nodules, of at least two types (Oakley, 1940, p.3), their dominant constituents variously sand, clay and lime, reflecting the sedimentary environments from which the pebbles were derived.

One type of 'nodule' is virtually identical to the enclosing rock matrix and consists of quartz 'grit' (with subordinate glauconite, orthoclase, microcline and chert) in a matrix of brown amorphous collophane; in the chert, tiny rhombs of dolomite appear to have been replaced by green isotropic greenalite. These 'nodules' are irregular in form, with corroded but glazed surfaces, against which the matrix is slightly pyritic and much darkened. In places contiguous 'nodules' coalesce. Authigenic glauconite patches in the matrix end sharply against the 'nodules' and commonly enclose detrital glauconite grains with some contact interaction.

The second type of 'nodule' consists largely of homogeneous amorphous collophane enclosing brown rock fragments of varying size. Some subangular fragments have glauconitic impregnations along their peripheries and a few have ball-like kernels of glauconitic 'grit'. Both types of 'nodule' show signs of solution but not of abrasion, and they are commonly bored.

It seems that only a single phosphogenic event is represented by the two types (Kelly, 1980, fig. 6). There were, however, several phases to this event, for some 'nodules' have been penecontemporaneously broken and then assimilated into matrix material. Homogeneous phosphate was also able to infiltrate shells and fill moulds after solution. Green glauconite of contemporaneous formation colours the phosphatic cement in patches, and forms irregular wisps within it. Localised calcite cement is also recognisable, but there is no indication that either of these cements has replaced the other.

The remainder of the Lower Spilsby Sandstone overlies the Basal Spilsby Nodule Bed with no detectable depositional break. It is a relatively homogeneous quartzose sandstone but varies in colour and in its content of phosphatic pebbles, silt and clay.

Figure 27 Generalised section of the Lower Cretaceous rocks and the lower part of the Spilsby Sandstone Formation

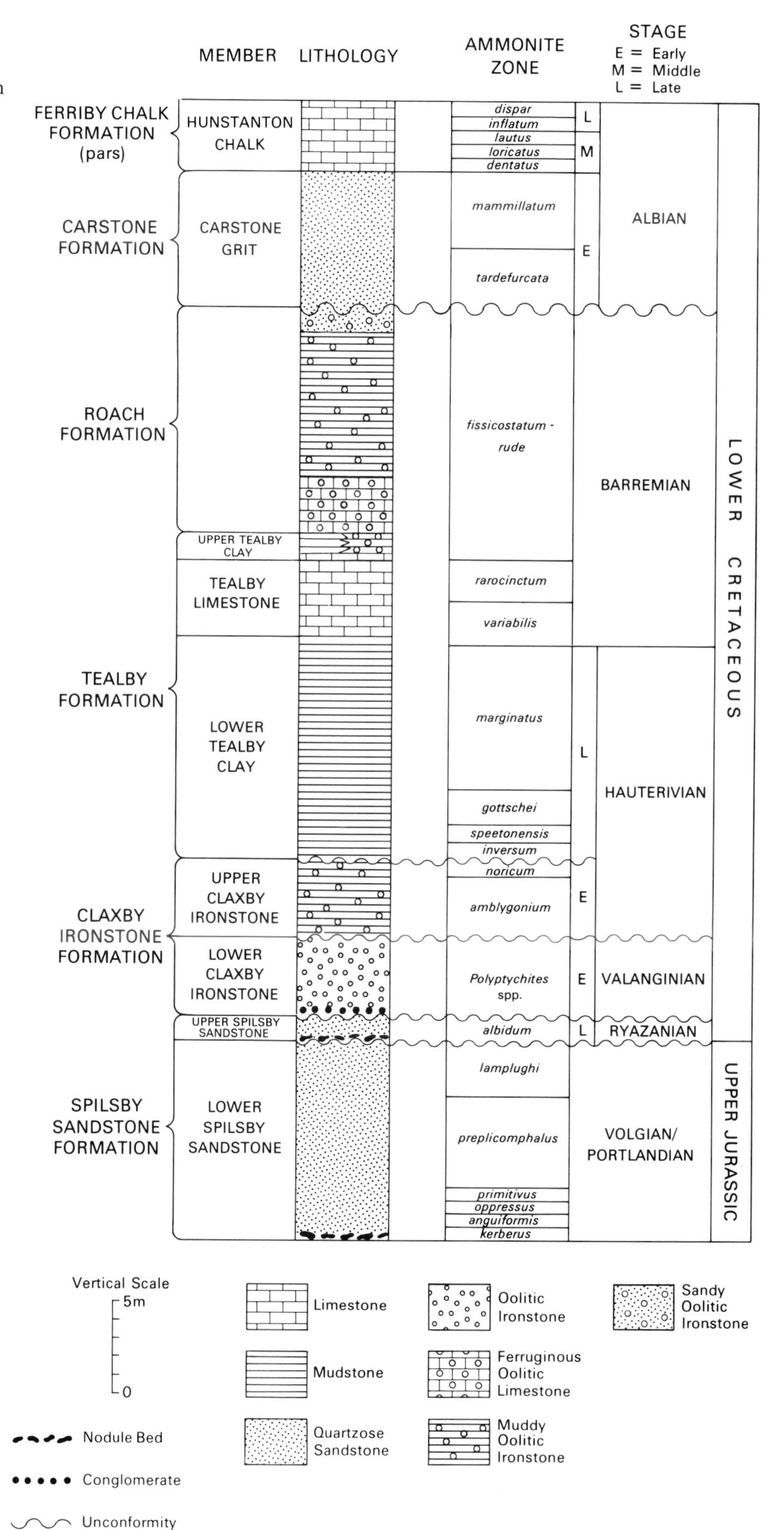

The sandstone varies from calcareous quartzose 'grit' to friable medium-grained quartz sand with silt and clay admixtures. The grains are mainly subangular and polished. Small pebbles are not uncommon and are concentrated into thin layers, emphasising the original bedding. Primary sedimentary structures are, however, rare because of intense bioturbation.

The unweathered rock is more or less colourless, but speckled with small grains of green glauconite, pink feldspar or blackish lydian chert fragments. The cementing matrix is of primary sparry calcite which shows no sign of late expansion. Where leached of carbonate, as is common, the rock is yellowish or reddish brown and green. Oxidation of glauconite is probably the cause of some of this staining, but the breakdown of sideritic cement and the percolation of iron-charged waters have also contributed. In thin section, quartz grains predominate and vary in size and form. Most are medium-sized, subangular and generally clear. Some contain minute inclusions of black dust, irregularly or in streaks; other inclusions include zircon, chlorite, tourmaline and rutile. The calcite matrix occurs as large irregular crystals of different optical orientations; it is also present as fossil remains and minute secondary veins. Thin goethite coatings on quartz grains possibly indicate some replacement of calcite by siderite, and subsequent oxidation and hydration of this mineral. Glauconite coatings are also present, commonly enclosing several grains.

The included pebbles are generally dark and up to 15 mm in diameter. They are well rounded and polished, and most are of chert or chalcedony stained by iron oxides. Phosphatic pebbles and fragments, similar to types in the Basal Spilsby Nodule Bed, are widespread but most abundant in the lower part of the sandstone.

In the Basal Spilsby Nodule Bed there is a mixed-age faunal assemblage of taxa, some derived and some coeval, from several Kimmeridgian and Portlandian zones. This concentration has been interpreted both as evidence for a long period of submergence without appreciable sedimentation (Lamplugh, 1896), and as an accumulation of reworked fossils in repeatedly winnowed sediment. Rolled fragments of *Pectinatites* sp. and *Pavlovia* sp. have been regarded as *remanié* material from late Kimmeridgian strata in the Ancholme Clay Group (Casey, 1973, p.206). More complete specimens of *Kerberites* cf. *kerberus*, *Epilaugeites* sp. and *Crendonites* sp. are lithologically identical to the enclosing matrix, and strongly suggest that initial sand deposition took place in Kerberus Zone times. There is no evidence of the Anguiformis and Oppressus zones within the district.

At Nettleton Top [TF 108 988] the Basal Spilsby Nodule Bed is overlain by 2 m of clayey sandstone containing *?Subcraspedites (Swinnertonia) primitivus* and *S. (Swinnertonia) cristatus*, indicative of the Primitivus Zone. Although Casey (1973, p.205) assigned over 4 m of the Nettleton Top section to the Preplicomphalus Zone, no diagnostic taxa of that zone are yet known hereabouts. The topmost 2.5 m of the member have yielded Lamplughi Zone ammonites, including the zone fossil and transitional species linking the subgenera *Subcraspedites* and *Volgidiscus*. The ammonites and other taxa occur at particular levels where the sandstone is unleached or where moulds are preserved in silty clay. Calcite shells are common, but nacreous specimens appear to be of phosphate rather than original aragonite; many moulds are strengthened by a coating of reddish iron oxide and filled with calcite spar or phosphate.

Bivalves dominate the macrofaunal benthos and represent several distinct environments. Most assemblages appear to have been transported, but each seems to represent a life community. In a recent study of the whole Spilsby Sandstone – Sandringham Sands outcrop, Kelly (1977) recognised five distinct 'groupings', and also noted a correlation between grain size and the proportion of epifauna. Only two of his 'groupings' are present within the district, a *Pleuromya – Grammatodon* Assemblage associated with the Basal Spilsby Nodule Bed and the strata immediately above, and an *Entolium* Assemblage in concretions of the Lamplughi Zone.

The glauconitic silty sand comprising the nuclei of the phosphatised 'nodules' contains moulds of a rich infauna. Following penecontemporaneous erosion, pebbles containing this infauna, and large shells such as ammonites, formed a quasi-hardground that was colonised by a rich byssate fauna of suspension feeders, encrusters and borers. The following taxa are listed by Kelly: serpulid, *Rouillieria ovoides*, pleurotomariid, *Antiquicyprina lincolnshirensis*, *Discoloripes fischerianus*, *Grammatodon compressiusculum*, *G. schourovskii*, *Hartwellia hartwellensis*, *Hiatella foetida*, *Neocrassina nummus*, ostreid, *Pleuromya* cf. *uniformis* and *Plicatula producta*. To this can be added the belemnites *Acroteuthis lindseyensis*, *A. partneyi* and *A. (Microbelus)* sp. The *Entolium* Assemblage at the top of the Lower Spilsby Sandstone comprises *D. fischerianus*, *E. orbiculare*, *Myophorella intermedia*, *Neocrassina asiatica*, *Pinna* aff. *subcuneata* and *Sowerbya longior*. *Entolium* 'pavements' with current-oriented valves, all convex upwards, are noted by Kelly, as also are deep burrowers such as *Pleuromya* and *Pholadomya* in life positions.

It appears that the Lower Spilsby Sandstone accumulated in Portlandian times as a coastal sand spread. Its uneven base suggests the presence of filled-up tidal channels, and rivers draining metamorphic rock terrains similar to those of the present Scottish Highlands (Ingham, 1929) were probably the source of detrital sediment that was then transported along the coast by longshore drift. Following initial transgression, tidal lagoons may have developed behind sand bars. This would explain the high organic activity with minimal sedimentation and the precipitation of phosphate and glauconite.

Upper Spilsby Sandstone Member

Only one exposure of Upper Spilsby Sandstone is known in the district, but a summary of the more important aspects of this member, derived from evidence farther south, is included in order to place this exposure in its regional stratigraphical context. The Upper Spilsby Sandstone is at least 11 m thick in places and rests on the eroded top of the Lower Spilsby Sandstone (Figure 27). Its basal layer, the Mid Spilsby Nodule Bed, closely resembles the Basal Spilsby Nodule Bed but its phosphatic cement is less well developed; it is equivalent to the Coprolite Bed at the base of the Speeton Clay in north-eastern Yorkshire.

The rest of the Upper Spilsby Sandstone consists of quartzose sandstone that is partly calcitic and partly decalcified and friable. Versey and Carter (1926, p.359) listed the heavy minerals. On faunal evidence the erosional phase that preceded deposition of the Upper Spilsby Sandstone is largely early Ryazanian in age, and it encompasses the two distinct erosive episodes that are discernible within the Sandringham Sands in Norfolk. The base of the Upper Spilsby Sandstone may be diachronous, for north of Winceby [TF 321 687] only the *P. albidum* Zone is unequivocally proven in the member.

The only exposure of Upper Spilsby Sandstone in the district is in an old trial pit [TA 1179 0252] for 'ironstone' near Hundon Manor, which reveals 1 m of sandstone, including a thin 'nodule' bed which may represent the Mid Spilsby Nodule Bed. The fauna in this 'nodule' layer includes *Acroteuthis (A.) explanatoides*, *A. (A.) paracmonoides arctica*, *Entolium* cf. *demissum* and *Procyprina* cf. *centralis* and indicates the late Ryazanian Belemnite Assemblage of Pinckney and Rawson (1974). Hard calcareous masses in the overlying, friable sandstone have yielded the ammonites *Bojarkia (B.) bodylevskii*, *Peregrinoceras albidum*, *P.* cf. *albidum*, *P.* aff. *bellum*, *P.* cf. *rosei*, *P.* cf. *subpressulum and P.* cf. *wrighti*, and the belemnite *Acroteuthis (A.)* cf. *explanatoides*, indicating a *P. albidum* Zone age. Other fossils collected from these beds include *Rouillieria* cf. *ovoides* (attached to *Camptonectes (Maclearnia* shells), *Camptonectes (C.)* sp., *C. (Mclearnia) cinctus*, *Corbicellopsis claxbiensis*, *Dicranodonta* sp., *E. demissum*, *E. nummulare*, *Lyapinella laevis*, *Camptonectes (M.)* sp., *Modiolus* sp., *Myophorella claxbiensis*, *Nanogyra* cf. *thurmanni*, *Nicaniella (Trautscholdia) claxbiensis*, *Panopea* cf. *neocomiensis*, *Pinna* sp., *Plagiostoma* cf. *subcardiiformis*, '*Pleuromya*' cf. *peregrina*, *Stegoconcha* aff. *plotii*, *Bathrotomaria* cf. *swinnertonia* and asteroid ossicles.

CHAPTER 5

Cretaceous

Cretaceous rocks are confined to the eastern part of the district, where they dip gently eastwards and form the Lincolnshire and Yorkshire Wolds. South of Audleby they crop out as a narrow, highly dissected west-facing scarp; to the north the outcrop widens to cover almost half the Kingston upon Hull district. East of a line from Kirmington to Swanland the Cretaceous rocks are concealed beneath Quaternary deposits. The Lower Cretaceous is represented dominantly by clastic sediments and the Upper Cretaceous by chalk, the boundary lying within the basal chalk beds.

LOWER CRETACEOUS

Lower Cretaceous rocks largely comprise marine quartz-rich sandstones, micritic limestones, oolitic muddy ironstones and grey silty mudstones, not unlike those deposited earlier in the Jurassic. Upper shoreface sandstones are overlain by iron-rich mudstones of the transition zone, which are in turn overlain by more open-shelf siltstones and mudstones. Clastic starvation in the late Hauterivian Stage allowed local transient carbonate sedimentation before a brief interval of erosion. After a further period of open-shelf and transition-zone deposition, during which mainly iron-rich sediments accumulated, the whole sequence was differentially eroded and in the north completely removed. In the early Albian Stage, renewed deposition of upper shoreface sands took place during a transgression from the south; these sands were subsequently succeeded by chalk sediment.

The Lower Cretaceous is represented in the district by the Claxby Ironstone, Tealby, Roach and Carstone formations, together with the Upper Spilsby Sandstone Member (see Chapter 4) at the base and the Hunstanton Chalk Member at the top (Figure 27). The most complete sequence is about 50 m thick and is preserved in the south, where several depositional breaks are recognisable. The extent of each nonsequence becomes greater northwards and the feather-edge of the Carstone gradually oversteps all older formations to rest directly upon the Jurassic Ancholme Clay Group at Grasby [TA 087 049]. Beyond this point, Ferriby Chalk overlies a very thin Carstone, the outcrop of which, north of Bonby [TA 008 150], is too narrow to show on the 1:50 000 map.

Spilsby Sandstone Formation

The upper part of the Spilsby Sandstone Formation, designated the Upper Spilsby Sandstone Member, was deposited in Cretaceous times; it is described along with the Jurassic Lower Spilsby Sandstone Member in Chapter 4.

Claxby Ironstone Formation

This formation is a muddy oolitic ironstone up to 8 m thick that sharply overlies the eroded Spilsby Sandstone, the change from sandstones to clay ironstones marking the basal Valanginian transgressive event (Rawson and Riley, 1982). In southern Lincolnshire the Claxby Ironstone grades upwards into the Tealby Formation, the boundary being taken at the upward disappearance of goethite ooliths and the simultaneous appearance of glauconite and dark phosphatic nodules (Swinnerton, 1935). Farther north in the present district, however, this definition no longer obtains, for both geothite ooliths and glauconite are present throughout the Tealby Formation.

The best section of ironstone in the district is at Nettleton Hill, where the top of the formation has been drawn at an indistinct erosion surface separating goethite-oolith wackestone below from paler and greyer oolitic mudstone above. In this section the formation is 5.74 m thick. A depositional break at the base of an upper current-bedded unit enables a division into two members. Elsewhere in the district, for example at Hundon Manor, this break is represented by thin lenses of conglomerate.

There are few other natural exposures of the Claxby Ironstone, but it forms reddish brown clayey soils with conspicuous shiny ooliths in ploughland. The northernmost occurrence is just north-east of Audleby, to the north of which it is overstepped by the Tealby Formation. South of Audleby, trial pits [TA 1182 0248; 1189 0237] near Hundon Manor proved the formation to be about 4 m thick, comprising a lower member 2.3 m thick, and an upper, brachiopod-rich member with a pebbly base. The ironstone has been worked in opencast and deep mines around Nettleton (Figure 45, p.134) but, except for two unrestored opencast faces [TF 1140 9868; 1164 9870] north of Nettleton Top Farm, sections are no longer accessible. South of Nettleton Top Mine, oolitic wackestones with pale yellowish grey calcitic patches [TF 1112 9778] have yielded numerous brachiopods (Wright, 1941), and lie above steep bluffs of sandstone. Minor exposures of the ironstone can also be seen in a gully [TF 1147 9526] near Normanby Lodge and along a track [TF 1196 9482] between Claxby and Normanby le Wold.

The formation was largely deposited below wave-base in an open marine environment, representing the lower reaches of the upper offshore zone (transition zone), to which faunas from both Boreal and Tethyan realms had ready access. Such an environment was inimical to iron precipitation and much of the ferruginous content seems to have been derived from landward marginal lagoons destroyed during a marine transgression. Sedimentation was interrupted by an erosive episode, and there is no evidence for the presence of late Valanginian deposits in the district. Bedding and grading are barely perceptible and are masked by intense bioturbation.

Lower Claxby Ironstone Member

This member is mainly an oolitic wackestone with small goethite ooliths set in a fine-grained groundmass of berthierine, goethite and siderite. A thin irregular conglomeratic grainstone at the base consists mainly of degraded and distorted ooliths with subordinate quartz grit, glauconite, flattish phosphatised pebbles of laminated calcite mudstone, and shell fragments largely derived from the underlying Spilsby Sandstone.

The overlying wackestone contains many small polished goethite ooliths; sand, characterising the lowest part, dies out rapidly upwards. A few thin lenses of bioclastic limestone also occur near the base, where calcitisation of the matrix has imparted an overall pale colour. Locally, there is an admixture of non-ferruginous silt and clay as well as packstone concentrations of calcite shells and ooliths. The highest part of the sequence also contains conspicuous angular chips, up to 8 mm long, of well polished hard goethitic mudstone. The ironstone is thoroughly bioturbated with several generations of burrows that have provided foci for the diagenetic precipitation of calcite, pyrite and phosphate, and which produce a pronounced olive and yellowish mottling. In thin section, the groundmass proves to be mainly a pale greenish, weakly birefringent berthierine with microcrystalline siderite rhombs having distinctive eye-like brown centres (Hallimond, 1925, p.82, plate VI, fig. 27). In places, pale brown patches of goethite replace the berthierine, and secondary calcite crystals occur in rounded cavities or in the matrix around shell fragments. Ooliths are virtually all of goethite and are predominantly spherical, highly polished and of equal size. Nuclei of quartz and glauconite grains are recognisable. In some of the oolith envelopes, the outermost layers, which are kaolinised, tend to adhere strongly to the clay matrix when the rock fractures. Some ooliths are bound together by a goethite coat to produce 'dumb-bell' or multi-oolith clots containing as many as eight ooliths.

The member is 4.2 m thick at Nettleton Hill and comprises 13 beds, poorly defined by separation planes along which irregular iron pans are developed. The beds, however, appear to be of only limited areal extent. Surface exposures provide poor sedimentological detail, and bioturbation and matrix diagenesis mask vague alternations of finer- and coarser-textured oolitic wackestone.

The basal conglomerate contains derived Spilsby Sandstone fossils, for example *Peregrinoceras* at Nettleton Hill (Kelly and Rawson, 1983). Limestone lenses from the lowest 0.5 m of the member at this locality yielded a rich and well-preserved fauna of brachiopods, gastropods, bivalves, belemnites and rare ammonites. The commonest fossils are bivalves, notably *Aetostreon subsinuata carinatoplicata*, *Dicranodonta benniworthensis*, *Lyapinella laevis* and *Myophorella (Pseudomyophorella?) rawsoni*. The brachiopods are represented by only two species, *Lamellaerhynchia rostriformis* and *Rouillieria tilbeyensis*. The belemnites, comprising predominantly *Acroteuthis explanatoides* and *A. subquadratus*, belong to *Acroteuthis* Assemblage 4 of Pinckney and Rawson (1974), of early Valanginian age. An earliest Valanginian age is indicated by the ammonites, which include both Boreal forms such as *Menjaites*—permitting correlation with Russian *Menjaites*-bearing sequences—and *Polyptychites*, and the Tethyan *Neocomites?* cf. *trezanensis* (see Kemper et al., 1981). The only ostracods found were *Protocythere hannoverana*, the zonal index for the early Valanginian ostracod zone, *Mandelstamia sexti*, and a single, poorly preserved specimen of the low-salinity (fresh to brackish water) *Fabanella* sp. cf. *F. boloniensis*.

The base of the member is diachronous, becoming progressively younger northwards, for Ryazanian (*albidum* Zone) ammonites occur within it in southern Lincolnshire, Valanginian (*Paratollia* Zone) ammonites at Benniworth Haven, and *Polyptychites* in this district.

The remainder of the member yields a less diverse fauna dominated by large bivalves such as *Aetostreon subsinuata carinatoplicata* and *Camptonectes (Mclearnia) cinctus*, with *Dicranodonta benniworthensis, Myophorella rawsoni* and *Plagiostoma planum*. The belemnites belong to *Acroteuthis* Assemblage 4, and the presence of *Polyptychites gravesiformis* at several levels indicates that the main ironstone is no younger than the early Valanginian *Polyptychites* Zone.

Upper Claxby Ironstone Member

This member is 1.54 m thick at Nettleton Hill, where its basal bed consists of cross-bedded packstone filling a channel. In the Hundon Manor succession, to the north, this horizon may be represented by a concentration of phosphatised pebbles resting on an erosion surface.

The upper member is similar to the more bioturbated parts of the lower member, except that calcite mudstone is here more obvious, both in the matrix and as clots up to 0.1 m in diameter. Polished angular goethitic mudstone clasts with reddish centres and narrow greyish pink rims also occur. Characteristically there are bands containing grey and pink calcareous patches, apparently the result of differential calcitisation and weak phosphatisation of bioturbated mud patches. Some patches have been so calcitised that small chalky cavities filled with friable, sugary calcite are developed. The overall calcite content of the top metre of this member has led to descriptive terms such as 'calcareous beds' (Kelly and Rawson, 1983). Angular and subrounded quartz and glauconite grains occur sporadically throughout and some are irregularly coated with blood-red goethite. Dumb-bell compound ooliths and goethite casts of foraminifera and ostracods are not uncommon.

At Nettleton Hill, the top of the member is marked by an inconspicuous erosion surface. In other sections, especially to the south and east, it has been drawn at the level at which thin lenses of grey non-ferruginous mud appear within the oolitic sequence.

There is no biostratigraphical evidence of a nonsequence at the base of the member at Nettleton Hill, but the presence of derived fossils such as Jurassic *Pleuromya* and pavloviid ammonites in the basal pebble bed at Hundon Manor suggest a period of erosion affecting beds as old as the Basal Spilsby Nodule Bed. The relationship of this erosive event to events in the North Sea Basin and adjacent areas, documented by Rawson and Riley (1982), is not clear, but it may correspond to their late Valanginian event. No macrofossil or microfaunal investigations were carried out on the basal 0.3 m of the member above the channel, but Penny et al. (1969) noted broken specimens of the long-ranging bivalve *Camptonectes (Mclearnia) cinctus* from the ironstone immediately beneath their 'top calcareous beds'. The diverse foraminiferal and ostracod fauna at the base of the 'top calcareous beds', including *Dorothia kummi*, *Lenticulina guttata*, *L.* cf. *schreiteri*, *Hechticythere frankei frankei*, *H. hechti* and *Protocythere triplicata* is, however, distinctly Hauterivian in aspect, and comparable with that in early Hauterivian strata of the Speeton Clay of Yorkshire (Neale, 1962; Fletcher, 1973).

The 'top calcareous beds' represent a condensed sequence which has yielded early Hauterivian ammonites, notably species of *Endemoceras*, with *Acanthodiscus*, indicative of the *Endemoceras amblygonium* and *E. noricum* zones, together with derived late Valanginian ammonites such as *Bochianites neocomiensis, Dichotomites (Prodichotomites)* cf. *grotriani, D. (P.)* cf. *obsoleticostatus, Karakaschiceras* cf. *heteroptychum* and *Neohoploceras submartini*. Belemnites present include *Acroteuthis acrei*, and thus belong at least in part to *Acroteuthis* Assemblage 5 of late Valanginian age, the other species present being *A. explanatoides* and *A. paracmonoides*, which range downwards and upwards respectively. In addition to ammonites and belemnites, the 'top calcareous beds' contain a rich brachiopod fauna comprising *Lamellaerhynchia rostriformis* and *Rouillieria tilbyensis* (which are also found in the basal Claxby Beds) together with *L. walkeri claxbiensis, Rugitela hippopus, R. rugosa, Rouillieria walkeri* and *Sellithyris sella lindensis*. Bivalves present are dominated by the long-ranging *Camptonectes (Mclearnia) cinctus*.

A conspicuous band of weathered concretions lies close to the top of the member. The ostracod fauna within and immediately below this band consists of typical early Hauterivian species, but also includes *Exophthalmocythere anterospinosa* and *Euritycythere parisiorum*, the latter being a Tethyan species apparently restricted to the *amblygonium* and *noricum* zones. In addition, numerous foraminifera, including *Dorothia kummi, Epistomina ornata* and *Lenticulina crepidularis*, were found with sponge spicules, bryozoa, numerous small gastropods, echinoid spines and fish remains.

Tealby Formation

In its most complete development, to the south of the present district, the Tealby Formation comprises three members: Lower Tealby Clay, Tealby Limestone and Upper Tealby Clay. The last is overlain by, and passes laterally into a complex sequence of argillaceous and calcareous ironstones, the Roach Formation, also known as the Fulletby Beds (Swinnerton, 1935). In the published 1:50 000 Brigg (89) Sheet, the beds between the Tealby Limestone and the transgressive base of the Carstone are all included in the Upper Tealby Clay. Since the publication of the map, however, it has been realised that the greater part of the Upper Tealby Clay as mapped should more properly by regarded as the Roach Formation, although it is not clear whether the Roach here is the time-equivalent of the Roach farther south, or whether it belongs wholly or in part to the 'ferruginous facies' of the Upper Tealby Clay described by Owen and Thurrell (1968). The Upper Tealby Clay *sensu stricto* is apparently a thin bed of mudstone locally underlying the Roach and restricted to the southern part of the Brigg district. It was seen, for example, in the former section at Normanby [TF 1205 9640] and recorded in Thoresway Borehole [TF 1646 9632), just beyond the eastern margin of the district.

The Tealby Formation is up to 28 m thick in the extreme south of the district, but thins northwards to about 4 m at Hundon Manor before being transgressed by the Carstone Grit. Its outcrop does not continue beyond Audleby [TA 112 040]. The strata range in age from late Hauterivian to early Barremian.

Lower Tealby Clay Member

The Lower Tealby Clay is between 3 and 13 m thick. Extensive landslips, cambers, mudflows and downwash complicate and conceal much of the outcrop, but some sections of the highest beds are exposed in slip-scars [e.g. TF 120 952], where selenite occurs in the weathering zone. Most of the better sections come from boreholes, especially Thoresway Borehole, and trial pits dug to the Claxby Ironstone [TA 1182 0253; 1189 0243; TF 1135 9852; 107 994]; a former quarry section [TF 1127 9581] was described by Lamplugh (1896, p.203).

The Lower Tealby Clay rests nonsequentially on the eroded top of the Claxby Ironstone, and although the *regale* Zone is missing in the district, Rawson (1971) recorded derived phosphatised *Endemoceras regale* from directly above the erosion surface at Nettleton. In Thoresway Borehole, at least the basal metre of the Lower Tealby Clay can be assigned to the *regale* Zone based on the first occurrence of the foraminifer *Pseudonodosaria vulgata*, the occurrence of the ammonite *Subastieria* cf. *sulcosa* and the last occurrence of the ostracod *Cythereis (Rehacythereis) senkenbergi*. Macrofossil and microfossil evidence from the borehole indicates that all zones from *regale* up to possibly the basal Barremian *variabilis* Zone are recognisable within the Lower Tealby Clay (p.74).

At Nettleton Hill, the basal 2.45 m are extremely oolitic and have much in common with the underlying calcareous Upper Claxby Ironstone. Weakly calcitised and phosphatised bioturbated mud patches are common, together with irregular concentrations of chalky and sugary calcite nodules. At Nettleton Hill, the beds overlying this basal ferruginous sequence contain bands of small septarian ironstone nodules or nodules with phosphatised crusts that superficially resemble 'potato stones' (Neale, 1974). The bulk of the main sequence consists of pale olive-grey mudstone with scattered quartz grains and a high content of goethite ooliths, locally sufficiently abundant to form thin beds of ironstone (packstone). The ooliths are superficially similar to those in the basal 2.45 m and in the underlying ironstone, but, except for the outermost lamina, are affected by decomposition, with irregular clusters of granular goethite replacing the concentric lamination. Disseminated goethite also occurs as angular clasts, as casts of foraminifera and ostracods and as remnant ooliths. Glauconite is present in the lower part, but is less common than goethite; it occurs as peloids and foraminiferal casts and is most obvious within broken ooliths, diagenetic growths fracturing granular goethite masses, and within burrowfills. Pyrite is rare and forms crystal aggregates on calcite shells. Bands of laminated and blocky calcareous silty mudstones containing powdery shell debris also occur. Tiny mica flakes, siderite rhombs and preferential calcitisation characterise these siltier bands, which are harder than the other sediments and have probably been misidentified as limestones in some borehole sections. In the upper part of the sequence, a red-brown powder produced by oxidation of iron imparts a speckled appearance to some beds. The Lower Tealby Clay appears to be a transitional facies between that of the Claxby Ironstone and a more open-shelf facies like that of the Speeton Clay.

All levels are thoroughly bioturbated, but many well preserved fossils of long-ranging but typically Hauterivian taxa occur. The fauna, based on data from Thoresway Borehole, is of high diversity, and corals (*Trochocyathus conulus*), serpulids (*Tetraditrupa*), gastropods, bivalves, echinoids, crustaceans and fish are well represented. The extensive molluscan fauna includes the gastropods *Bathrotomaria* cf. *speetonensis* and *Nerineopsis aculeatum* and the bivalves *Acesta longa*, *Corbula isocardaeformis*, *Entolium orbiculare*, *Goniomya* sp., '*Limea*' *granulatissima*, '*Lucina*' cf. *hauchecornei*, *Panopea* sp., and *Resatrix* cf. *neocomiensis*. The occurrence of the echinoid *Toxaster retusus* within a 0.2 m interval suggests a possible correlation with Bed C3 (the 'Echinospatangus Bed') of the Speeton Clay in the lowest part of the *marginatus* Zone. The belemnite *Hibolites jaculoides* occurs throughout, and allows a broad correlation with the C Beds at Speeton. Few ammonites have been recorded; *Subastieria* cf. *sulcosa* (*regale* Zone), *Aegocrioceras* cf. *bicarinatum* (*inversum* Zone), *Simbirskites (Milanowskia) concinnus* and *Simbirskites* sp. (*concinnus* Subzone) have been recognised in Thoresway Borehole [TF 1646 9632], and *M. speetonensis* (*speetonensis* Subzone), *Crioceratites* (Rawson in Rawson et al., 1978), *Endemoceras*, *Simbirskites* aff. *toensbergensis* and *S. (S.)* cf. *virgifer* have been collected from the Nettleton area.

Within the lower, oolitic part of the member (*inversum* Zone) at Nettleton Hill Quarry, diverse ostracod faunas include *Cytherella fragilis*, *Parexophthalmocythere rodewaldensis*, *Quasihermanites bicarinata* and numerous Cytherurinae, together with *Amphicytherura bartensteini* and *Paranotacythere (P.) anglica* near the top. Above the basal 2.45 m *Apatocythere simulans* is found together with *Neocythere (N.) protovanveenae*. *Amphicytherura roemeri* and *Cytherelloidea pulchra*, present 3.45 m above the base of the Lower Tealby Clay, are used to infer the presence of the *gottschei* to *marginatus* zones. It seems likely that this is approximately the same level as that from which Bartenstein (1956) figured eight species of ostracods. A similar distribution of ostracods is seen in Thoresway Borehole, where a diverse foraminiferal assemblage includes *Lenticulina ouachensis wisselmanmi*. Dinoflagellate cysts also occur, and include *Batioladinium longicornutum*, *Chlamydophorella trabeculosa*, *Cribroperidinium sepimentum*, *Hystrichodinium furcatum*, *H. voigtii*, *Isthmocystis distincta*, *Muderongia simplex*, *M. tetracantha*, *Phoberocysta neocomica* and *Pseudoceratium pelliferum*.

Tealby Limestone Member

The Tealby Limestone ('greystone' of Judd, 1867) is a lenticular unit, generally less than 4 m thick, which conformably overlies the Lower Tealby Clay. The lower boundary is gradational in some of the ironstone boreholes, but elsewhere, for example in Thoresway Borehole, it is quite sharp. The upper limit is taken here at the top of a bed of friable ferruginous limestone, which is rich in large oysters and belemnites near its base and distinct from the underlying limestones. The sequence, totalling 3.46 m in Thoresway Borehole, comprises alternations of hard micritised bioclastic limestones (grainstones and packstones), and thin shaly beds (wackestones). Locally about ten hard beds are present, but the development of the thicker limestones appears to be extremely variable. The Tealby Limestone contains belemnites and rare ammonites which indicate an early Barremian age, ranging from the *variabilis* Zone to the *fissicostatum* Zone. The only exposures are at Oxgangs [TF 1122 9745] and Normanby Lodge [TF 1126 9584].

The wackestones tend to be wispy bedded, with a matrix of ferruginous mud containing sporadic lenses of medium grey mudstone. Geothite peloids are numerous, but limestone ooliths are rare. Glauconite and quartz grains persist from the Lower Tealby Clay, and muscovite flakes and shiny carbonaceous laths occur in the youngest beds. The cement is of calcite. The main characteristic of the limestone is the micritisation; almost all the calcitic elements are affected and the resultant pale olive meshwork (orange-brown where oxidised) shows much of the micrite to be ferruginous and siliceous. The peloids are generally rounded with a thin hard shiny outer lamina enveloping soft honeycombs of powdery to granular goethite or crystalline meshes. Goethite cases of foraminiferal chambers or echinoid spines are common and similar in structure to the peloids. Glauconite is patchily distributed as small hard peloids; it also replaces goethite and fills foraminiferal chambers. Well rounded to sharply angular quartz grains are also present. Small siderite rhombs occur in some calcareous bands, but pyrite is virtually absent. Bioturbation is common in the form of long meandering horizontal burrows, the rounded cross-sections of which are emphasised by concentric zones of green and brown which become grey on weathering. Halo-like concentric bands associated with weak phosphatisation around large disarticulated crustacean fragments are distinctive, especially near the base, where the decapod *Meyeria rapax* is common. Bivalves and echinoids are recognisable in the abundant shell debris, especially in grainstones high in the sequence.

The Tealby Limestone is only locally and patchily fossiliferous, the commonest fossils being long-ranging bivalves such as *Acesta longa*, *Camptonectes* (*Mclearnia*) *cinctus*, *Entolium orbiculare* and *Rastellum macropterum*. Particularly characteristic bivalves are the oyster *Gyrostrea osmana* (previously misdetermined as *Exogyra sinuata*) and the trigoniids *Iotrigonia scapha* (which is abundant at Hundon Manor) and *Linotrigonia (Oistotrigonia) ornata*. The belemnites *Hibolites jaculoides* and *Acroteuthis* (*Boreioteuthis*) *rawsoni*, characterising *Acroteuthis* Assemblage 7 (Pinckney and Rawson, 1974; Mutterlose et al., 1987), occur in the lower part of the limestone and permit correlation with the *variabilis* Zone. Ammonites are rare, and there are no records of the exact stratigraphical position from which specimens in museums came. However, specimens of *Crioceratites hildesiensis* sensu Immel and *C. woekeneri*, collected loose, but probably from low in the limestone, at Hundon Manor, indicate the *variabilis* Zone, as do specimens of *Simbirskites* (*Craspedodiscus*) spp., including *S. (C.) discofalcatus* and *S. (C.) juddii* from other localities.

Ostracods, investigated in three samples from Thoresway Borehole, proved to be common and diverse. They include *Euryitycythere dorsicristata*, restricted to the Lower and Middle Barremian of Lincolnshire; *Paranotacythere (P.) blanda*, which has been recorded from the *variabilis* Zone and up into the Middle Barremian in Yorkshire (Speeton Clay) and Germany; and *Protocythere strigosa*, known only from the Lower Barremian of the Paris Basin. Among the common foraminifera are *Citharina discors*, which is not known above the lower part of the *fissicostatum* Zone, *Dorothia kummi*, which disappears from the record in the early *variabilis* Zone, and

Vaginulopsis humilis cf. *praecursoria*, which is recorded for the last time in the *rarocinctum* Zone. Dinoflagellate cysts are represented by *Canningia* cf. *reticulata*, *Ctenidodinium elegantulum*, *Muderongia tetracantha* and *Pseudoceratium pelliferum*.

The top bed of the Tealby Limestone is much more fossiliferous than the underlying massive limestones. It is exposed at Oxgangs and Normanby Lodge, and was formerly seen in a pit at Normanby [TF 1205 9640] (Lamplugh, 1896; Ussher et al., 1888) and on Nettleton Hill, where it was regarded as part of the Tealby Limestone 'Greystone' (Dikes and Lee, 1837). The fauna is characterised by the large oysters *Aetostreon subsinuata*, *Gyrostrea osmana* and *Rastellum macropterum* associated with the long-ranging bivalves *Acesta longa*, *Camptonectes* (*Mclearnia*) *cinctus* and *Entolium orbiculare*. This assemblage is comparable with that sparsely represented in the underlying limestones, but is distinguished by the abundance of serpulids such as *Glomerula gordialis* and *Mucroserpula* sp, as well as by the belemnite assemblage comprising common *Praeoxyteuthis* spp. (*P. jasikofiana* and *P. pugio*) with rarer *Aulacoteuthis* cf. *speetonensis*. Other fossils include the brachiopods *Cyrtothyris cyrta arminiae*, *Lamellaerhynchia rawsoni* and *Rugitela* cf. *rugosa*. No ammonites have been recorded, but the co-occurrence of *Aulacoteuthis* and *Praeoxyteuthis* indicates the junction between the *Praeoxyteuthis pugio* and *Aulacoteuthis* spp. belemnite zones, which falls within the *fissicostatum* ammonite Zone in the Speeton Clay (Rawson and Mutterlose, 1983, fig. 7).

This oyster-belemnite bed constitutes a significant bioevent, which is represented in southern Lincolnshire by a 3 m interval within the lowest third of the type Upper Tealby Clay in the Alford and Skegness boreholes. The same assemblage was recovered from Roach ironstone facies near Kirmond Le Mire [TF 1996 9140], in the adjacent Grimsby district, demonstrating the complexity of lateral facies variation in this part of the succession.

Upper Tealby Clay

As noted previously (p.73), Upper Tealby Clay *sensu stricto* is present only as a feather edge in the southern part of the Brigg district, where it is represented by the 0.66 m (2 ft. 2 in.) bed of yellowish brown clay overlying the top oyster-rich bed of the Tealby Limestone in the former pit near Normanby (Ussher et al., 1888, p.101). It is also present immediately to the east of the district in Thoresway Borehole, where it comprises 0.52 m of silty mudstone with fragmentary bivalves (*Acesta sp.* and *Procyprina?*). Elsewhere in the Brigg district, i.e. to the north of Normanby, the Tealby Limestone is succeeded directly by the Roach Formation.

Roach Formation

This formation, depicted on the published map as Upper Tealby Clay (see p.73), is a sequence of highly ferruginous limestones, mudstones and sandstones. It is up to 11.5 m thick near Walesby, but is gradually overstepped by the Carstone northwards, being absent at outcrop north of Nettleton Bottom [TF 1257 9856] and north-west of Acre House [TF 1117 9701]. Exposures are meagre and near-surface oxidation and leaching are intense. Borehole records are poor, but generally show a lower unit of ferruginous limestones some 3 m thick, overlain by a variable thickness of muddy and sandy ironstones. This twofold division appears to correspond with the Roach Stone and Upper Roach respectively of southern Lincolnshire (Swinnerton, 1935) where, however, they may be younger (Owen and Thurrell, 1968).

The basal carbonate unit within the district is between 1.8 and 3 m thick. In the slightly thicker Thoresway Borehole succession it is 3.16 m thick (between 42.31 and 45.47 m). There, the basal 0.41 m is a calcitic biomicrite and contains numerous degraded goethite ooliths and peloids in addition to scattered quartz grains. It is thoroughly bioturbated, with large irregular patches of pale greenish grey mud containing most of the ooliths. Its top is marked by large calcite and partly goethitic shell fragments.

The succeeding 1.46 m are distinctive for their high percentage of disseminated speckled pyrite within a greenish grey calcitic biomicrite. Goethite is absent and tiny glauconite and quartz grains are conspicuous in the lowest 0.86 m. In the top 0.25 m there is a decrease in calcite and pale greyish olive berthierinitic mud forms the matrix. Quartz grains reappear near the top which is delineated by a 'flood' of relatively fresh goethite ooliths and goethitic shell fragments. The whole sequence is bioturbated. Burrows generally display concentric mud laminae associated with disseminated fine shell debris and micritic mud. Near the top specular pyrite with pale brown oxidised margins fills large vertical burrows.

The overlying 0.3 m are of hard bioturbated goethite-oolith packstone with rare quartz grains and cements of calcite biomicrite or, more rarely, biosparite. A little over 0.6 m of soft bioturbated goethite-oolitic wackestone and mudstone intervenes between this and an uppermost 0.39 m-thick hard calcitic bed in the basal 0.2 m of which goethite ooliths and smooth shiny angular clasts are concentrated to form packstone patches. Scattered small quartz grains are conspicuous as are partly goethitic shell fragments, among which prismatic calcite shells have yellowish orange goethite rods between prisms.

The carbonate-rich Roach Stone is overlain by a sequence of muddy ironstones about 8 m thick. The basal few centimetres are slightly indurated by calcite, but otherwise the junction is sharply defined. The ironstones are dominated by large fresh shiny goethite ooliths, peloids and flattish, sharp-cornered rafts, all being grain types present in the Claxby Ironstone. These are embedded in olive-grey berthierinitic mudstone, but are randomly distributed within grainstone-wackestone patches. Subrounded grains and rare small pebbles of clear quartz are scattered throughout. All levels are bioturbated and burrows are marked as grain-free paler greyish mudstone containing siderite. Very few shells are present.

The muddy ironstone is succeeded by a more fossiliferous sandy ironstone, but because of the Carstone overstep it occurs only to the south of Normanby le Wold, whence it gradually thickens towards Tealby. As much as 0.4 m is known to be preserved in the district, but in borehole records it has not been differentiated from the overlying sandy Carstone. It is distinguished from the underlying muddy facies by the presence of a considerable amount of quartz sand and pebbles of quartz and phosphate. The pebbles are mainly concentrated at the base where glauconite is also conspicuous. Higher up, the sequence is essentially a sandy, pebbly packstone with goethite ooliths and sand grains as the main constituents.

Little is known of the Roach macrofaunas in the present district, except in Thoresway Borehole. The top bed of the 'Roach Stone' yielded the long-ranging bivalves *Camptonectes* (*Mclearnia*) *cinctus* and *Entolium orbiculare*. Shelly beds near the base of the 'Upper Roach' contain *E. orbiculare*, *Rastellum macropterum*, a single *Oxyteuthis* sp. juv. tentatively assigned to the *O. germanica* group and *Ditrupa* (*Tetraditrupa*) sp. The belemnite suggests that the highest part of the Roach is of late Barremian, *germanica* Belemnite Zone age, though it should be noted that *O. germanica* in southern Lincolnshire is believed to be restricted to the Lower Roach (Rawson and Mutterlose, 1983). The topmost beds of the 'Roach' in the borehole contain *Plagiostoma ferdinandi*, in addition to *E. orbiculare*, also indicative of a late Barremian age, but in England known only from the upper part of the Lower Roach of southern Lincolnshire. The Upper Roach in Thoresway Borehole may therefore be coeval with the Lower Roach of the southern area, a possibility reinforced by the fact that *C.* (*M.*) *cinctus* and *Rastellum macropterum* are known only from the Lower Roach in the type area. If this proved to be the case it would be reasonable to view the Roach-type sediments of the present district as an anomalous development of the 'ferruginous facies' of the Upper Tealby Clay, as posulated by Owen and Thurrell (1968).

Ostracod faunas in the Roach are sparse, of low diversity, and commonly poorly preserved, partly as a result of reworking. The long-ranging zonal index *Protocythere triplicata* has been found throughout much of the formation in Thoresway Borehole. Other species present include *Amphicytherura bartensteini*, *Cytheropteron nova*, *Haplocytheridea parallela*, *Paranotacythere* (*Paranotocythere*) *blenda*, *Schuleridea hammi* and *S. rhomboidalis*, together with long-ranging forms such as *Acrocythere hauteriviana*. In contrast to the ostracods, dinoflagellate cysts are quite numerous, with common *Muderongia tetracantha* and *Pseudoceratium pelliferum*.

Carstone Formation

The name Carstone first appeared on William Smith's 1815 Geological Map of England and Wales, and was formally applied to pebbly sands and clays at the base of the Chalk by Strahan (1886, p.486). These largely arenaceous sediments grade upwards into the Chalk and may span the Aptian–Albian boundary. In southern Lincolnshire two members are recognisable: the Carstone sands and clays and the overlying Carstone grit (Swinnerton, 1935). Only the upper member is preserved in the present district.

The Carstone Grit (Thoresway Sand of Dikes and Lee, 1837, p.563) is a gritty basement bed to the Chalk, and is the facies generally associated with the term Carstone. Its base is transgressive, and northwards through Lincolnshire and Humberside it gradually cuts across the underlying beds. Along the outcrop the Carstone rests on Roach to the south of Nettleton Bottom, on Tealby Limestone between Nettleton Top and Audleby, on Lower Tealby Clay between Audleby and Clixby, on Spilsby Sandstone between Clixby and Grasby, and on progressively older levels of the Ancholme Clay Group farther north. It does not form a distinct topographical feature and north of Melton Gallows is barely traceable.

North of Grasby downwash obscures the outcrop. Kent (1937, p.167) found 0.3–0.5 m of sand with rolled nodules on Ancholme Group mudstones in a cutting [TA 074 059] near Searby, and pebbly sand lies at the base of red chalk to the east [TA 0866 0493]. A small pit [TA 1025 0440] near Clixby exposes 0.9 m of pebbly sandstone resting on the Spilsby Sandstone, and there are similar sections in nearby crags [TA 1054 0427]. South of Grasby the thickness varies from 2.29 to 8.53 m. Ussher (1890, p.111) recorded 2.5–3.7 m in a small quarry at Nettleton Grange Barns [TF 1205 9942], and Nettleton Bottom Quarry [TF 1249 9823] reveals 4.58 m of highly weathered orange, gritty sands. The Carstone is coarsest and thickest in the south-eastern part of the district, where 8.53 m are known just north of Normanby le Wold. Boreholes to the east of the outcrop in the Melton Ross–Kirmington area have shown up to 3.7 m [TA 0640 1030] of sand possibly referable to the Carstone. There are 1.5 m of gritty sands in the Caistor Waterworks Borehole [TA 1211 0137], and 4.5 m in a borehole at Whitegate Hill Watermill [TA 1255 0067].

Between Melton Gallows [TA 051 109] and Hundon [TA 119 023] the formation is barely preserved, being represented only by pebbles at several localities, for example near Barnetby le Wold [TA 063 103], and this area appears to separate two distinct facies. To the south the coarse sands have a cementing matrix of goethitic/berthierinitic clay, with glauconite as a minor component; to the north the sands are finer grained and weakly cemented by highly calcareous clay.

The southern facies is a friable gritty sandstone. The main component is subangular quartz, but the cement gives the rock a pale green colour where fresh, and a pale orange or dusky red colour where weathered. The quartz grains commonly contain inclusions, and have strain shadows. Many have pale to dark yellowish films or hardened ferruginous clay adhesions. Sand-size grains of degraded glauconite, orthoclase, sodic plagioclase and microcline also occur. Vein quartz and quartzite pebbles showing signs of metamorphism and mylonitisation are abundant in thin persistent bands at several levels. They are accompanied by rounded fragments of dusky brown hematite-veined 'lydian' chert, phosphatic nodules, calcareous siltstone and boxstone chips. Virtually all these fragments can be matched to rocks in the underlying sequence, and support Strahan's (1886, p.490) view that the Carstone sediments largely derive from that source. The basal 0.3 m contains many smooth, irregularly etched, fossiliferous phosphatic nodules ('potato stones') similar to those in both the Spilsby Sandstone and the Lower Tealby Clay. Phosphatic nodules also occur in higher pebbly bands, which probably represent erosional lag deposits rather than horizons of phosphatisation at times of slow deposition as was thought possible by Versey and Carter (1926, pp.349–350). The heavy mineral suite separated from an exposure at Thoresway (Versey and Carter, 1926, p.352) is similar to that recorded by Ingham (1929) from the Spilsby Sandstone of the same area, which thus seems to have been the principal contributor to this part of the Carstone.

The northern facies is usually less than 1 m thick. It resembles the upper part of the southern succession but is much finer grained, with rare calcite ooliths. Pebbles of phosphate, polished quartz and calcareous siltstone again lie at the base, together with fragments of clayey sandstone and oyster-bearing siltstone. North of the Humber the matrix contains fresh goethite ooliths, probably locally generated, together with dark bluish and purplish green weathering clay (Kaye, 1964). Bedding, where not disrupted by weathering, is planar-laminated. Large boudin-like goethitic boxstones, with irregular hard purplish brown Liesegang rings and goethite-filled pipes and joints are characteristic along the whole outcrop, particularly in the middle of the exposures at Nettleton Bottom Quarry. The Carstone is highly porous and a significant aquifer, soft enough to be dug out by spade, but with a few cemented patches.

The Carstone includes both derived and indigenous faunas. By far the greatest proportion of the fossils occur in the reworked nodules, pebbles and clasts. Four main *remanié* suites are recognisable, indicating provenances from the Basal Spilsby Nodule Bed, Mid Spilsby Nodule Bed, Lower Tealby Clay and Tealby Limestone. The fossils are predominantly ammonites and bivalves. The commonest ammonites, notably *Subcraspedites (S.) claxbiensis*, *S. (Volgidiscus) lamplughi*, in addition to rarer *Paracraspedites* sp. and *Subcraspedites (Swinnertonea)* sp., are of latest Jurassic Portlandian age, but *Surites* is of earliest Cretaceous (Ryazanian) age. The bivalves include *Hiatella (Pseudosaxicava) foetida*, suggesting derivation from the Basal Spilsby Nodule Bed, and *Gyrostrea osmana*, which is common in the Tealby Limestone. Dinoflagellate cysts are sparse, poorly preserved and dominated by the long-ranging *Odontochitina operculata*, many specimens of which are corroded.

It is only in recent years that indigenous fossils have been proved in the Carstone. From the top 0.15 m at Melton Bottoms [SE 973 273] in North Humberside, Owen et al. (1968) recorded the brachiopods *Burrirhynchia leightonensis*, *Cyclothyris mirabilis* and *Modestella festiva*, associated with *Aetostreon latissimum*, *Neithea* sp. and *Rastellum colubrinum*, and drew a comparison with the Lower Albian *tardefurcata* Zone Shenley Limestone fauna of Bedfordshire. A foraminiferal fauna from the same horizon (Dilley, 1969) included *Osangularia schloenbachi*, which is restricted at Speeton to the 'Greensand Streak' and overlying basal A3 marls, thus supporting the *tardefurcata* Zone age indicated by the brachiopods. *Burrirhynchia leightonensis* and *Neithea* aff. *quinquecostata* were found near the top of the Carstone at South Ferriby Quarry (Smart and Wood, 1976) while, farther south, the motorway excavation in the Carstone at Melton Gallows [TA 0498 1102] yielded the same fauna as at Melton Bottoms, with the addition of *Entolium orbiculare*, *'Exogyra' conica* (striated variety of Keeping, 1883, pl.4, fig. 3c), *Oxytoma* ex gr. *pectinatum*, *Rastellum macropterum* of possible *mammillatum* Zone age, and corroded belemnites including *Neohibolites minimus* and *N.* cf. *minimus pinguis*. Microfaunas from the top part of the Carstone at this locality include the foraminifer *Arenobulimina macfadyeni* and *Osangularia schloenbachi* (recorded previously by Dilley (1969) from Melton Bottoms), associated with *Marginulinopsis cephalotes*, *Saracenaria bononiensis* and long-ranging Albian ostracods..

The indigenous faunas indicate an open-sea environment, and the deposits suggest upper shore-face sedimentation similar to that pertaining during Lower Spilsby Sandstone times.

The Carstone is conformably overlain by Ferriby Chalk with an apparently gradational junction (see p.81). The lowest part of the chalk is generally red and of Albian age. However, since it is lithologically an integral part of the mainly Upper Cretaceous Ferriby Chalk Formation, it is described with the latter in the next section.

UPPER CRETACEOUS (CHALK)

The Chalk, 250 + m thick, crops out in the eastern part of the district, though the outcrops of the higher parts of the sequence are extensively obscured by superficial deposits. It dips generally east-north-eastwards at 1½ to 2 degrees and, from the northern margin of the district to near Grasby, the lower part of the Chalk forms a prominent scarp with a north-west to south-east trend, modified only by a swing into the Humber, accentuated by a minor anticline, and another swing into a valley at Melton. South-east of Grasby there is a sudden change in the alignment of the scarp to approximately north–south, the result of the east-north-east-trending 'Caistor Monocline', originally postulated by Versey (1931).

The Chalk of the district falls within a loosely defined 'Northern Province' that extends from Flamborough Head to northern Norfolk and continues eastwards beneath the North Sea. The northern succession has much in common with correlatives in Germany and areas farther east, and contrasts strongly with that in southern England, which constitutes a 'Southern' or 'Anglo-Paris Basin' Province. In general, the Northern Province succession is characterised by chalks that are hard and thinly bedded compared with the relatively soft massive chalk to the south; the flints are predominantly of tabular rather than burrow-form types. Faunally, too, there are differences between the two provinces, notably in the echinoids and brachiopods; there is significantly less faunal diversity in the north, possibly because the water was deeper there.

Table 3 shows the various classifications that have been applied to the Chalk of the Northern Province. The classification used in this account is that of Wood and Smith (1978), who also formally established the lithostratigraphical marker horizons in the Welton and Burnham Chalk formations. Figure 28 shows the correlation between the Northern and Southern provinces based on marl bands, and the relationship between the Wood and Smith classification and that established by Mortimore (1986) for Sussex and adjacent areas. Some of the evidence for this correlation is discussed by Mortimore and Wood (1986) in greater detail than is possible in this memoir.

The Chalk of the Northern Province is divided into four formations—in ascending order the Ferriby, Welton, Burnham and Flamborough Chalk formations (Wood and Smith, 1978) respectively some 26, 53, 150 and 20 + m thick in this district. The Ferriby Chalk thins northwards towards and over the Market Weighton Structure (p.106), but above this

Table 3 Classification of the Northern Province Chalk

Initial classification used in Yorkshire	Jukes-Browne and Hill, 1903; 1904	Wood and Smith, 1978	Macrofossil zones	Stages
			Sphenoceramus lingua	Campanian
Upper Chalk without Flints		Flamborough Chalk Formation	*Marsupites testudinarius*	
			Uintacrinus socialis	Santonian
	Upper Chalk		*Micraster coranguinum*	
		Burnham Chalk Formation	*Micraster cortestudinarium*	Coniacian
Middle Chalk with Flints			*Sternotaxis planus*	
			Terebratulina lata	Turonian
	Middle Chalk	Welton Chalk Formation	*Mytiloides labiatus* (*s.l.*)	
			Neocardioceras	
			Metoicoceras	Cenomanian
Lower Chalk = Grey Chalk without Flints	Lower Chalk	Ferriby Chalk Formation	*Holaster trecensis*	
			Holaster subglobosus	
			Stoliczkaia dispar	
Red Chalk	Red Chalk	Hunstanton Chalk Member	*Mortoniceras inflatum*	Albian
			Condensed Middle Albian	

stratigraphical level there is no evidence that the structure had any control on sedimentation. There is a general increase in thickness eastwards, presumably into the major north-west-trending plunging syncline shown on structural maps of Yorkshire (e.g. Neale, 1974, fig. 56). South of the Humber, the Ferriby and Welton chalks thicken southwards with, to the south of the Brigg district, an accompanying reduction in hardness and a significant increase in faunal diversity. Part at least of the Burnham Chalk actually thins away from the Market Weighton Structure, a phenomenon that can be demonstrated throughout the Kingston upon Hull district from Willerby in the north to Ulceby on the eastern margin. It is not clear, however, whether this represents a regional trend, or whether the thinning is controlled by structures in the vicinity of the Humber that did not become active until after the beginning of Burnham Chalk deposition.

The sediments are predominantly chalks, i.e. extremely pure, fine-grained (micritic), pelagic limestones of biogenic origin mainly composed of debris derived from the calcitic plates (coccoliths) of a group of unicellular algae (Coccolithophoridae). In addition to coccolith debris, there are bioclastic components comprising spherical bodies (calcispheres), 50 μm in diameter, that are the calcitic tests of a specialised group of dinoflagellates, planktonic and benthonic foraminifera (typically less than 5 per cent), and sparse larger bioclasts such as molluscan and echinoderm fragments. At some horizons, notably in the Ferriby Chalk, the chalks are coarse grained, being composed largely of shelly debris with little or no coccolith component, the original coccolith 'flour' having been winnowed away by current action in relatively shallow water. The calcisphere component is highly variable, but at some horizons assumes almost rock-forming proportions.

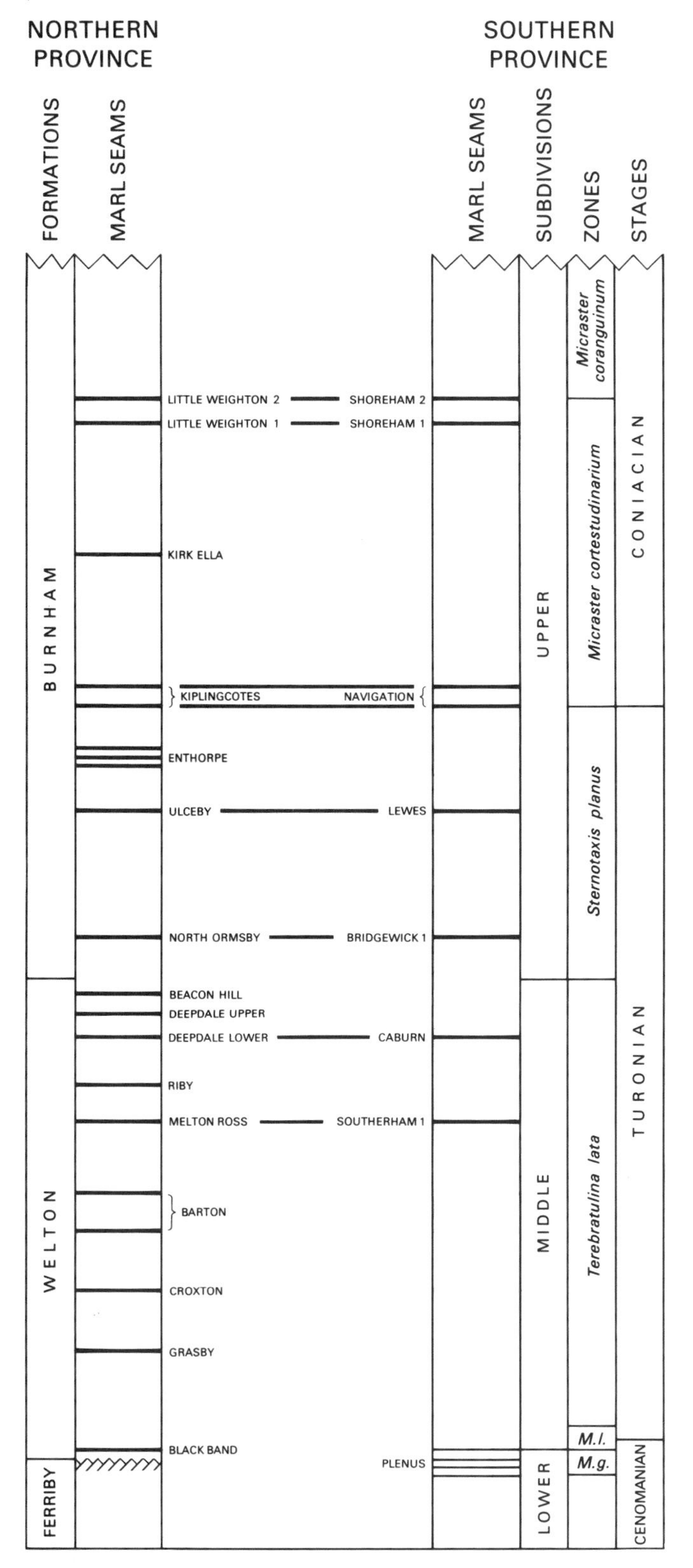

Figure 28 Correlation of the Northern Province Chalk, as seen in the district, with the Chalk of the Southern Province by means of marl seams. (Correlation of the marl seams below the Melton Ross Marl with those below the Southerham 1 Marl is at present uncertain).

The chalks of the Ferriby Chalk Formation are relatively impure, containing clay minerals such as montmorillonite and illite, together with small amounts of quartz and feldspar. At some horizons, and particularly in the basal few metres comprising the so-called Red Chalk, the rock is red or pink due to the presence of finely divided hematite, but the Ferriby sediments are otherwise predominantly grey or off-white. The succeeding Welton and Burnham chalks are extremely pure, white to off-white, and characterised by insoluble residues of the order of 2 per cent. A higher clay-mineral content is found in the more marly chalks of the Flamborough Chalk, but this lies outside the scope of the present memoir.

In addition to small quantities of silica in the clay minerals, silica in the form of cryptocrystalline quartz is found throughout the greater part of the Chalk Group of the district as nodules and tabular masses of flint. There are no flints in the Ferriby Chalk or in the basal few metres of the Welton Chalk, but above this level they occur, more or less conspicuously, up to the top of the Burnham Chalk. In the overlying Flamborough Chalk flints are again absent. In contrast to the black or otherwise dark-coloured flints of southern England, the flints of the Northern Province are predominantly grey with ill-defined margins, and commonly appear to merge with the surrounding chalk; thus they are difficult to see in weathered sections. At certain horizons the tabular flints are carious, presenting a texture described by the early workers as 'intermingled chalk with flints'. In general, nodular burrow-form flints are succeeded upwards by tabular flints, both types being found in courses parallel to the bedding. Giant cylindrical vertical flints known as paramoudras are common at some levels in the Burnham Chalk. They may be several metres long and over 1 metre in diameter, one of the best known examples being that illustrated by Hancock (1975, pl.1, fig. D) from the Ashby Hill Quarry in the adjacent Grimsby district. In contrast to nodular and tabular flints, both of which are related in some way to the fill of crustacean burrows of *Thalassinoides* type, a paramoudra forms around, either close to or at some distance from, an extremely thin vertical burrow (*Bathichnus paramoudra*) only a few millimetres in diameter (Bromley et al., 1975).

Marl bands are another significant feature of the succession, particularly in the Welton Chalk and the lower part of the Burnham Chalk. They are from 1 to 10 cm thick and contain 30 to 65 per cent of insoluble noncarbonate material; many are of wide lateral extent. Some marls are of volcanogenic origin, as indicated by their clay-mineral and trace-element chemistry, and the presence of pyroclasts (Pacey, 1984). In addition to clay minerals the marls contain minute quantities of a pelletal, uraniferous calcium apatite (francolite) which possesses a strong gamma radiation. It is possible to identify the various marls in wells and boreholes from the high peaks they produce in gamma logs. The marls also have low electrical resistivity, and are correspondingly expressed as low 'spikes' on resistivity logs, particularly 16 inch normal and single point. Using data on thickness and relative spacing of marls from measured sections, it has proved possible to relate each gamma or resistivity spike to the marl that caused it (Figure 31), and in this way to interpret uncored borehole sections with a high degree of accuracy (Barker et al., 1984, fig. 4). Extending this technique outside the Northern Province, it can now be demonstrated (Mortimore and Wood, 1986; Murray, 1986) that many of the marls of the northern succession can be correlated with marls in East Anglia and southern England (Figure 28), although a volcanogenic origin for the latter remains to be proved. Even farther afield, it is highly likely that some of the marls of the Northern Province have their correlatives on the other side of the North Sea in the Münster 'Basin' and Lower Saxony Basin successions of Germany (Wood et al., 1984).

The chalks of the Northern Province are generally much harder than those of southern England because of secondary deposition of calcite cement in the pore spaces of the original sediment following pressure-solution under overburden and tectonic stress (Scholle, 1974; Mimran, 1977). The northern chalks are also characterised by the development of stylolites, another manifestation of pressure-solution acting upon an already lithified sediment. In this district, the degree of induration varies both vertically and horizontally. Thus, chalk that is no harder than most chalks in southern England may be intercalated with beds of relatively indurated and cemented chalk, while successions of relatively soft chalks may be found well to the north of correlative, indurated successions. The picture is obviously extremely complex, and may have as much to do with the character of the original sediment as with a supposedly regional tectonic stress-induced overprint.

Seen under the scanning electron microscope, these hard chalks exhibit relatively well-preserved coccoliths embedded in a matrix of secondary euhedral calcite crystals with partially interlocking textures (e.g. Scholle, 1974, figs. 11, 12), although in some cases the coccoliths are partly obscured by syntaxial overgrowths of secondary calcite (Mimran, 1977). In these compacted and cemented chalks there is an intimate relationship between the crystals of the matrix and those comprising the skeletons of macrofossils. Hence, specimens are difficult to extract except where a stylolitic surface has

Plate 2 Ancholme Clay Group mudstones (dark grey) overlain by Carstone and Hunstanton Chalk Member (brown) and Chalk at South Ferriby Quarry. The Chalk is clearly divided by the Black Band into Welton Chalk above and Ferriby Chalk below. The Chalk and mudstones are worked for cement manufacture. A 14379

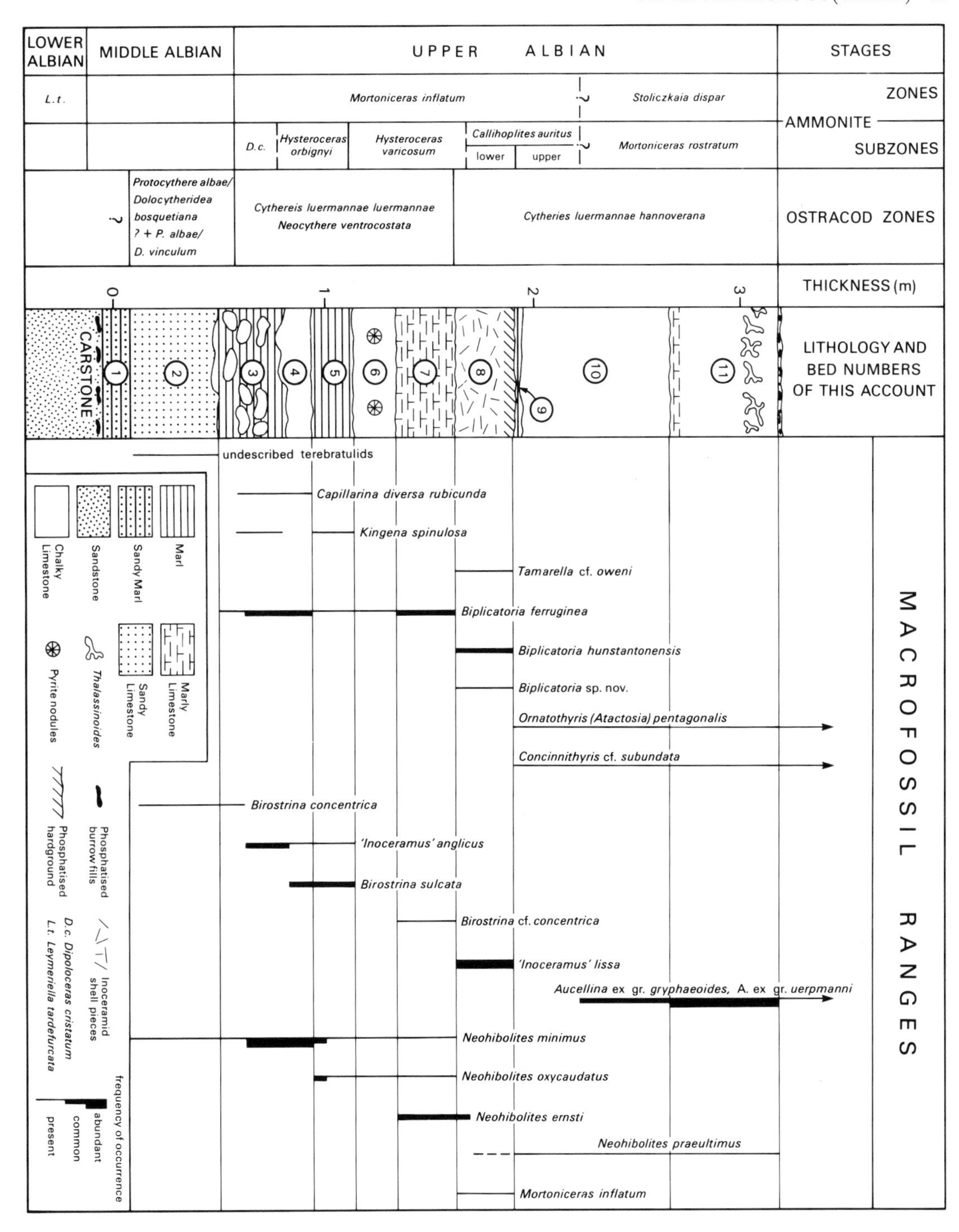

Figure 29 Stratigraphy of the Hunstanton Chalk Member (Red Chalk), with ranges of significant macrofossils

developed between them and the surrounding matrix, or where they have been gradually exposed through the natural processes of weathering.

Much of the Northern Province succession is relatively barren of macrofossils, except at certain horizons in which they are locally abundant. The faunas in the present district are generally of very low diversity, being restricted largely to brachiopods, inoceramid bivalves and echinoids. However, there seems to be an increase in diversity from thinner into thicker successions, this being particularly noticeable in the lower part of the Burnham Chalk as it is traced southwards. The macrofaunas indicate that the sediments of the Chalk Group were deposited in a fully marine environment. The depth of deposition is a subject of controversy, but the extreme rarity of original aragonite-shelled fossils, coupled with the general absence (except in the Ferriby Chalk and at some horizons in the Burnham Chalk) of phosphate and glauconite-impregnated hardgrounds, indicates deposition near the maximum of 300 m normally suggested for white chalks.

Ferriby Chalk Formation

Hunstanton Chalk Member

The Hunstanton Chalk Member or Red Chalk *sensu* Jeans (1973, 1980) comprises a complex succession of marls and both nodular and massive indurated chalky limestones of mid to late Albian age. The sequence is much condensed, with several nonsequences, and is typically brick-red in colour, particularly in its lower part. A high content of detrital quartz and feldspar grains is present in the lower part, and many of the more calcareous beds are rich in calcispheres and foraminifera. The thickness is variable, the maximum recorded within the district being just over 3 m in the working quarry at South Ferriby [SE 9915 2045]. In most localities, the member appears to grade down into the Carstone Formation, though the contact can be identified by an horizon of small phosphatised burrow-fills. In some places, however, the Hunstanton Chalk apparently oversteps the Carstone and rests either on older Cretaceous sediments (e.g. near Melton Ross) or on Upper Jurassic mudstones (Bigby Quarry [TA 0594 0782]). The upper limit of the member, as already noted by Jeans, is marked by a clearly defined, locally bored and stromatolite-encrusted erosion surface. This surface does not mark the upper limit of redness, which is variable and can extend to the top of the overlying Sponge/Paradoxica Bed (Wood and Smith, 1978, fig. 3). The 0.025 m rusty red, silty marl overlying the erosion surface, regarded by Andrews (1983) as the topmost bed (HF) of the Hunstanton Red Rock at Hunstanton, is treated here as an independent unnamed unit intercalated between the Hunstanton Chalk and the Paradoxica Bed (p.84).

The best and thickest exposed section at South Ferriby Quarry (Plate 2) is taken as a standard for the member in Humberside, although the much condensed sequence at Hunstanton (Andrews, 1983, and references therein) is provisionally accepted as the stratotype. Mr A A Morter recognised eleven distinct beds (Figure 29), but their relationship to the beds recognised by Andrews (1983) in the stratotype is not clear.

	Thickness m
Bed 11 Brick-red, massive, indurated, chalky limestone with abundant large *Aucellina* and small *Neohibolites*; top third of bed contains a *Thalassinoides paradoxica* burrow system; *Concinnithyris subundata*, *Ornatothyris (Actactosia) pentagonalis*, *Aucellina* ex gr. *gryphaeoides*, *A.* ex gr. *uerpmanni*, *Neohibolites praeultimus*, *N.* sp. [small, 15–20 mm long], *Holaster* sp. and cidarid radioles.	0.50
Bed 10 Brick-red, massive, rubbly, indurated, chalky limestone with some clayey wisps and partings; manganese on joints; common small *Neohibolites* and less common terebratulids; top marked by weak separation plane; *Concinnithyris* cf. *subundata*, *Ornatothyris (A.) pentagonalis*, *Aucellina* ex gr. *gryphaeoides*, *A.* ex gr. *uerpmanni* (both with smooth shell ornament), *Plicatula minuta* and *Neohibolites praeultimus*.	0.68–0.69
Bed 9 Red marl with chalky pebbles; *Biplicatoria hunstantonensis*.	0.03–0.04
Bed 8 Pale yellow and rusty, indurated, nodular, chalky limestone terminated above by phosphatised hardground; manganese on joints; abundant fragments of '*Inoceramus*' *lissa* throughout; concentration of *Neohibolites ernsti* at base; *Biplicatoria hunstantonensis*, *B.* sp. nov., *Tamarella* cf. *oweni*, '*Inoceramus*' *lissa*, *Pycnodonte (Phygraea)* aff. *vesicularis*, *Neohibolites ernsti*, *N.* transitional between *N. ernsti* and *N. praeultimus*, *Hemicrinus canon* and *Nielsenicrinus* aff. *cretaceus*; probably source of the ammonite *Mortoniceras (M.) inflatum* figured by Kent (1980b, pl.21)	0.30
Bed 7 Dark brick-red, marly limestone with abundant *Neohibolites*; common brachiopods and inoceramid shell fragments; *Biplicatoria ferruginea*, *Birostrina* cf. *concentrica*, *N.* spp. including *N. ernsti*, *N. ernsti* transitional to *N. oxycaudatus*, *N.* cf. *oxycaudatus*, *N. minimus*, and a cirripede valve.	0.24
Bed 6 Yellow, gritty, indurated, massive chalky limestone with pyrite nodules and inoceramid shell fragments; strong separation plane at top; limited fauna comprising *Biplicatoria ferruginea*, *Neohibolites minimus* and *N. oxycaudatus*	0.20
Bed 5 Brick-red marl with inoceramid debris; greenish at top; common *Birostrina sulcata* and *Neohibolites*; '*Rotularia*' cf. *umbonata*, *Kingena spinulosa*, *Biplicatoria ferruginea*, *Terebratulina martiniana* sensu Schmid *non* d'Orbigny, *Eopecten studeri*, '*Inoceramus*' *anglicus* [fragments], *Pycnodonte?* sp., *Neohibolites* spp., including *N. minimus* and *N. oxycaudatus*.	0.14–0.18
Bed 4 Pale brick-red, chalky limestone; continuous, with irregular base; rich macrofauna with common *Neohibolites*; *Biplicatoria ferruginea*, *Capillarina diversa rubicunda*, *Terebratulina* cf. *martiniana*, *Birostrina sulcata*, '*Inoceramus*' *anglicus* [fragments], '*I.*' cf. *anglicus*, *Plicatula minuta*, *Neohibolites minimus* and fish remains including sharks' teeth.	0.13
Bed 3 Dark red-brown marl with chalky limestone cobbles, phosphatic nodules and polished pebbles; *Neohibolites* abundant, particularly at base; '*Rotularia*' cf. *umbonata*, *B. ferruginea*, *Capillarina diversa rubicunda*, *Kingena spinulosa*, *Atreta* sp., *Birostrina concentrica* [derived, presumably ex Middle Albian], '*I.*' *anglicus*, *Plicatula* spp. including *P. minuta*, *N. minimus*, *Nielsenicrinus cretaceus* and indet. fish debris; fauna from base of bed comprises *B. ferruginea*, *B. concentrica*, *Gryphaeostrea canaliculata* and *N. minimus*.	0.20–0.26
Bed 2 Pale red-brown, silty, rubbly, chalky limestone, irregularly blocky with marl envelopes around blocks, becoming massive on south side of quarry; undescribed	

	Thickness m
terebratulids, *B. concentrica* and *N. minimus*.	0.30–0.35
Bed 1 Red-brown, sandy marl with common *Neohibolites*; horizon of pebbles and phosphatised burrow-fills at base; *N. minimus*.	0.15

The lower part of the Hunstanton Chalk (Beds 1–8) includes most of the argillaceous beds and contains detrital grains throughout. It typically retains the red coloration in those localities where discoloration due to sulphidisation has occurred. The top of Bed 8 is cemented, with a well-defined phosphatised hardground.

The upper part of the Hunstanton Chalk (Beds 9–11) has far fewer detrital grains, is much more calcareous, and tends to form massive, well-jointed units in exposures, in contrast to the more rubbly appearance of the underlying beds. It tends to be paler coloured, and is prone to discoloration through sulphidisation.

Jeans (1980, fig. 3) also recognised two broad divisions (the Goulceby Member and the overlying Brinkhill Member) within his 'Red Chalk' Formation, but treated the inoceramid shell-rich Bed 8 as the basal unit of an 'upward-fining cycle'. Whilst his classification has some merit, it is clear that there is a significant nonsequence (see below) between Bed 8 and the overlying beds. Furthermore, Beds 9 and 10 could themselves be regarded as successive parts of a fining-upward rhythm, a second such rhythm perhaps beginning at the base of Bed 11.

Ammonites are extremely rare in the Hunstanton Chalk, probably mainly due to non-preservation as a result of sea-floor dissolution of their aragonite shells. The classification of the succession within the context of the standard ammonite-based zonal scheme for the Albian is based on indirect correlation with ammonite-bearing Gault (mudstone) successions of eastern and southern England using other faunal elements, notably bivalves (*Aucellina*, *Birostrina* and '*Inoceramus*', belemnites (*Neohibolites*) and ostracods. Some of this evidence is unpublished, but the biostratigraphical significance of *Neohibolites* and inoceramid bivalves in the Gault of East Anglia has been reviewed by Gallois and Morter (1982), and the use of *Aucellina* and ostracods has been described by Morter and Wood (1983) and by Wilkinson and Morter (1981) respectively.

Figure 29 shows the ranges and relative frequencies of selected macrofossil taxa plotted against the stratigraphy. Beds 1–8 inclusive are characterised by inoceramids and by species of the biplicate terebratulid genus *Biplicatoria* [commonly referred to *Moutonithyris dutempleana*]; in the higher sequence, both inoceramids and *Biplicatoria* are absent [*Biplicatoria* just reaches Bed 9], *Aucellina* ex gr. *gryphaeoides* and *A.* ex gr. *uerpmanni* are relatively common, particularly at the top, and the terebratulids are represented by *Ornatothyris (Actactosia)* spp. and smooth non-plicate forms tentatively assigned to Cenomanian species of *Concinnithyris*.

The following biostratigraphically significant faunal changes have been identified by A A Morter:

1 In Bed 8, *Biplicatoria hunstantonensis* and *B.* sp. nov. replace the *B. ferruginea* of Beds 3–7.

2 Beds 1–7 inclusive are characterised by species of the thin-shelled inoceramid *Birostrina*, with *B. concentrica concentrica* ranging up to the basal part of Bed 3, *B. sulcata* restricted to beds 4 and 5, and *B.* cf. *concentrica* (including subspecies C and D of Kauffman, 1978) occurring in Bed 7. *Birostrina* is absent from Bed 8, which yields abundant thick-shelled fragments of the large '*Inoceramus*' *lissa*. This occurrence of '*I*' *lissa* in flood abundance constitutes a major bioevent, which can be identified in Gault successions in East Anglia (Gallois and Morter, 1982) and southern England, as well as in northern Germany.

3 The successive entries of *Neohibolites oxycaudatus*, *N. ernsti* and *N. praeultimus* at the bases of beds 5 and 7, and within Bed 8 respectively are significant. Above Bed 8, only *N. praeultimus* is present, but it is replaced by *N. ultimus* in the Paradoxica Bed and its underlying red marl. The common occurrence of *Neohibolites*, particularly *N. ernsti* and *N. oxycaudatus* in the marly chalks of Bed 7 constitutes a widespread bioevent.

Beds 1, 2 and the basal part of Bed 3 can be assigned to the Middle Albian, although correlation with the standard ammonite zonation is possible only on a broad basis. Bed 1, with *Neohibolites minimus*, is of uncertain age, but presumably belongs to the *dentatus* Zone. Bed 2 yielded *Matronella corrigenda* (Kaye) at Elsham Interchange cutting [TA 052 111]. This species is restricted elsewhere in Britain to the *dentatus* and basal *loricatus* zones, and forms an important element of the *albae/vinculum* ostracod zonal assemblage (Wilkinson and Morter, 1981). The succeeding *albae/bosquetiana* ostracod Zone was recognised in Bed 2 at South Ferriby, indicating, by analogy with data from the Gault of East Anglia, a range from the upper part of the *intermedius* to the *subdelaruei* subzones of the *loricatus* Zone. It would appear that Bed 2 is extremely condensed and comprises part of the *dentatus* Zone and most if not all of the *loricatus* Zone.

Bed 3 is complex, and consists of a dark red-brown marl with a central concentration of cobbles of chalky limestone, and similar cobbles preserved in irregular pockets in the basal part. Macrofossils collected specifically from the basal cobbles include *Biplicatoria ferruginea* and *Birostrina concentrica concentrica*; *Neohibolites minimus* occurs in the surrounding marl. The presence of *B. concentrica concentrica* indicates that the cobbles belong to the Middle Albian, since this species (in Gault facies) does not range above the *lautus* Zone at the top of the substage. It is probable, therefore, that the whole of the *lautus* Zone and, possibly, the uppermost part of the underlying *loricatus* Zone are represented in extremely condensed form at the base of Bed 3.

The central cobbles in Bed 3 yield '*Inoceramus*' *anglicus*. The occurrence of this species and the absence of *Birostrina sulcata* suggests that the Upper Albian part of this bed could be assigned to the *inflatum* Zone, *cristatum* Subzone, at the base of the substage. A sample taken from the marl between the central cobbles at Elsham Interchange belonged to the *luermannae/ventrocostata* ostracod Zone.

The abundance of *Birostrina sulcata* in Beds 4 and 5 indicates the *orbignyi* Subzone. This is supported by the entry of *Neohibolites oxycaudatus* at the base of Bed 5, correlating with the sudden appearance of this belemnite in the higher part of the subzone in the Gault of East Anglia (Gallois and

Morter, 1982, fig. 3). *B. sulcata* is absent from Beds 6 and 7, permitting them to be assigned to the *varicosum* Subzone, the flood occurrence of *Neohibolites* spp. in Bed 7 correlating with the belemnite bioevent (Gault Bed 14) in the higher part of the subzone in East Anglia. A sample from Bed 7 at Elsham Interchange belongs to the *luermannae/ventrocostata* ostracod Zone.

Bed 8, constituting the *'Inoceramus' lissa* bioevent, correlates with the *'I'. lissa*-rich Bed 15 of the East Anglian Gault. The occurrence of *Neohibolites ernsti* at the base of Bed 8 suggests that the boundary between the *varicosum* and *auritus* subzones should perhaps be drawn within this bed (as in Figure 29). Bed 9, which overlies the hardground at the top of Bed 8, contains the highest known occurrence of *Biplicatoria* and, by analogy with the Gault, may represent the Milton Brachiopod Bed at the base of the upper part of the *auritus* Subzone.

The entry of *Aucellina* with smooth shell ornament 0.29 m above the base of Bed 10 (at South Ferriby) is taken to mark the base of the *dispar* Zone, by analogy with the *Aucellina*-rich Bed 17 in the Gault of East Anglia and the condensed equivalent (Bed XII) in the Gault of southern England (Morter and Wood, 1983). *Aucellina* is present in strength in the overlying Bed 11, also with smooth ornament. Both Beds 10 and 11 belong to the *hannoverana* ostracod Zone which, in ammonite-bearing Gault successions, ranges no higher than the top of the *rostratum* Subzone of the *dispar* Zone. The higher *perinflatum* Subzone of the *dispar* Zone equates with the *steghausi/bemerodensis* ostracod Zone. The latter zone has not been recognised in Lincolnshire or East Anglia, and present evidence suggests that the highest Albian has been removed by erosion in both areas.

From a maximum thickness of just over 3 m at South Ferriby, the succession thins northwards across the Humber to 2 m in Melton Bottom Quarry [SE 973 273], the maximum reduction apparently occurring in the post-*'Inoceramus' lissa* bed sequence. The succession in Elsham Interchange Cutting is essentially similar to that at South Ferriby, although there is some expansion and modification of Beds 3 to 5, perhaps at the expense of a thin Carstone.

At Bigby Quarry [TA 0594 0782] the red colour is completely lost, this being typical of the successions in the area between Barnetby and Searby (Kent, 1937; Wood and Smith, 1978, fig. 3), where it is exceedingly difficult to distinguish the Hunstanton Chalk from the overlying beds. The Bigby section differs from that given by Wood and Smith in that it is now known that the Hunstanton Chalk at this locality rests directly on Jurassic mudstones.

Another thin Hunstanton Chalk succession is found at Nettleton Bottom Quarry [TA 1252 9810]. The upper part of this section matches that in Elsham Interchange Cutting, but the lower part, comprising Beds 3 to 5, is much reduced, with Bed 3 almost wedging out in places and Bed 5 being represented by a marly parting.

There is an apparent geographical coincidence between the discoloration of the Hunstanton Chalk and the wedging out of the Carstone, which may also be related to the reduction in thickness of the former in this area.

Ferriby Chalk above the Hunstanton Chalk Member

This succession maintains a thickness of 21–24 m throughout the district, and comprises a variety of lithologies including porcellanous chalks with mineralised erosion surfaces, argillaceous chalks, discrete marl bands, coarse-grained chalks rich in large fragments of *Inoceramus*, and sand-grade chalks with much comminuted shell debris. It is flintless throughout. The sediments are generally grey, weathering to buff, but at two levels in the middle and at the top of the succession (the Lower and Upper Pink Bands respectively) they are pink or brick red and superficially similar to the underlying Hunstanton Chalk. This red coloration is present, however, only locally within the district due to subsequent sulphidisation. A generalised section is given in Figure 30.

Figure 30 Generalised section of the Ferriby Chalk Formation above the Hunstanton Chalk Member. The section is based on South Ferriby Quarry, Conoco No. 2 Borehole [TA 1681 1613] and the resistivity log of a well at Burnham [TA 0603 1685].

The succession is characterised by several clearly defined marker beds including, in ascending sequence, the porcellanous Paradoxica or Sponge Bed with its underlying marl, the coarse-grained, shell-fragmental Inoceramus beds, the sand-grade Totternhoe Stone and Nettleton Stone beds, and the terminal convolute hardground with its erosional upper surface. Jeans (1980) regarded the basal porcellanous chalk and each of the shell-rich and/or sand-grade chalks as marking the base of an 'upward-fining cycle', continuing the sequence of sedimentary rhythms that began with the *'Inoceramus' lissa* Bed in the Hunstanton Chalk. In addition to recognising five sedimentary phases (II–VI inclusive of Jeans, 1980, fig. 3), he subdivided the post-'Red Chalk' succession into seven members, but in a somewhat inconsistent manner; he divided his third rhythm (IV) into two members, treated the extremely complex succession comprising his fourth rhythm (V) as a single member, and gave member status to the coarse basal portion of his fifth rhythm (VI). The following outline account of the lithostratigraphy and biostratigraphy has been compiled using all available information from within and immediately outside the district.

Unnamed marl (HF of Andrews' 1983 stratotype succession)

The erosion surface at the top of the Hunstanton Chalk Member is overlain by a 0.025 m bed of iron-stained, silty marl. In the present district, this marl has yielded no macrofossils, but collections from Hunstanton include terebratulid brachiopods having affinities with species from the Belgian 'Tourtia', as well as *Neohibolites ultimus* d'Orbigny. The brachiopod assemblage differs significantly from that of the top beds of the underlying Hunstanton Chalk as well as from that of the overlying Paradoxica Bed. Offshore, in the western part of the Central North Sea Basin, this thin marl expands to a 0.4 m bed of pink marly chalk that yields *Aucellina gryphaeoides* and *A. uerpmanni*, and is overlain by a 0.35 m Paradoxica Bed (Lott et al., 1985; Morter, unpublished BGS report). The pink bed is also present in the submarine succession of Heligoland, where it is characterised by the same species of *Aucellina* and by *Neohibolites ultimus*, and is taken to be of earliest Cenomanian age (Wood and Schmid, in preparation). Present evidence suggests that this thin marl of condensed onshore sequences

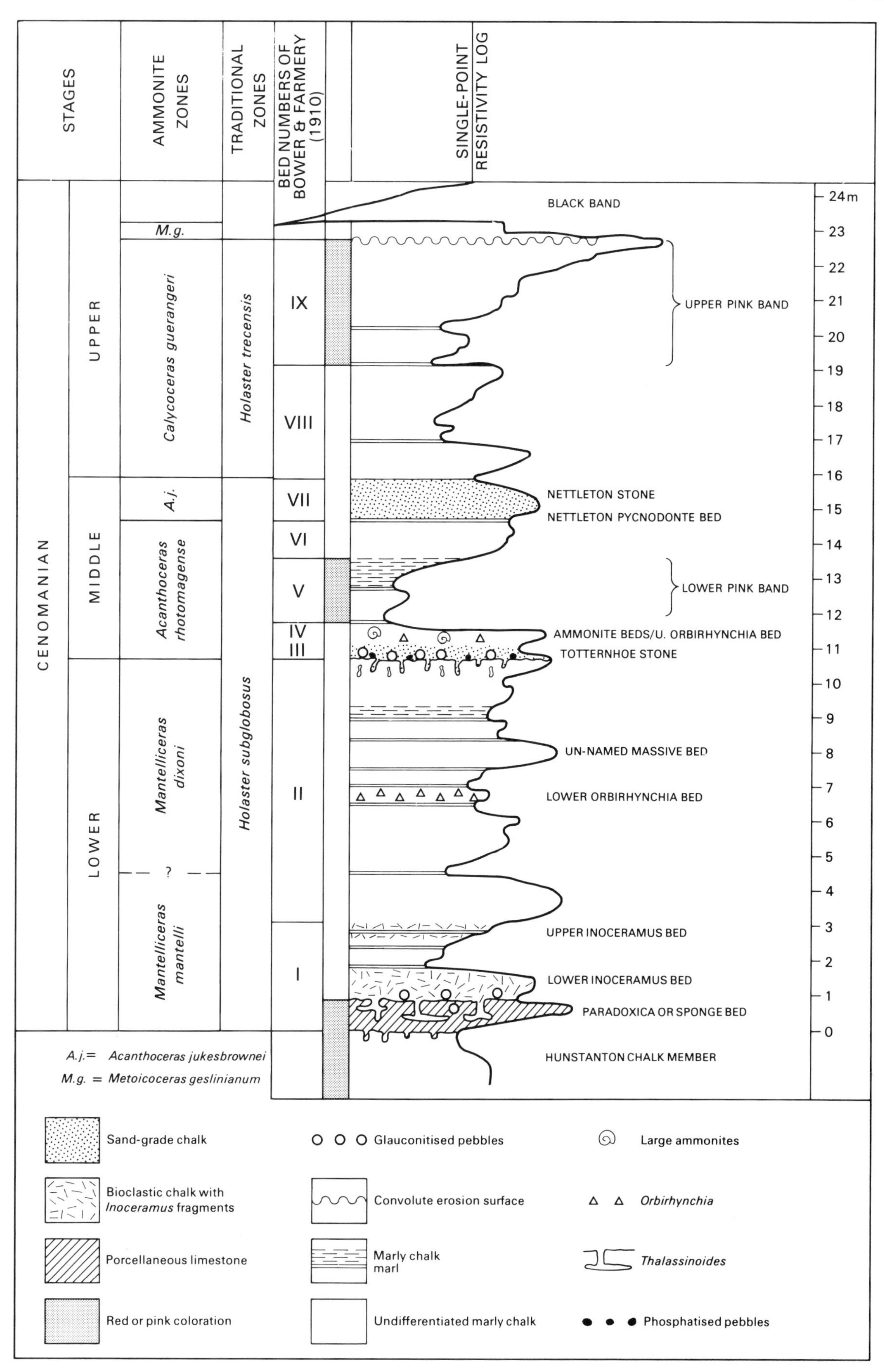
STAGES
AMMONITE ZONES
TRADITIONAL ZONES
BED NUMBERS OF BOWER & FARMERY (1910)
SINGLE-POINT RESISTIVITY LOG
CENOMANIAN
UPPER
MIDDLE
LOWER
M.g.
Calycoceras guerangeri
A.j.
Acanthoceras rhotomagense
Mantelliceras dixoni
?
Mantelliceras mantelli
Holaster trecensis
Holaster subglobosus
IX
VIII
VII
VI
V
IV
III
II
I
BLACK BAND
UPPER PINK BAND
NETTLETON STONE
NETTLETON PYCNODONTE BED
LOWER PINK BAND
AMMONITE BEDS/U. ORBIRHYNCHIA BED
TOTTERNHOE STONE
UN-NAMED MASSIVE BED
LOWER ORBIRHYNCHIA BED
UPPER INOCERAMUS BED
LOWER INOCERAMUS BED
PARADOXICA OR SPONGE BED
HUNSTANTON CHALK MEMBER
24m
23
22
21
20
19
18
17
16
15
14
13
12
11
10
9
8
7
6
5
4
3
2
1
0
A.j.= Acanthoceras jukesbrownei
M.g. = Metoicoceras geslinianum
Sand-grade chalk
Bioclastic chalk with Inoceramus fragments
Porcellaneous limestone
Red or pink coloration
Glauconitised pebbles
Convolute erosion surface
Marly chalk
marl
Undifferentiated marly chalk
Large ammonites
Orbirhynchia
Thalassinoides
Phosphatised pebbles

represents the earliest (non-chalky) phase of Cenomanian deposition, and that it rests nonsequentially on eroded chalky limestones of late Albian *dispar* Zone, *perinflatum* Subzone, age.

Paradoxica or Sponge Bed

This unit, regarded by Jeans (1980) as a sedimentary 'cycle' and attributed to a 'Belchford Member', comprises cemented porcellanous chalks, typically pink or yellow, penetrated by an anastomosing *Thalassinoides paradoxica* burrow network. There is a reddish brown marl 10–20 cm above the base and, higher in the bed, a planar erosion surface followed by a glauconitised convolute erosion surface. The latter is overlain by greyish white chalk which is classified with the Paradoxica Bed for convenience, although it does not exhibit the characteristic lithology. The Paradoxica Bed is rich in brachiopods, notably *Monticlarella carteri* and terebratulids including *Concinnithyris* aff. *subundata*, *Harmatosia crassa*, *Ornatothyris* (*Actactosia*) cf. *obtusus* and *O.* (*A.*) sp. It is also characterised by the thin-shelled bivalves *Aucellina gryphaeoides* and *A. uerpmanni*, which range up to the convolute erosion surface. The belemnite *Neohibolites ultimus* is known from the base of the bed in localities outside the district. The association of *A. gryphaeoides* and *N. ultimus* places the Paradoxica Bed within the *carcitanense* Subzone of the Lower Cenomanian *mantelli* Zone (Morter and Wood, 1983).

Lower and Upper Inoceramus Beds

These are coarse-grained chalks rich in large fragments and complete valves of '*Inoceramus*' *crippsi*. The Lower Inoceramus Bed yields a large fauna of serpulids, for example *Parsimonia antiquata*, brachiopods including *Ornatothyris* (*Ornatothyris?*) spp., bivalves such as *Pseudolimea* cf. *echinata* and echinoids including *Holaster* cf. *subglobosus*, *Hemiaster morrisii* and *Hyposalenia* sp. nov. At several localities there is a concentration of glauconitised pebbles associated with numerous *Holaster* at the base of the bed. The Upper Inoceramus Bed is less fossiliferous, and the inoceramid fragments are generally thinner and smaller; the upper limit is diffuse. Immediately above or within the bed there is a glauconitised surface with large, poorly preserved glauconitised moulds of turrilitid ammonites, probably mainly *Hypoturrilites* spp. (the 'turrilitoid plane' of Jeans, 1980). Contrary to the statement by Wood (*in* Kent, 1980b), it is now believed, on the basis of inoceramid and brachiopod evidence from higher parts of the Northern Province Cenomanian succession, that the Inoceramus Beds belong to the Lower Cenomanian and that the Lower Inoceramus Bed can be attributed to the higher part of the *carcitanense* Subzone.

Lower Orbirhynchia Bed

This is a thin ill-defined but laterally continuous bed characterised by a relative abundance of an *Orbirhynchia* near *O. mantelliana*. It was taken by Jeans (1980) to mark the base of his Bigby Member. Current unpublished work on southern England successions suggests that this bed belongs in the topmost Lower Cenomanian *Mantelliceras dixoni* Zone.

Totternhoe Stone

The Totternhoe Stone, formerly known in Yorkshire and Lincolnshire as the Grey Bed (Hill, 1888), comprises dark grey or brown sand-grade chalk largely composed of comminuted inoceramid shell debris. This coarse sediment rests with a sharp contact on an irregular, locally cemented erosion surface at the top of the underlying argillaceous chalks, and is piped down over a metre below the erosion surface in *Thalassinoides* burrows. In some places, for example the Somerby Pit [TA 0648 0656] and Bigby Quarry, a true hardground is present beneath the Totternhoe Stone. The erosion surface is overlain by a concentration of pale green, weakly glauconitised pebbles.

The Totternhoe Stone is very fossiliferous, particularly at the base, but the macrofossil assemblage varies as the unit is traced southwards towards Hunstanton and then south-westwards towards the type-locality in the Chiltern Hills. In the present district extensive collections made during the survey from Mansgate Quarry, Caistor [TA 1234 0023], include poorly preserved glauconitised composite moulds of the serpulid *Genicularia* (*Glandifera*) *rustica*, the bivalves *Entolium laminosum*, *Oxytoma seminudum* and *Plicatula inflata*, and the ammonites *Acanthoceras* cf. *rhotomagense confusum* and *Schloenbachia* sp., in addition to a diverse and apparently largely undescribed assemblage of terebratulid brachiopods including *Capillithyris* sp. nov., '*Rectithyris*'? sp. nov. and *Concinnithyris* aff. *burhamensis*. *Camerogalerus cylindricus*, *Echinocorys sphaerica* and *Holaster subglobosus* were collected from other localities within the district. The minute and enigmatic belemnite *Belemnocamax boweri* is known from adjacent areas, and was originally described from this horizon in a quarry at Louth. The Totternhoe Stone is assigned on ammonite and other evidence to the Middle Cenomanian *rhotomagense* Zone, *costatus* Subzone.

Overlying the Totternhoe Stone there is a unit of pale nodular limestones (the Ammonite Beds or Bed IV of Bower and Farmery, 1910) yielding *Orbirhynchia mantelliana* and very large ammonites (*Austiniceras*). This unit, Jean's upper *Orbirhynchia* Band, is the correlative of the *Orbirhynchia mantelliana* Band described by Kennedy (1969) from southern England successions, and its top equates with the mid-Cenomanian nonsequence of Carter and Hart (1977), i.e. the level at which a sudden deepening of the sea was accompanied by an incursion of large numbers of planktonic foraminifera of the genus *Rotalipora*.

The Lower Pink Band comprises marly nodular chalk overlain by marly chalk including one or more well-defined marl bands. The typical red or pink colour is generally absent in the Kingston upon Hull/Brigg district. The relatively high marl content is expressed in resistivity logs as a broad low resistivity spike (Barker et al., 1984, fig. 3). The Lower Pink Band is rich in brachiopods, notably *Concinnithyris subundata* and *Kingena concinna*, and a small *Terebratulina* which occurs in such profusion in the pits near Louth that it was used by Bower and Farmery (1910) to characterise a 'Subzone of *Terebratulina ornata*'. Small *Holaster subglobosus* and *Camerogalerus cylindricus* are not uncommon.

Nettleton Stone and Nettleton Pycnodonte Bed

The Lower Pink Band passes up into more calcareous off-white chalks terminating in a cemented erosion surface. This surface is overlain by a dark grey silty marl, the Nettleton Pycnodonte Bed, which was termed by Bower and Farmery the 'Gryphaea Bed' from the profusion within it of isolated and associated valves of a small *Pycnodonte* (*Gryphaea*). Associated with this oyster are fragments and sparse, more complete specimens of '*Inoceramus*' *atlanticus*. The Pycnodonte Bed grades up into the Nettleton Stone, just over a metre of hard grey sand-grade chalk that projects in quarry faces and is topped by a marl. This bed, the 'Nettleton Member' of Jeans (1980), marks a widespread but short-lived regressive phase that is expressed in northern Germany by the so-called Pycnodonte Event (Ernst et al., 1983) with its overlying massive chalk bed, and in southern England by a unit of heavily bioturbated, coarse-grained chalk with laminated scour structures (Jukes-Browne Bed 7 of Dover and correlative successions). Using ammonite evidence from Germany and southern England, the Nettleton Stone can be attributed to the highest Middle Cenomanian *jukesbrownei* Zone. In all areas it marks the upper limit of *Holaster subglobosus* and thick-tested *Echinocorys sphaerica* as well as of '*Inoceramus*' ex gr. *atlanticus*. The Nettleton Stone is expressed in resistivity logs as a high-resistivity spike (the 'Jukes-Browne Band' of Murray, 1986).

UPPER PINK BAND TO EROSION SURFACE BENEATH BLACK BAND

The Nettleton Stone is overlain by marly chalks with thin marl bands and beds of relatively pure chalks. The upper half of the sequence is red or pink-stained outside the present district, constituting the Upper Pink Band of Bower and Farmery. The succession has a low-diversity fauna of thin-tested echinoids, notably *Camerogalerus cylindricus*, *Echinocorys sphaerica*, and a holasterid conventionally, but probably incorrectly, referred to *Holaster trecensis* and used as the index of the local *trecensis* Zone. These echinoids are commonly flattened parallel to the bedding. Other important elements of the fauna are oysters and *Inoceramus pictus*. One or more horizons rich in the exogyrine oyster genus *Amphidonte* occur in the sequence below the Upper Pink Band, for example in Barnetby Pit [TA 0660 0994], these horizons correlating with the *Amphidonte* events of northern Germany (Ernst et al., 1983). The whole succession can be attributed to the Upper Cenomanian *guerangeri* Zone, although no zonally significant ammonites have been found at this level within the Northern Province. The terminal erosion surface is co-extensive with the sub-Plenus Marls erosion surface of southern England, and marks the extinction level of *Rotalipora*. Giant poorly preserved ammonites (*Austiniceras?*) occur sporadically in the hardened chalk beneath this surface at South Ferriby.

Welton Chalk Formation

This formation extends from the erosion surface below the Black Band to the bedding plane defining the base of the Burnham Chalk (Wood and Smith, 1978; fig. 3). The stratotype is the upper quarry [SE 970 280] at Melton Bottom, near Welton, where the greater part of the succession can be seen in the working face, and the lowest beds have been proved in boreholes in the quarry floor. The basal boundary stratotype is taken in the abandoned lower pit at Elsham [TA 038 131]. The succession at Melton Bottom is relatively inaccessible, but was the best available section when the nomenclature was established. Today the Welton Chalk is best exposed in the western (working) Melton Ross Quarry [TA 082 112], the abandoned eastern quarry [TA 086 115] and (the highest beds only) in a number of small pits around Burnham, for example Deepdale [TA 0455 1820] and Burnham cross-roads [TA 059 169], as well as in the Humber Bridge Approach Road cutting [TA 025 197] (Figures 32 and 33). The formation is approximately 44 m thick in the stratotype section, and typically 53 m in the Burnham–Melton Ross area. There is some evidence of thinning immediately to the north of the district (Weedley Dale railway cutting [SE 952 335]), although it is believed that this reduction is related to the general trend towards thin successions in the western part of the Yorkshire Wolds, and is unconnected with proximity to the Market Weighton Structure.

The Welton Chalk, except for the basal beds, comprises massive or relatively thick-bedded chalk characterised by thalassinoidean burrow-form flints. It should be noted that laminate chalk, a characteristic lithology at several levels in the Burnham Chalk, is virtually absent. The basal beds are different in lithology from the main mass of the formation, but are grouped in the Welton Chalk for convenience. The base of the Welton Chalk is marked by a topographical slack caused by the soft marly chalks, including the Black Band, that rest on the hard chalks at the top of the underlying Ferriby Formation. A strong positive feature, caused by relatively hard chalk immediately below and above the Grasby Marl, forms the top of the main Chalk scarp over much of the district.

BASAL BEDS, INCLUDING THE BLACK BAND

The basal 5 m of the Welton Chalk comprise marly sediments overlain by coarse-grained bioclastic and, at some levels, pebbly chalks (Figure 33). Comminuted and incomplete shells of the inoceramid bivalve *Mytiloides* are a characteristic feature. The succession is flintless, although there are unconfirmed reports (Rowe, 1929) of flints at Claxby in southern Lincolnshire. There are numerous thin (less than 1 cm) marls, some of which are greenish and comparable in their 'buttery' texture with the thick marls of the main mass of the formation, while others are more silty with a high bioclastic content. These basal beds represent an extremely condensed sequence, and the marls, including the Black Band, present a distinctive signature on geophysical logs (Barker et al., 1984, fig. 4).

The basal erosion surface is orange-stained and highly convolute. At Mansgate Quarry and Elsham Lower Pit, pockets within this surface are filled with green marl containing abundant brachiopods and sparse echinoids. Some of these fossils are filled with green marl and are largely crushed, while others are filled with hard chalk and are mainly uncrushed. Collections made during the survey from these localities include *Monticlarella jefferiesi*, *Ornatothyris* (*Actactosia*) spp., *Orbirhynchia multicostata*, a rhynchonellid of uncertain affinities previously incorrectly attributed to *O. wiesti*, and small flattened mushroom-shaped variants of *Camerogalerus cylindricus*. Jefferies (1963) recorded *Actinocamax plenus* and *Oxytoma seminudum* from this horizon.

The basal erosion surface is overlain by a 10 cm bed of hard chalk terminating in a low-amplitude black (MnO_2?)-stained erosion surface, this being succeeded by a 5–7 cm bed of buff-coloured silty sediment, locally containing poorly preserved, crushed and largely decalcified *Orbirhynchia* cf. *multicostata* together with *Inoceramus pictus*. This succession is particularly well seen at the top of Bigby Quarry. At this locality a large, poorly preserved, smooth ammonite (*Austiniceras?*) was collected from the hard chalk between the two erosion surfaces, in addition to two specimens of *Actinocamax plenus*, one from immediately above the black-stained surface and the other resting on top of the silty bed. These belemnite occurrences are of importance in that they are stratigraphically *higher* than the specimens recorded by Jefferies (1963).

The silty bed is overlain by a complex sequence of variegated marls up to 0.5 m thick, ranging in colour from buff to brown, grey, purple and khaki. In the middle there is an irregular bed of dark grey carbonaceous marl, the Black Band, which tends to split near its top along a silty surface covered with fish-scales; it was seen to contain a large piece of bituminous wood at Melton Ross Quarry. The Black Band is bioturbated, the type and intensity of bioturbation varying from locality to locality: *Chondrites* and *Planolites?* are the most commonly encountered ichnofossils. The most complex bioturbation was seen in South Ferriby Quarry, where lamination is only weakly developed; this contrasts with the poorly bioturbated and strongly laminated Black Band at Elsham. Traced to the south, the dark grey colour of the Black Band changes to dark purple.

The grey and khaki marls overlying the Black Band terminate in a green sticky marl and are succeeded by a 0.4 m-thick bed of off-white silty chalk (up to 53 cm thick at South Ferriby) containing the aberrant cylindrical bivalve '*Turnus*' *amphisboena*, sparse poorly preserved inoceramids (possibly of earliest Turonian age) and the small ichnofossil *Gyrolithes*. This bed is overlain by three marl seams within 30 cm. Above the second of these there is a significant lithological change from marly or silty chalk to pebbly chalk with inoceramid fragments. The succeeding 0.35 m-thick bed, which is pebbly at the top, is characterised by *Lewesiceras* sp. and entire valves of *Mytiloides* aff. *labiatus*, the latter being particularly common at this horizon at Mansgate Quarry. This represents the lowest unequivocal Turonian assemblage, although microfaunal and other evidence suggests that the base of the Turonian Stage must be drawn significantly lower, close to or possibly even within the Black Band itself (see discussion below). There is a profusion of crushed *M. labiatus* in the overlying 10 cm bed of pebbly chalk, which is overlain in turn by an orange, silty marl; these two beds together provide a distinctive lithostratigraphical marker above the Black Band. It is highly probable that the orange marl correlates with the 'Violet Marl' marker horizon of the 'Rotpläner' facies successions in northern Germany (Ernst et al., 1984).

Above the orange marl the remainder of the basal sequence comprises shell-detrital chalks with *Mytiloides* spp. and several discrete marl seams. These shell-debris rich chalks probably correspond to the post-Melbourn Rock *labiatus* Zone of the standard macrofossil zonal scheme. Extraction of macrofossils from these hard chalks is very difficult, but limited collections from the Elsham Lower Pit and Grasby Quarry comprise a low-diversity fauna of *Concinnithyris?* sp., *C. protobesa*, *Monticlarella* sp., *Orbirhynchia* cf. *compta*, *Mytiloides* spp. including *M. mytiloides* and *Sciponoceras bohemicum anterius*. *Conulus subrotundus* and large specimens of *Mammites nodosoides* occur in the lower part of this succession in Elsham Lower Pit.

The top of this sequence is marked by three marl seams within 25 cm, the lowest of which is brown and silty, and the upper two black. Rare small 1 cm nodular flints occur between the two black marls, but have been observed only at Elsham lower pit and the Melton Ross working quarry. These marls, which are herein named the Chalk Hill Marls (see Appendix 3), probably correlate with the highest of the four Gun Gardens Marls in Sussex (Mortimore, 1986; Lake et al., 1987). Above the Chalk Hill Marls there is a pronounced change to pure, white, poorly fossiliferous chalks with flints.

The Black Band is virtually devoid of macrofossils apart from debris and complete skeletons of fish; but exceedingly rare occurrences of oysters, terebratulid brachiopods and a lobster are also known. It is characterised by a low-diversity dinoflagellate cyst assemblage of Turonian aspect dominated by *Cyclonephelium compactum* together with common smooth forms of *Deflandrea* (R J Davey: unpublished BGS report); the comparative lack of long-spined taxa points to reduced water temperatures. The foraminiferal assemblage is entirely composed of simple agglutinating forms of indeterminate age. Bralower (1988) provided nannofossil data for the Black Band and the associated marls from the Elsham and South Ferriby quarries, and noted that *Eprolithus floralis* reached up to 47 per cent of the non-*Watznaueria barnesae* fraction at the latter locality.

The microfaunal evidence indicates a near-anoxic bottom environment with restricted water circulation, perhaps in a silled basin. However, foraminiferal assemblages from below and above the Black Band are dominated by planktonic forms such as '*Hedbergella*', suggesting a deep-water environment (Hart and Bigg, 1981). Those authors regarded the Black Band succession as the local expression of the global Cenomanian–Turonian high sea-level stand with its associated depleted oxygen levels (the so-called Oceanic Anoxic Event) and suggested (largely on the basis of the degree of etching of the microfossils) that the Black Band itself correlated with bed 6 of the Plenus Marls.

The Black Band was formerly considered to be the Northern Province equivalent of the Plenus Marls of southern England, particularly in view of the unsubstantiated record (Hill, 1888) of the eponymous *Actinocamax plenus*. However, Jefferies (1963) observed that at South Ferriby Quarry *A. plenus* occurred not in the Black Band itself, but in a conglomerate immediately overlying the basal erosion surface of the Welton Chalk Formation, together with fossils such as *Oxytoma seminudum* and *Orbirhynchia wiesti*, indicative of the higher part of the Plenus Marls, i.e. beds 4–6 and bed 7 respectively of his standard succession. He suggested that the Black Band and the underlying and overlying variegated marls probably postdated the Plenus Marls and, by implication, correlated them with the basal part of the Melbourn Rock.

This interpretation that only the upper part of the Plenus Marls is represented in northern England is confirmed from foraminiferal evidence by Hart and Bigg (1981), who showed that *Rotalipora cushmani* became extinct in the white chalks immediately beneath the basal erosion surface. Since

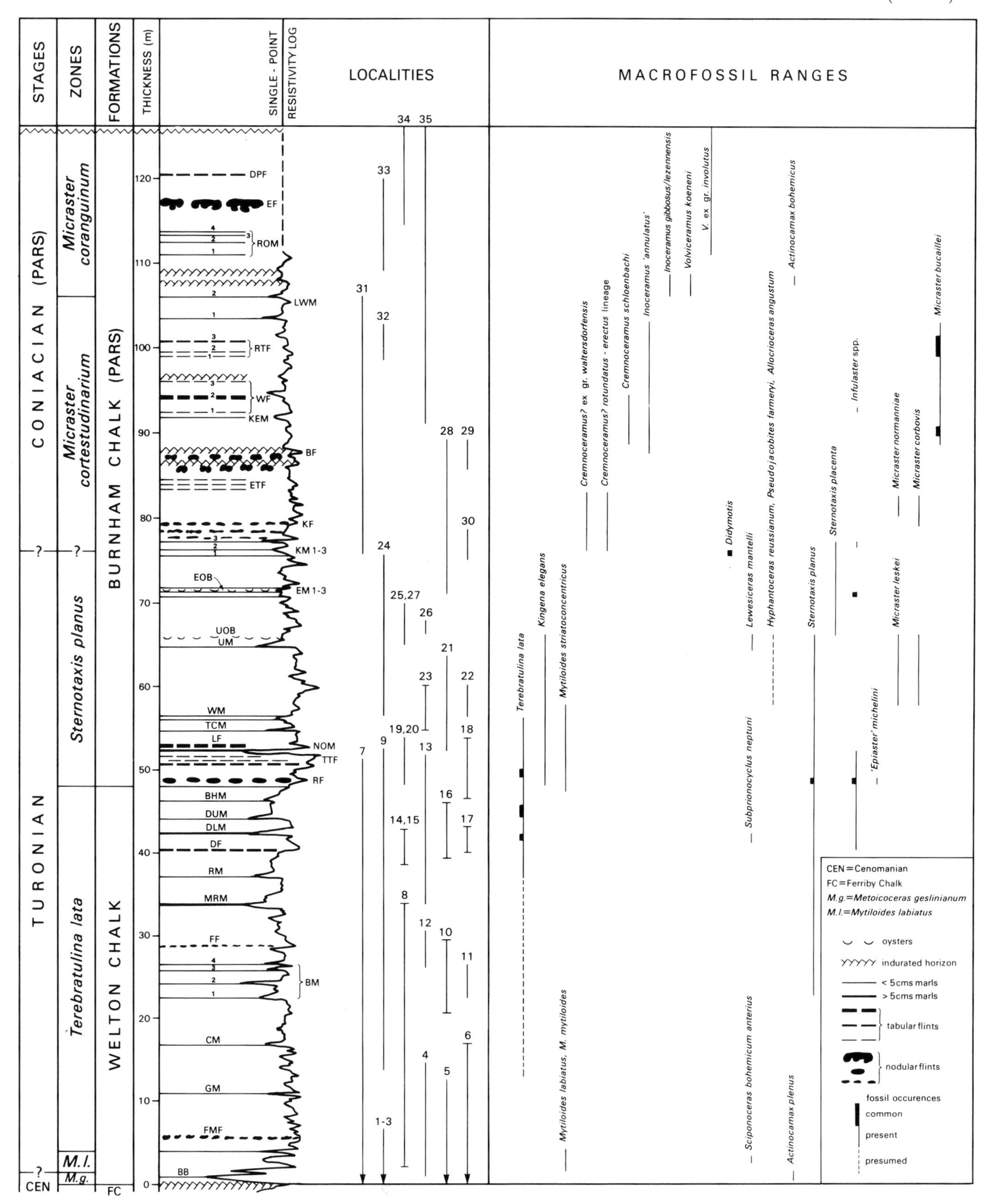

Figure 31 Stratigraphy of the Welton and Burnham (pars) Chalk formations, with ranges of significant macrofossils. For names of marker horizons see Figures 33 and 34, and for localities Figure 32

this extinction datum (marking the top of the *R. cushmani* Total Range Zone) occurs at the top of bed 3 of the Plenus Marls there is clearly a significant nonsequence at the erosion surface, involving beds 1–3 (Hart and Leary, 1989, fig. 4). Hart and Bigg also showed that *Praeglobotrunca praehelvetica* and *P. helvetica*, marking the base of the (Turonian) *P. helvetica* Partial Range Zone, entered in the marls immediately overlying the Black Band. The succession from the basal erosion surface up to the top of the Black Band thus falls into the late Cenomanian–early Turonian '*Hedbergella*' *archaeocretacea* Interval Zone. Since *P. helvetica* in southern England first enters many metres above the Plenus Marls, it follows that not only is the succession in the north extremely attenuated, but also that the base of the Turonian Stage probably falls within the Black Band itself or in the underlying marls.

Supporting evidence for this last interpretation is provided from the carbon stable isotope data. The Cenomanian–Turonian boundary succession in most areas of the world is characterised by a significant increase in $\delta^{13}C$ values followed by an equally dramatic decrease (the so-called $\delta^{13}C$ excursion or spike). Many workers believe this excursion to be globally isochronous and consequently directly applicable to the resolution of problems of stratigraphical correlation. Carbon isotope data for South Ferriby (Schlanger et al., 1987, fig. 5) show that there is an extremely marked increase in values across the basal erosion surface corresponding to the major nonsequence demonstrable from foraminiferal and macrofaunal evidence (see Hart and Leary, 1989, fig. 4). The $\delta^{13}C$ spike corresponds to the succession from the basal surface up to the top of the Black Band and thus equates with the '*Hedbergella*' *archaeocretacea* Interval Zone. The excursion comprises two peaks below the Black Band and a minor peak at the base of the Black Band, followed by a sharp drop in values within the Black Band itself. At Dover, the carbon isotope excursion extends from the base of the Plenus Marls into the basal part of the Melbourn Rock to just above the level taken on macrofaunal evidence as the Cenomanian–Turonian boundary (Hart and Leary, 1989, fig. 4). The main peaks correspond to Beds 4 and 6 of the Plenus Marls, with the sharpest drop in carbon isotope values taking place within the Melbourn Rock, across the Cenomanian–Turonian boundary.

The isotope evidence from northern England strongly suggests that the stage boundary falls within the Black Band itself (Jarvis et al., 1988, p.80), that at least part of the Black Band correlates with the basal part of the Melbourn Rock, and that the change to chalk facies sedimentation occurred significantly later in northern England than in the south. The attenuated Northern Province succession would thus compare with the succession at Rheine, Germany, where a thin black shale overlies beds which correlate with the higher part of the Plenus Marls (Ernst et al., 1984, fig. 3).

Beds above the shell-detrital chalks

Above the basal beds, the bulk of the Welton Chalk Formation is composed of extremely pure chalks—99 per cent $CaCO_3$ in a sample from Melton Ross Quarry (Hancock, 1975). Wood and Smith (1978) noted that the Welton Chalk is relatively soft compared with the overlying Burnham Chalk, but this difference in hardness varies both laterally and vertically. The flints of the formation are for the most part relatively small and inconspicuous, and are rare in some parts of the succession. Certain flint bands, however, are strongly developed and provide important marker horizons. The First Main Flint, near the base of the flinty succession, comprises closely spaced nodular flints, typically contained within a bed defined by stylolitic surfaces. The description (Wood and Smith, 1978, p.277) of these flints as being 'dumb-bell shaped' is inaccurate; their shape is complex and may include vertical elements. Locally, for example in some faces of the western Melton Ross Quarry, the flint may appear semicontinuous over more than a metre laterally. The beds between Barton Marl 1 and Barton Marl 2 contain abundant, closely spaced, large burrow-form flints with vertical extensions downwards and upwards (see Wood and Smith, 1978, pl. 12, fig. 2). The frequency of these flints imparts a distinctive rubbly nature to this interval. The Ferruginous Flint is a semi-tabular flint with pyrite-coated fracture surfaces; oxidation of the pyrite to limonite or hematite is responsible for the distinctive reddish colour. This flint lies just above a conspicuous 25 cm marl complex, named herein the Yarborough Marl (see Appendix 3) after the Yarborough Camp earthworks which overlook the type-locality, the eastern Melton Ross Quarry. The Yarborough Marl forms a marked recess in quarry faces. The Deepdale Flint, which overlies a rare horizon of laminate chalk, is semitabular and of Burnham rather than Welton type. At many localities it is closely overlain by a line of nodular flints. Paramoudra flints are rare or absent in the Welton Chalk Formation, but similar cylindrical, subvertical flints occur sparsely in the western Melton Ross Quarry.

Discrete marl bands, a conspicuous feature of the Welton Chalk, typically have sharply defined bases and grade upwards into overlying chalks. Greenish in the unweathered state, they weather rapidly to orange-brown. Thicknesses vary from 1 to 10 cm, the thickest bands being the Grasby, Melton Ross and Deepdale Lower marls. Variation in thickness combined with the relative spacing of the marls provides a distinctive geophysical signature for the Welton Chalk, particularly in the natural gamma and resistivity logs (Figure 31). The tentative correlation between the marls and those of the southern England succession was discussed by Mortimore and Wood (1986), the identification (on both geophysical and lithostratigraphical grounds) of the Melton Ross Marl as the Southerham Marl of Sussex being critical to the correlation (Figure 28).

Lithostratigraphical details of the succession are given in Figure 33. In addition to the marker horizons named by Wood and Smith (1978), and the Chalk Hill Marls and Yarborough Marl discussed above, it has proved necessary to name two further marker horizons. The Hall Farm Marl (Appendix 3, p.143) is a thin, less than 0.5 cm marl, which is found about 2.5 m above the First Main Flint, and is underlain by a 0.25 m bed with a basal 0.10 m concentration of 'rubbly' indurated chalk with marl envelopes. The combination of the thin marl and the rubbly chalk is very conspicuous in quarry faces. Although the Hall Farm Marl can be observed over several hundred metres of section at the type-locality, it is rather inaccessible, and can be best studied at the Elsham upper pit.

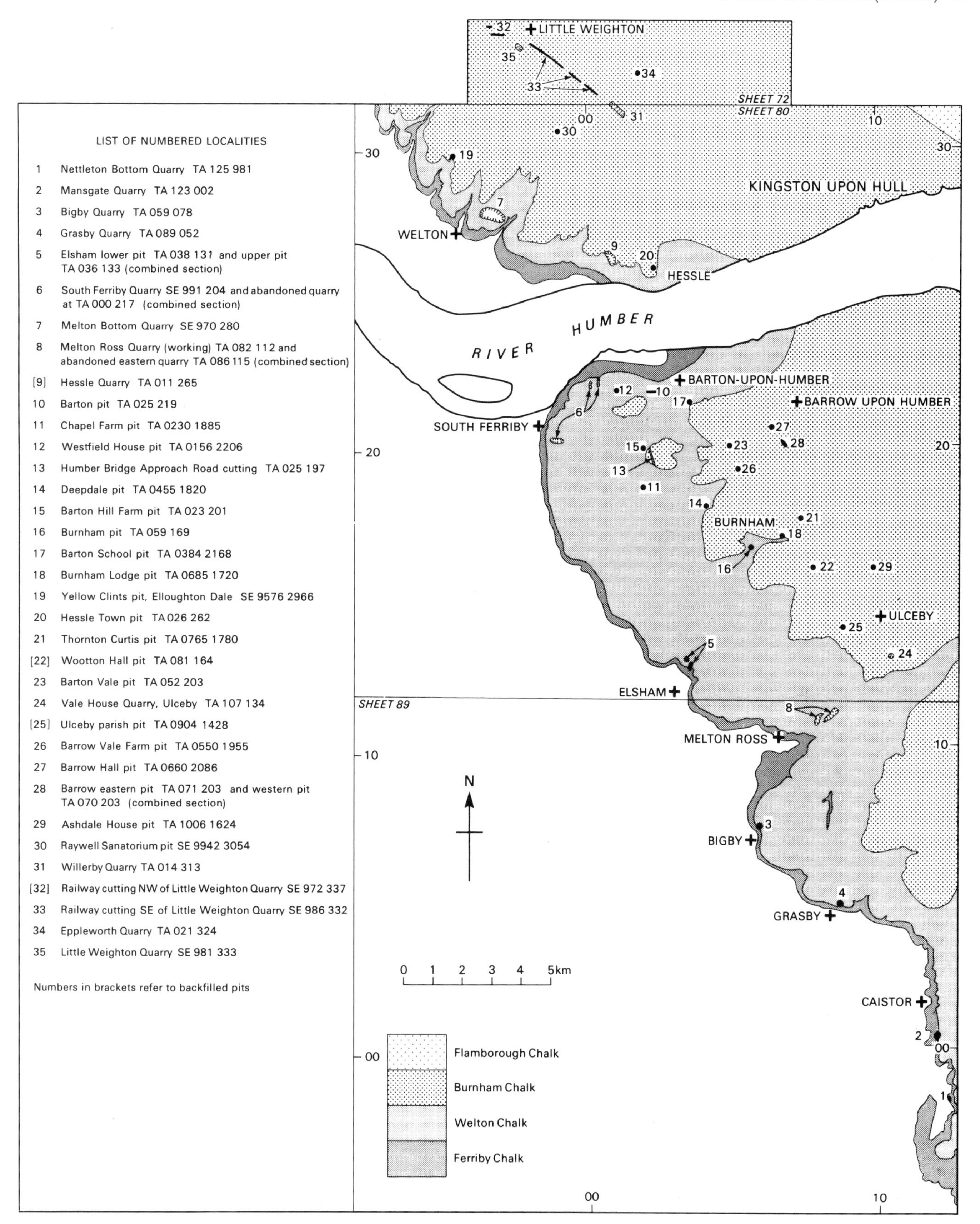

Figure 32 Map of the outcrops of the Chalk formations. Localities in the Welton and Burnham Chalk formations referred to in the text and in Figures 31, 33 and 34 are shown

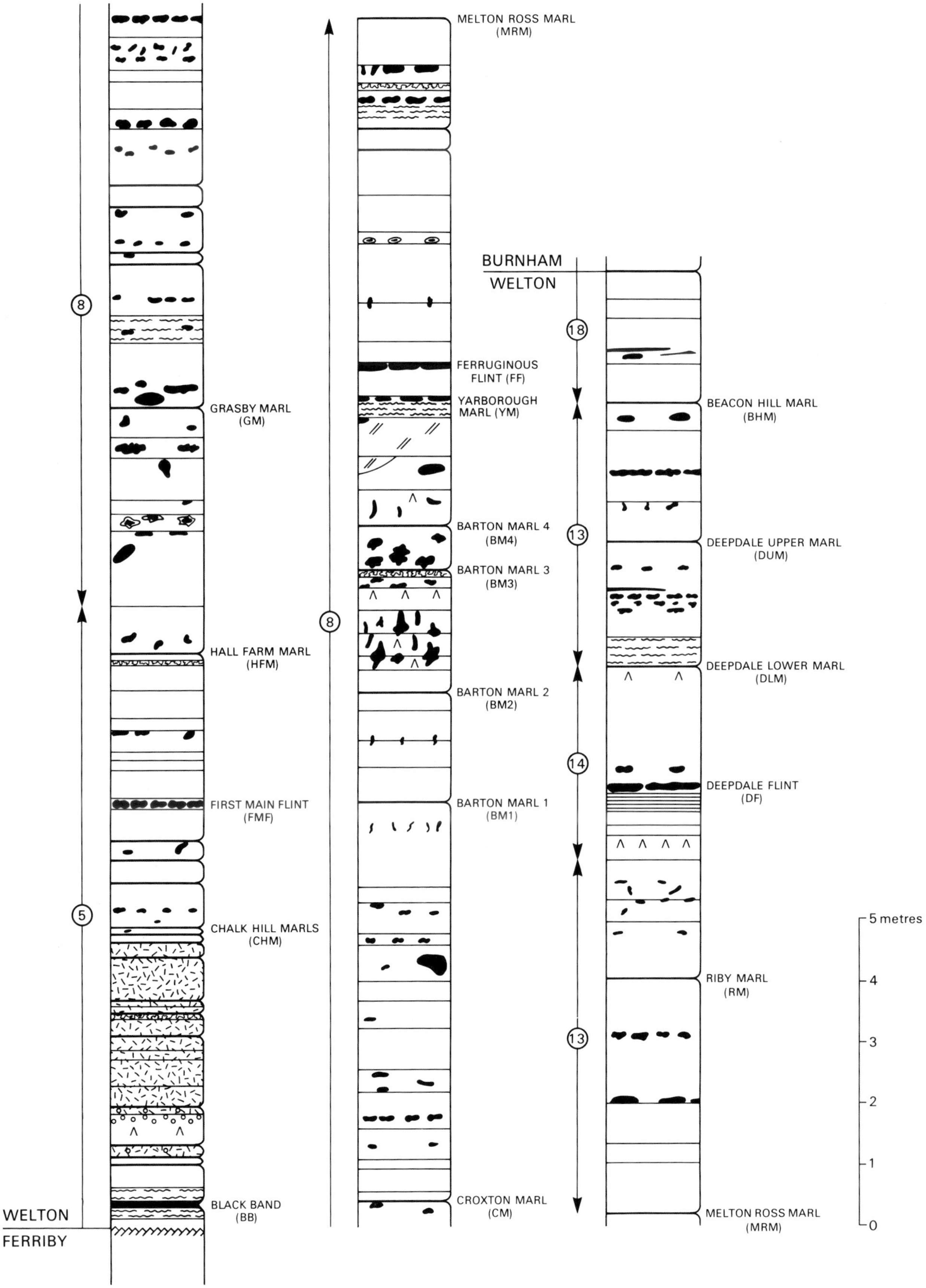

GRASBY MARL (GM)
HALL FARM MARL (HFM)
FIRST MAIN FLINT (FMF)
CHALK HILL MARLS (CHM)
BLACK BAND (BB)
WELTON
FERRIBY
8
5

MELTON ROSS MARL (MRM)
FERRUGINOUS FLINT (FF)
YARBOROUGH MARL (YM)
BARTON MARL 4 (BM4)
BARTON MARL 3 (BM3)
BARTON MARL 2 (BM2)
BARTON MARL 1 (BM1)
CROXTON MARL (CM)
8

BURNHAM
WELTON
BEACON HILL MARL (BHM)
DEEPDALE UPPER MARL (DUM)
DEEPDALE LOWER MARL (DLM)
DEEPDALE FLINT (DF)
RIBY MARL (RM)
MELTON ROSS MARL (MRM)
18
13
14
13

5 metres
4
3
2
1
0

Figure 33 Detailed section of the Welton Chalk Formation. For localities see Figure 32 and for symbols Figure 34

The Croxton Marl is the thin (up to 1 cm) but conspicuous marl midway between the Grasby Marl and the lowest of the Barton Marls. It is best developed in the now inaccessible underground sections of the LPG storage caverns near Immingham in the adjacent Grimsby (Sheet 90) district, where it is underlain by small, closely spaced burrow-form flints. It is, however, best studied at the type locality, the western Melton Ross Quarry, and in a pit at Swallow Vale Farm [TA 175 043] (Sheet 90). In both these exposures the flints underlying the marl are inconspicuous and widely spaced, locally failing entirely. The Croxton Marl is marked by an overhang in quarry faces, and is expressed as a minor low-resistivity spike in resistivity logs. The Croxton Marl at the type locality can be readily identified from the tendency for preferential solution to take place at and immediately above the level of the marl; the cavities so formed become filled with a variety of sediments of uncertain (?Pleistocene) age, including brown quartz sands and indurated green mudstones, which extend down from the surface in enlarged fissures corresponding to high-angle conjugate joint sets.

The beds up to the Grasby Marl have yielded a low-diversity fauna comprising *Concinnithyris* cf. *albensis*, small *Orbirhynchia*, including *O.* cf. *compta* and *O. heberti*, rare *Inoceramus* sp. and *Conulus subrotundus*. This fauna is not zonally diagnostic, and its attribution to the *lata* Zone is questionable in view of the fact that the lowest record of the zonal index in the district is 2.3 m above the Grasby Marl, although the apparent absence of *Mytiloides* spp. might be taken to support a *T. lata* rather than a *M. labiatus* zonal position. Current work in southern England suggests that the base of the *lata* Zone in the Northern Province falls between the top of the basal *Mytiloides*-rich beds and the Grasby Marl. Thus, in Trunch Borehole [TG 2933 3455] near North Walsham, Norfolk, *T. lata* extends up to 2 m below the presumed equivalent of the Grasby Marl (Wood *in* Arthurton et al., in preparation).

The succession above the Grasby Marl belongs unequivocally to the *T. lata* Zone and is richer in fossils, but the fauna is still of low diversity, comprising brachiopods, inoceramids and thin-tested echinoids such as *Infulaster* and *Sternotaxis*. Fossils tend to be concentrated at particular horizons, the beds in between being more or less barren. The commonest fossils are inoceramids, but these are difficult to extract and are rarely complete, so that identifications present problems. Most of the inoceramids belong to the *I. cuvierii*, *I. lamarcki* and *I. inaequivalvis* lineages. Inoceramid shell beds occur within the Barton Marl sequence and beneath the Deepdale Lower and Upper marls. At Deepdale Pit, a hard iron-stained bed 2.7 m below the Deepdale Lower Marl yields abundant small well-preserved thin-shelled *Inoceramus* ex gr. *inaequivalvis* and *I.* aff. *lusatiae*. *Sternotaxis* spp., including forms questionably attributable to *S. planus*, are first found below the third Barton Marl and beneath the Ferruginous Flint; above this there are no records until immediately above the Deepdale Flints, where they are relatively common together with the first *Infulaster excentricus*. A little higher, *S. planus* is not uncommon between the Deepdale Marls. *Terebratulina lata* is extremely rare up to the echinoid horizon over the Deepdale Flints, above which it occurs regularly, and even abundantly at some levels, notably immediately beneath the Deepdale Lower Marl and up to 0.75 m above the Deepdale Upper Marl: at the latter horizon in Burnham Pit the shells are dissolved away leaving hollow limonite-stained external moulds. The only ammonite known is a poorly preserved specimen of the Upper Turonian international zonal index *Subprionocyclus neptuni* from 1 m beneath the Deepdale Lower Marl in the railway-side pit at Kiplingcotes [SE 9156 4350] to the north of the present district. At this locality an *Inoceramus* close to *I. lamarcki geinitzi* occurs commonly below the Deepdale Upper Marl.

Burnham Chalk Formation

This formation is named after Burnham, South Humberside, and extends from the bedding plane beneath the Ravendale Flint to the top of the flinty chalk. The basal boundary stratotype is the disused Burnham Lodge Quarry [TA 0685 1720] (Plate 3). No stratotype has been proposed for the formation itself because the only place where a complete succession is exposed is in the inaccessible sea cliffs between North Landing and Flamborough Head. A composite thickness of 130–150 m was given by Wood and Smith (1978), based on measurements from discontinuous exposures and boreholes in Humberside. Only the lower part of the succession is exposed in the present district.

The Burnham Chalk is predominantly thin bedded and characterised by continuous tabular and lenticular flints. The tabular flints range in thickness from less than 1 cm to over 0.3 m, and may be solid or carious. Paramoudra flints are common at some levels, particularly in the lower part of the succession. Burrow-form flints are relatively uncommon, and tend to occur in the more thickly bedded, less indurated chalks and also towards the top of the formation. Belts of laminate chalk, in which the typically marl-coated bedding planes are only a few millimetres apart, occur at several levels and can be used for long-range correlation. As in the Welton Chalk, discrete marl bands are a characteristic feature. These generally range up to 5 cm, but the North Ormsby Marl is locally 11 cm thick. The main marls were named by Wood and Smith (1978), but subsequent work in the Willerby–Little Weighton area has revealed further important marls in the succession above the Kiplingcotes Marls. These additional marls, as well as some particularly conspicuous flint bands, are here given formal names (Figure 34 and Appendix 3).

No hardgrounds have been identified in the Burnham Chalk, but there are several laterally continuous horizons of intensely indurated orange or yellow chalk, notably below the Ulceby Marl and in the succession above the Kiplingcotes Marls in Willerby Quarry [TA 0148 3118] and correlative localities. The topmost of the three yellow hard beds at Willerby, within which there is some evidence of glauconitisation of burrows, floors a wide and shallow sedimentary channel. The absence of hardgrounds and associated phosphate and glauconite mineralisation suggests that the Burnham Chalk was deposited in relatively deep water, an indication that is supported by the tabular form of the flints (Clayton, 1984). The indurated chalk below the Ulceby Marl forms a conspicuous topographical feature, which can be mapped throughout the drift-free parts of the district and is readily recognised in other districts to both north and south.

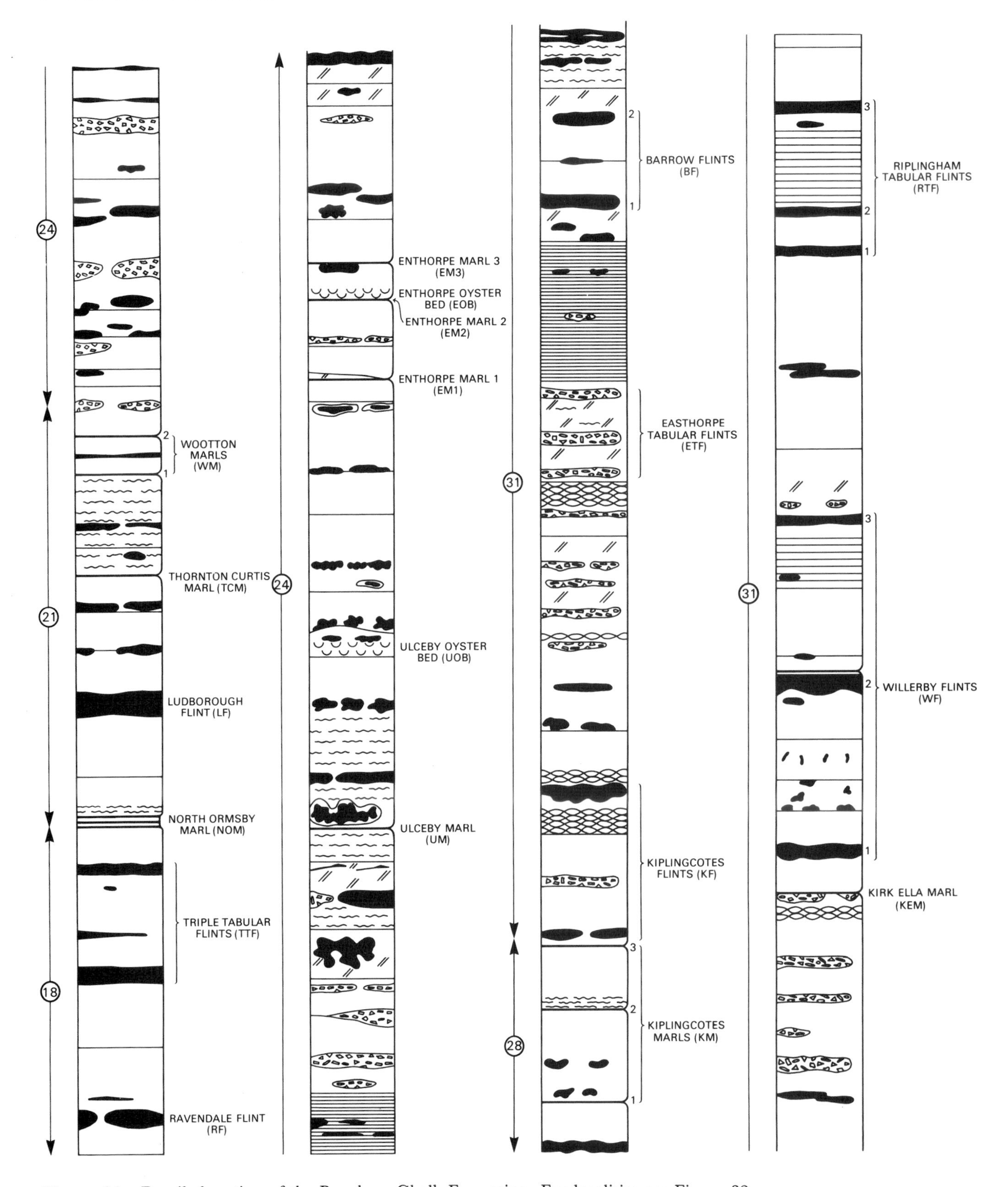

Figure 34 Detailed section of the Burnham Chalk Formation. For localities see Figure 32

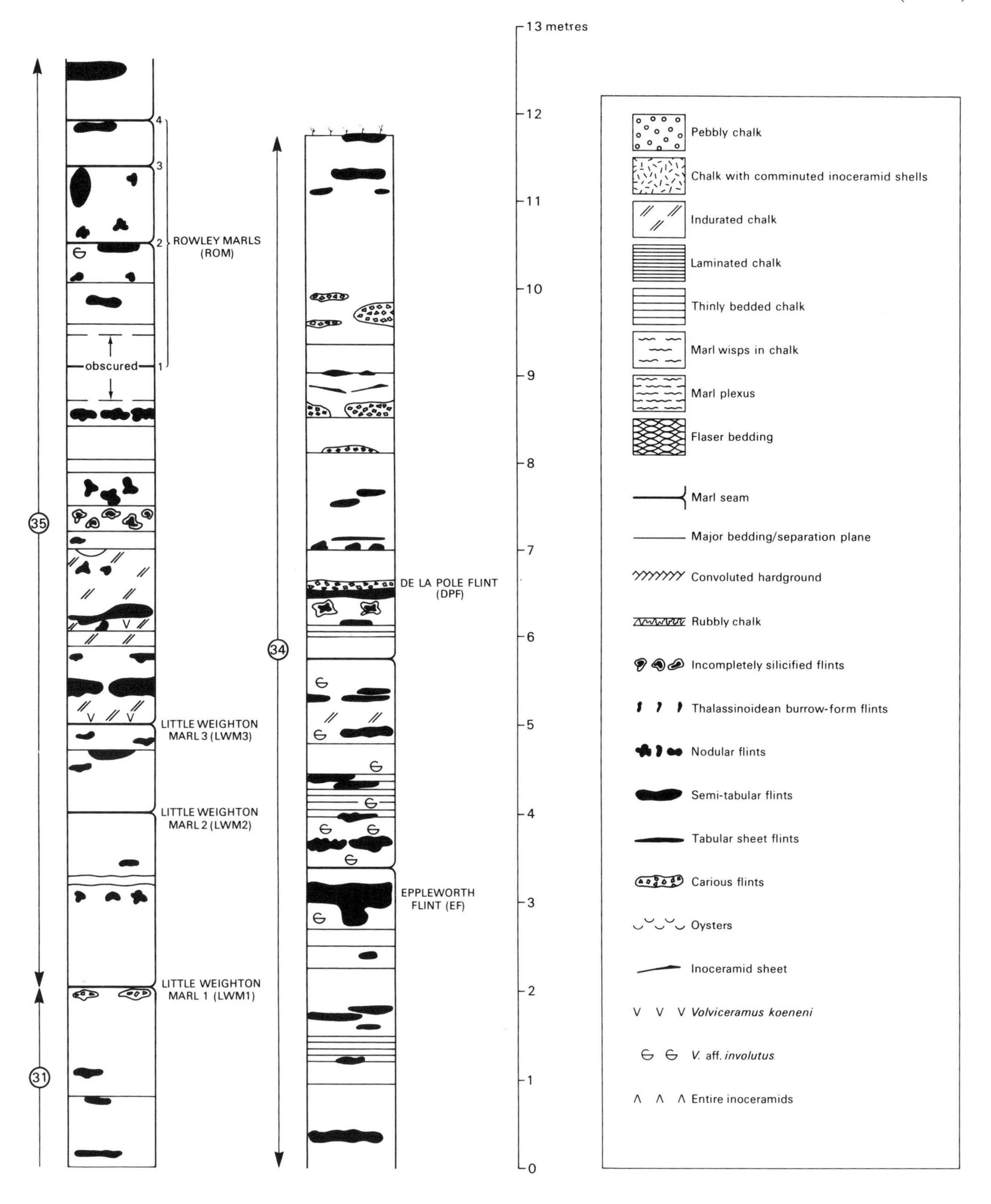

13 metres
12
11
10
9
8
7
6
5
4
3
2
1
0
4
3
2
1
ROWLEY MARLS (ROM)
obscured
35
31
34
LITTLE WEIGHTON MARL 3 (LWM3)
LITTLE WEIGHTON MARL 2 (LWM2)
LITTLE WEIGHTON MARL 1 (LWM1)
DE LA POLE FLINT (DPF)
EPPLEWORTH FLINT (EF)
Pebbly chalk
Chalk with comminuted inoceramid shells
Indurated chalk
Laminated chalk
Thinly bedded chalk
Marl wisps in chalk
Marl plexus
Flaser bedding
Marl seam
Major bedding/separation plane
Convoluted hardground
Rubbly chalk
Incompletely silicified flints
Thalassinoidean burrow-form flints
Nodular flints
Semi-tabular flints
Tabular sheet flints
Carious flints
Oysters
Inoceramid sheet
V V V Volviceramus koeneni
V. aff. involutus
Λ Λ Λ Entire inoceramids

Compared with that of the Welton Chalk, the macrofauna of the Burnham Chalk is more abundant and of higher diversity. There is an apparent increase in both abundance and diversity in the relatively thicker and less cemented successions to the south of the present district, this change being particularly conspicuous immediately to the north of Louth in the type area of the North Ormsby Marl. It is uncertain, however, to what extent this apparent increase is attributable to improved collecting conditions in these relatively soft chalks rather than to a more favourable ecological environment.

Excellent exposures of the lower part of the Burnham Chalk are provided by Burnham Lodge Quarry (Plate 3), Thornton Curtis Pit [TA 0765 1780], the two disused quarries at Barrow upon Humber [TA 0710 2034; 0700 2034], Eppleworth Pit [TA 021 324] and, in particular, the Willerby and Little Weighton [SE 981 333] quarries and the railway cuttings between. The type section of the Ulceby Marl (Wood and Smith, 1978) at Vale House Quarry, Ulceby [TA 107 134], is now poorly exposed.

Details of the lithostratigraphy of the exposed Burnham Chalk are given in Figure 34. The higher part of the succession was penetrated by boreholes in the adjoining Patrington (81) district, e.g. at East Halton [TA 1396 2260], for which excellent geophysical logs are available (Barker et al., 1984, fig. 6). The deep cored boreholes drilled at North Killingholme for the NIREX site investigation (1986–87) have enabled satisfying interpretation of the logs from East Halton.

The exposed Burnham Chalk can be conveniently divided into four parts for descriptive purposes.

Base of formation to Ulceby Marl

These beds are characterised by closely spaced, thick, continuous, tabular flints and paramoudras. They were originally identified as marking the *Sternotaxis planus* Zone on the Yorkshire coast (Barrois, 1876; Rowe, 1904) and subsequently yielded macrofossil evidence of this zone in the Louth area of Lincolnshire (Hill, 1902). The distinctive basal sequence, comprising the giant Ravendale Flint

Plate 3 Burnham Chalk on Welton Chalk at Burnham Lodge Quarry. The Burnham Chalk, containing tabular flints, is more resistant to weathering. The boundary between the two formations is the marked bedding plane below the prominent flint (the Ravendale Flint) at the right of the picture. A 14381

overlain by the Triple Tabular Flints, North Ormsby Marl and Ludborough Flint, produces a marked topographical feature that enables the base of the formation to be mapped in unexposed drift-free areas. The beds correlate with the 'Brandon Flint Series' of East Anglia, the Ludborough Flint being the lateral extension of the 'Floorstone' Flint at the base of the Brandon flint-mines. They are the northern expression of the flint maximum taken by the Geological Survey to mark the base of the Upper Chalk in the North Downs area of southern England (see Mortimore and Wood, 1986). The North Ormsby Marl, which is the thickest and best developed of all the Turonian marls, appears to correlate with a thick marl (M_E) used as a major lithostratigraphical marker throughout northern Germany (Wood et al., 1984). It provides a key low-resistivity and high gamma spike in geophysical logs.

The beds from the base of the formation up to the North Ormsby Marl are locally rich in echinoids, notably *Infulaster hagenowi*, *Sternotaxis planus* and, less commonly, *Echinocorys* sp. They also contain the brachiopods *Kingena elegans*, *Orbirhynchia?* spp. (mainly undescribed forms), *O. dispansa* and *Terebratulina lata* (abundant between the Ravendale Flint and the basal Triple Tabular Flint). Inoceramids are also relatively common, but are usually crushed, incomplete or otherwise poorly preserved. Small *Inoceramus* ex gr. *inaequivalis* and large sheet-like forms that are probably *I. lamarcki stuemckei* are typical in most localities, and a thin-shelled form, possibly *I. glatziae*, is known from the hard bed immediately beneath the North Ormsby Marl. Of particular importance is the occurrence of the characteristically Southern Province echinoid '*Epiaster*' *michelini*, known as a single specimen from the echinoid-rich horizon at the base of the formation in the Newbald Wold Pit [SE 9342 3772] immediately to the north of the district. In marked contrast to southern England, where '*E*'. *michelini* and *Micraster corbovis* of *lata* Zone type are relatively common at this level associated with *S. planus*, the lowest specimens of true *Micraster* from northern England enter higher, some metres above the North Ormsby Marl.

The succession from the North Ormsby Marl to the Wootton Marls is poorly fossiliferous, the macrofauna comprising mainly *Orbirhynchia* spp., *Kingena elegans*, *Terebratulina lata* and rare *Echinocorys* sp., and the flints are solid (i.e. non-carious). The Wootton Marls mark the upper limit of *T. lata* and also the level at which solid tabular flints are replaced by carious tabulars. Some metres higher, the carious flints are replaced by solid continuous tabulars again, above which there is a belt of laminate chalk with extremely thin discontinuous tabular flints. The beds above the Wootton Marls are rich in echinoids, the assemblage being dominated by *S. planus* and *Micraster corbovis* with subordinate *M. leskei* and rare *Echinocorys* sp. This is the fauna and horizon of the famous Boswell Farm Pit [TF 2796 9062] a little to the south of the present district, where Hill (1902) first identified the *planus* Zone in Lincolnshire and where the holotype of the ammonite *Pseudojacobites farmeryi* was collected. *Infulaster* is rare in these beds in most localities, and the *Micraster–Sternotaxis* echinoid assemblage contrasts strongly with the *Infulaster–Sternotaxis* assemblage below the North Ormsby Marl. The hardness of the beds makes it very difficult to extract and develop the echinoids without 'airbrasive' and ultrasonic cleaning.

Apart from bedding-parallel sheets of thick-shelled inoceramids (? including crushed large *I. lamarcki stuemckei*), inoceramids are uncommon between the Wootton Marls and the Ulceby Marl and are restricted to small thin-shelled *Mytiloides striatoconcentricus*. Crushed brachiopods including *Kingena elegans*, *Terebratulina striatula* and *Orbirhynchia* spp. are common at several horizons, notably in a thin bed with carious flints immediately above the Wootton Marls. Rare limonitic moulds of heteromorph ammonites such as *Allocrioceras* sp. and *Hyphantoceras reussianum* provide evidence of the Chalk Rock '*reussianum* fauna' which is such a distinctive feature of this part of the *planus* Zone in southern England, though the characteristic Chalk Rock brachiopod and inoceramid assemblages are apparently unrepresented. The poor preservation and rarity of the ammonites suggests that even at this time of maximum regression and condensation elsewhere, sedimentation in the northern area took place in relatively deep water. The only real evidence of condensation is found in the orange, intensively indurated chalks beneath the Ulceby Marl, which in this district are unfossiliferous. The Ulceby Marl itself is a friable, bioclastic, marly chalk, rich in crinoid debris like its southern correlative the Lewes Marl, and very different in character from the 'buttery' marls (argillised ashes?) of other parts of the succession.

Ulceby Marl to Kiplingcotes Marls

The beds above the Ulceby Marl are generally less indurated and much more massive than the underlying Burnham Chalk. Large nodular and tabular flints are found, but the latter tend to be discontinuous (semitabulars or ellipsoidal masses). There are two beds of marly chalk with abundant pycnodonteine oysters, the Ulceby Oyster Bed and the Enthorpe Oyster Bed, and two sequences of discrete 'buttery' marls. This part of the sequence corresponds to the post-Chalk Rock *planus* Zone succession of southern England.

The beds between the Ulceby Marl and the Ulceby Oyster Bed were formerly exposed in Ulceby Parish Pit [TA 0904 1428] are currently visible in Vale House Quarry. They consist of relatively marly chalks with numerous wispy marl streaks, and yield *Inoceramus* near *I. modestus*, a giant *Gibbithyris?*, *Kingena elegans* and *Orbirhynchia* sp. nov. A poorly preserved mould of the ammonite *Lewesiceras* sp. was collected at Vale House, and the sponge-bed immediately beneath the oyster bed is rich in the sponge *Cystispongia bursa*.

In addition to the oysters, which occur in profusion, the Ulceby Oyster Bed contains a diverse terebratulid and rhynchonellid fauna including *Gibbithyris subrotunda*, a possible *Najdinothyris* or related form, *Orbirhynchia* sp. and sparse *Cretirhynchia cuneiformis*. This assemblage is similar to that of the Chalk Rock of southern England, but differs in the absence of '*Cretirhynchia*' *minor*, and is stratigraphically younger. A juvenile *Micraster leskei* was also collected from this bed. Immediately above the oyster bed is a weakly developed sponge-bearing horizon which has yielded *Sternotaxis placenta* and a large *Micraster* with a retracted posterior that is transitional in profile between the so-called 'large *M. leskei*' of southern England (Stokes, 1975) and the distinctive *Micraster bucaillei* of the higher part of the succession above the Kiplingcotes Marls.

The ichnofossil *Zoophycos* characterises the beds from just beneath Enthorpe Marl 1 to the top of the Kiplingcotes Marls, being particularly conspicuous in the Enthorpe Oyster Bed and beneath Kiplingcotes Marl 1. Occurrences of *Zoophycos* in the higher part of the *planus* Zone above the Chalk Rock are also a feature of the southern England succession. Giant *Infulaster* sp. occur between Enthorpe Marls 1 and 2 at Enthorpe and at a pit near Hatcliffe [TA 2278 0214] in the adjoining Beverley (72) and Grimsby (90) districts respectively. The Enthorpe Oyster Bed is much less well developed than the Ulceby Oyster Bed and the oysters occur more sparsely, particularly at Vale House Quarry, Ulceby, where the bed is only 0.18 m thick. Small brachiopods including *Orbirhynchia* sp. are also present.

The beds above the Enthorpe Oyster Bed are relatively barren, but small thin-shelled inoceramids including *Cremnoceramus?* cf. *waltersdorfensis* and *Mytiloides?* cf. *dresdensis* were collected to the north of the district in a roadside pit [SE 9285 4320] near Kiplingcotes. The Kiplingcotes Marls sequence by contrast is highly fossiliferous. *Sternotaxis placenta* and *Infulaster* sp. have been found at Barrow upon Humber [TA 070 203] and in Raywell Sanatorium Pit [SE 9942 3054]. There is a dramatic faunal change between Kiplingcotes Marls 1 and 2, with the entry of a basal Coniacian inoceramid assemblage comprising *C.? waltersdorfensis*, *C.? rotundatus* and *Inoceramus glatziae?*, associated with the thin-shelled bivalve *Didymotis*, the last being very poorly preserved and restricted to an horizon 0.15 m beneath Kiplingcotes Marl 2, where it locally occurs in flood abundance. The occurrence of this inoceramid–*Didymotis* association at this horizon is of great significance, for it may equate with the younger of the two *Didymotis* events of northern Germany (Wood et al., 1984). The Kiplingcotes Marls may correlate with the Navigation Marls of southern England near the base of the *Micraster cortestudinarium* Zone as currently defined. The Raywell Sanatorium section shows a similar succession to that in the south, in that Kiplingcotes Marl 3 rests on intensely hard nodular spongiferous chalk, analogous to the top Navigation Hardground.

The total or virtual absence of *Micraster* and the rarity of inoceramids between the sponge bed above the Ulceby Oyster Bed and the base of the Kiplingcotes Marls is particularly noteworthy. Comparatively few inoceramids are found in the equivalent beds in southern England; they do not become a significant element of the fauna until the horizon of the Navigation Marls is reached.

Kiplingcotes Marls to Little Weighton Marls

These beds correspond to the greater part of the succession shown as 'c.35 m' (without further details) by Wood and Smith (1978, fig. 2). New lithostratigraphical marker horizons are here designated; they comprise, in ascending order, the Easthorpe Tabular Flints, the Barrow Flints (and associated nodular chalks), the Kirk Ella Marl, the Willerby Flints and the Riplingham Tabular Flints. Biostratigraphical data are taken mainly from the Barrow upon Humber, Willerby and Little Weighton quarries, supplemented in the lower part of the succession by records from the Enthorpe Railway Cutting section in the Beverley (72) district.

The Kiplingcotes Flints maintain the characteristic sequence of the type area (a lower line of smooth solid lenticular flints, a central continuous carious tabular flint and an upper giant nodular flint with downward projections) throughout the district except at Great Limber [TA 141 082], where the upper flint is apparently represented by a belt of small lenticular flints. *Micraster corbovis* is found just below the top Kiplingcotes Flint (at Kiplingcotes) and *M. normanniae* occurs above this flint at Barrow upon Humber, associated with inoceramids. In the expanded succession at Enthorpe this latter horizon has yielded common *Cremnoceramus? waltersdorfensis* and large *C? w. hannovrensis*.

The succession above the Kiplingcotes Flints shows lateral variation in thickness and in the number and type of flints present. At Willerby Quarry, these beds are characterised by carious semitabular and tabular flints, the uppermost three (the Easthorpe Tabular Flints) being generally thicker and more continuous than those below, as in the Enthorpe type-section. In Barrow upon Humber Quarry, however, the Easthorpe Tabular Flints are thin, solid and lenticular, and it is only the underlying flints that retain the carious character. Patchy induration emphasised by pale yellow coloration is present throughout much of the succession, with the exception of the beds immediately beneath the Easthorpe Tabular Flints, which are flaser-bedded chalks, locally (Enthorpe, and possibly elsewhere) with abundant *Zoophycos*. The succession is highly fossiliferous, and is particularly rich in echinoids at two horizons. The lower horizon, about 1 m beneath the Easthorpe Tabular Flints at Willerby and Barrow upon Humber, has yielded abundant small *Echinocorys*. Large *Infulaster* sp. and *Sternotaxis* aff. *placenta* occur at approximately this horizon in the Enthorpe Cutting (F Witham Coll.). The higher horizon, corresponding to the flaser-bedded and/or *Zoophycos* chalks, has a much more diverse fauna including *Cretirhynchia* sp., *Echinocorys* sp., *Hemiaster* sp., *Micraster corbovis* and *M. normanniae*. *Echinocorys* and large *Infulaster* sp. are also found in the indurated chalks between the Easthorpe Tabular Flints. *Cremnoceramus?* of the *rotundatus–erectus* lineage occur throughout, most specimens being transitional forms between *C? rotundatus* and *C? erectus*, but *C?* cf. *erectus* occurs between the middle and top Easthorpe Tabulars and *C?* aff. *waltersdorfensis hannovrensis* immediately over the top tabular. The association of these inoceramids with *Micraster normanniae* allows these beds to be correlated with the lower part of the *Micraster cortestudinarium* Zone of southern England, i.e. the succession from the Navigation Marls to the Hope Gap Hardground of Mortimore's (1986) classification.

The Easthorpe Tabular Flints are overlain by laminate chalks with small, discontinuous, carious and solid flattened nodular flints. In contrast to the underlying beds, these laminate chalks are poorly fossiliferous, only small, indeterminate terebratulid brachiopods and *Echinocorys* sp. having been noted. Above are two beds of yellow, massive, strongly indurated chalks with giant flints (the Barrow Flints) separated by further laminate chalks with thin lenticular flints. The yellow coloration and degree of induration of the massive beds is most strongly developed at Willerby, where the lithology approaches that of a chalkstone, but without any obvious hardground. The development of the Barrow Flints is markedly different at Willerby and Barrow upon Humber: at the former locality both flints are solid and semitabular, whereas at Barrow upon Humber the lower flint is composed of overlapping and superimposed lenticular flints and the upper flint is a giant overgrown thalassinoid burrow-form flint, similar in shape to the top Kiplingcotes Flint. These indurated chalks with giant flints are exposed in a small pit near Ashdale House [TA 1006 1624], where they have yielded large inflated inoceramids, including *Cremnoceramus deformis* and forms transitional between *C. deformis* and *C. schloenbachi*, and a poorly preserved mould of a large gastropod (*Bathrotomaria* sp.). The Barrow Flints with their associated indurated chalks and *Cremnoceramus* concentration probably equate with the giant Stroud Flints, strongly developed hardgrounds and one or more horizons of *Cremnoceramus* shell fragments in the middle of the *M. cortestudinarium* Zone of the North Downs succession (Robinson, 1986). It has even been suggested on both lithostratigraphical and biostratigraphical grounds that the Barrow Flints equate with two horizons of flint ('Hornstein') in the otherwise flintless Lower Coniacian limestones of northern Germany (Wood et al., 1984).

The Barrow indurated chalks are succeeded by up to 0.5 m of relatively soft marly chalk with marl wisps and (at Willerby) thin green-stained vertical burrows. The beds above have incompletely silicified lenticular and semitabular flints, these being the highest beds seen at outcrop on the south side of the Humber, with the possible exception of the top of Ashby Hill Quarry [TA 2405 0060] in the adjacent Grimsby district (now backfilled). These beds are overlain by chalk with carious and solid lenticular and semitabular flints, above which is an irregularly shaped, extremely carious and/or incompletely silicified flint overlain by the Kirk Ella Marl (up to 1 cm thick). The combination of the brown marl and the underlying cream and brown-coloured flint provides an excellent marker horizon, but one that could be confused with the lowest Little Weighton Marl with its carious flint (p.100), were it not for the distinctive sequence of Willerby Flints in the beds above the former. The succession from the Barrow Flints to the Kirk Ella Marl yields *Cremnoceramus deformis*, *C. schloenbachi*, large non-sulcate *Inoceramus* ex gr. *lamarcki* (*I. annulatus*, in part?) and *Micraster bucaillei* sensu Stokes (1975).

The Kirk Ella Marl is followed by the Willerby Flints, comprising three conspicuous flints within some 4 m. Willerby Flint 1 is a solid lenticular to semitabular flint, up to 15 cm thick. It is the highest of a succession of dark grey flints, all the flints above being pale grey, a colour contrast which is conspicuous in fresh quarry faces. Willerby Flint 1 is overlain by massive chalks with small thalassinoid burrow-form flints (an exceptional occurrence in this part of the succession). Willerby Flint 2 is a giant (up to 25 cm) tabular flint with an undulating base and a flat top, followed 0.18 m above by a thin marl and lenticular flint. The overlying chalk is relatively thick bedded in the lower part, but towards the top becomes thinly bedded beneath Willerby Flint 3, a tabular flint up to 10 cm thick. *Inoceramus websteri* and *I. annulatus?* were found above Willerby Flint 1 at Willerby and in the western of the two abandoned railway cuttings at Little Weighton [SE 9722 3370] respectively. The massive chalks with burrow-form flints produce a conspicuous pale-coloured band in the quarry face at Willerby and are rich in echinoids (*Echinocorys*, *Micraster bucaillei* and rare *Conulus* sp.). *I. annulatus?* and *Cremnoceramus schloenbachi* have been collected from just below Willerby Flint 2 outside the present district in a small quarry [SE 9196 4640] north-east of the Enthorpe cutting; the same horizon at Little Weighton Quarry has yielded a small *Echinocorys* and several *M. bucaillei* (F Witham Coll.), and *Infulaster* sp. was found in the adjacent railway cutting. The thin-bedded chalks beneath Willerby Flint 3 are rich in *C. schloenbachi* and *I. annulatus?* which are flattened parallel to the bedding. No definite records of *C. schloenbachi* are known from above this horizon.

In Figure 34, Willerby Flint 3 is shown to be overlain by a bed of yellow, intensely indurated chalk with a burrow-form flint and a sharply defined upper limit. Indurated chalk was seen at this horizon in Little Weighton Quarry and the adjacent railway cutting, and in the former and now backfilled excavation at Willerby (to the north-east of the present quarry). In the present Willerby Quarry, the indurated chalk with its underlying flint exhibits a discordant relationship to the underlying succession, and appears to represent the floor of a broad shallow channel. The axial alignment of this channel cannot be determined, but the extent of downcutting differs significantly between the south-western and north-eastern faces of the quarry. On the former face, the channel begins over 2 m above Willerby Flint 3 and cuts down to just below its horizon. At the bottom of the channel the burrow-form flints are particularly large and conspicuous. On the north-eastern face, what appears to be the same channel cuts down some 3 m from an horizon between Willerby Flints 2 and 3, reaching to just above the level of Willerby Flint 1 (which itself disappears) at its lowest point. The succession above the channel as seen on the south-western face is apparently normal, though there is marked attenuation of the lower beds towards the sides of the channel to judge from the distance between the channel floor and the first conspicuous line of flints above it.

Overlying the Willerby Flints is a succession of relatively thin-bedded chalks with solid lenticular and continuous tabular flints, terminating in three prominent tabular flints within 1.75 m (the Riplingham Tabular Flints), which recall the Triple Tabular Flints near the base of the Burnham Chalk (p.97). In this case, however, it is the top rather than the bottom flint that is the thickest and most continuous. *Micraster* aff. *bucaillei* and a specimen of *M.* cf. *gibbus* were collected from the base of the succession at Willerby about 1 m above the yellow indurated chalk, the latter being the only record of this essentially Anglo-Paris Basin species in the Northern Province. *M. bucaillei* is common between the Riplingham Tabulars, and *I. annulatus?* occurs just beneath the top flint.

The succession above the Riplingham Tabulars comprises a bed of hard massive chalk overlain by 0.2 m of marly chalk between thin marls, followed by flaggy chalk with flattened burrow-form and discontinuous semitabular flints. The marly bed yielded *Micraster* aff. *normanniae* rather than the *M. bucaillei* which appears to characterise most of the succession above the Barrow Flints. The overlying, flaggy bed at Little Weighton is rich in small fossils including *Orbirhynchia* sp., *Terebratulina striatula*, *Spondylus* ex gr. *dutempleanus*, *Bourgueticrinus* sp. columnals, asteroid marginals and radioles of *Stereocidaris sceptifera*. The succession correlates with the top of the *Micraster cortestudinarium* Zone.

LITTLE WEIGHTON MARLS TO TOP OF EXPOSED SUCCESSION

The Little Weighton Marls are the first thick marl bands above the Kiplingcotes Marls, and the lower two (each 2 cm thick) provide marked low-resistivity 'spikes' on resistivity logs—the 'Conoco Marls' of Barker et al. (1984). The lowest marl, Little Weighton Marl 1, recalls the Kirk Ella Marl in that it rests on an incompletely silicified flint; at Willerby, a flint of this type is also locally present immediately above the marl. Little Weighton Marls 1 and 2 are separated by a bed of massive chalk with a continuous (up to 14 cm) tabular flint. No fossils other than traces of sponges have been noted from this bed. These marls are almost cetainly the correlatives of the East Cliff Marls and Shoreham Marls of the Kent and Sussex successions respectively, the upper and more persistent of which is conventionally taken to mark the base of the *Micraster coranguinum* Zone.

Overlying Little Weighton Marl 2 is a bed with lenticular flints in its higher part. At Little Weighton a local fossil accumulation some 30 cm above the base yielded a rich fauna comprising *Synolinthia?*, fragments of *Inoceramus* ex gr. *gibbosus/lezennensis*, asteroid marginals including *Metopaster* sp., *Echinocorys* sp., numerous (mainly crushed) *Micraster* cf. *coranguinum* and a tooth of *Scapanorhynchus raphiodon*. It is noteworthy that all the specimens of *Micraster* above Little Weighton Marl 3 are *M.* cf. *coranguinum* (the Southern Province species) rather than derivatives of the essentially northern *M. bucaillei*.

Little Weighton Marl 3 is thinner (up to 1 cm) than the lower marls, and is overlain by a concentration of valves of *Volviceramus koeneni* in an indurated yellow ferruginous chalk, marking the entry of *Volviceramus* in the Northern Province succession and permitting correlation with the *V. koeneni* Zone of the northern European inoceramid zonation. This horizon presumably equates with the calcarenitic chalk bed with fragments of *V. koeneni* and *Platyceramus* occurring just over 1 m above the higher East Cliff Marl at Dover (Bailey et al., 1983).

Little Weighton Marl 3 with its associated *V. koeneni* shell-bed is overlain by a complex 4 m succession including thin marls and, in the lower part, several horizons of yellow, patchily to intensely indurated chalks. There is a great diversity of flint type, including semitabulars, thin tabulars locally passing into or fusing with lenticular flints, and both ramifying and flattened thalassinoid burrow-form flints. Exposures in Little Weighton Quarry, in the two railway cuttings to the south-east and at the entrance to Willerby Quarry, show considerable lateral variation in the development of flint and indurated chalk. There appears to be a general condensation from the Eppleworth Wood Farm cutting [TA 002 321] to Little Weighton Quarry [SE 981 333]. The section used in the preparation of Figure 34 is of necessity the most condensed, for it is only at Little Weighton that the complex of indurated chalks overlying the Little Weighton Marls is demonstrable. The succession is characterised by *Inoceramus gibbosus/lezennensis*, *Volviceramus koeneni* (including forms transitional to *V. involutus* at the top), *Echinocorys* sp. and *Micraster* cf. *coranguinum*. The assemblage can be assigned to the *V. koeneni* inoceramid zone, and the indurated chalks are presumably an expression of the so-called '*koeneni* regression' (Ernst et al., 1983) of Germany and adjacent areas. An anomalous and slightly expanded succession near the entrance to Little Weighton Quarry yielded the only specimen [BGS CJW 8205] known from Britain of the extremely rare belemnite *Actinocamax bohemicus* (Christensen, 1982, p.72, fig. 1).

This complex sequence is overlain by beds of white, massive, relatively inoceramid-rich chalk with lenticular and burrow-form flints, with the Rowley Marls 1–4 (within 3 m and each up to 2 cm thick) and several stylolitic surfaces with marl films. There is some lateral variation in the thickness and persistence of the marl seams. Rowley Marl 3 at Little Weighton Quarry rests on vertical 'potstone'-type (paramoudra?) flints. It is replete with inoceramid shell fragments, as is Rowley Marl 4. The latter normally rests on a lenticular flint, e.g. in Little Weighton Quarry and in the Eppleworth Wood Farm cutting, but in the cutting south-east of the quarry the solid lenticular flint is large and carious, and the marl is absent. Fossils collected from these beds include *Volviceramus* ex gr. *involutus*, *Platyceramus mantelli* and *Echinocorys* sp., and can be assigned to the base of the *V. involutus* inoceramid zone.

The Rowley Marls are followed by 3 to 4 m of hard, relatively massive, poorly fossiliferous chalks with stylolitic partings. There is a great diversity of flint type, including lenticular and semitabular flints up to 20 cm thick near the base, flattened semi-continuous thalassinoid burrow-form flints, thin overlapping lenticular flints and, at the top, the giant (up to 30 cm) overgrown thalassinoid burrow-form Eppleworth Flint. The last is one of the largest flints in the Northern Province and occupies a variable position within a 75 cm bed of massive chalk, being thus reminiscent of the Ravendale Flint at the base of the Burnham Chalk (p.96). It is particularly well seen in Eppleworth Pit [TA 021 324] and towards the top of the railway cutting south-east of Little Weighton Quarry, where it forms a prominent ledge. The only fossils noted are isolated valves of *Volviceramus* sp., which occur sporadically below the Eppleworth Flint and, more commonly, at the base of the bed that includes the flint. There is considerable lateral variation in flint development, particularly at the base of the succession, where reduction in the number of flints from the Little Weighton cutting to the quarry provides some evidence of condensation. The massive chalks associated with the Eppleworth Flint, and the flint itself, provide a conspicuous peak on resistivity logs.

The 3 m of chalk above the Eppleworth Flint comprise massive beds delimited by marl-coated bedding planes or thin (up to 10 cm) marls. Flints are small (up to 10 cm) and of flattened thalassinoid burrow-form or overlapping lenticular type. The bottom 1.5 m is relatively fossiliferous, yielding a low-diversity fauna comprising *Gibbithyris* sp. and isolated valves of *Volviceramus* cf. *involutus* s.s. and *V.* sp. nov. aff. *involutus*. The beds above are replete with inoceramid shell fragments, but no complete specimens have been noted. The succession terminates in a flat-topped, carious tabular flint, the De La Pole Flint, beneath which an incompletely silicified rounded lenticular flint (up to 15 cm) is developed in Eppleworth Pit.

The 5 m of beds above the De La Pole Flint are exposed only in Eppleworth Pit in the present district. The bottom 2.5 m comprise massive beds delimited by thin marls as in the underlying succession. Flints are predominantly small and lenticular at the base, replaced upwards by irregular carious tabulate flints. One flint of the latter type is up to 18 cm thick, and is overlain by massive chalk with large pieces of *Platyceramus* shell lying parallel to the bedding. The top 2.5 m consist of chalk which is thin-bedded, particularly in the topmost metre. The lower flints are irregular, lenticular and carious, and are replaced in the highest beds by solid discontinuous semitabular flints. No macrofossils have been observed.

The giant Eppleworth Flint has been tentatively correlated with the East Cliff Semitabular and Seven Sisters flints of Kent and Sussex respectively, and also with an enlarged thalassinoid burrow-form flint found in the Willowmere Spinney Pit at Euston near Bury St Edmunds, on the basis that all these named flints represent a flint optimum within the lower part of the *Micraster coranguinum* Zone (Mortimore and Wood, 1986). On palaeontological grounds, however, the relative abundance of *Volviceramus* spp. below the De La Pole Flint, contrasted with its apparent absence above, could be taken to support correlation of this flint with the East Cliff Semitabular/Seven Sisters flints rather than the Eppleworth Flint, particularly in view of the appearance of *Platyceramus* sheets in the beds above in all areas.

CHAPTER 6

Structure

DEEP STRUCTURE

The pre-Carboniferous rocks and, to a large extent, the Lower Carboniferous rocks lie below the levels reached by boreholes and resolved by seismic reflection surveys. The only pertinent data are summarised on the regional Bouguer gravity anomaly and aeromagnetic maps (Institute of Geological Sciences, 1977a; 1977b). The contours on the Bouguer anomaly map of the district (Figure 35i) have been smoothed slightly to reduce the effects of near-surface density variations, but they still reflect the effects of thickness changes in the relatively low density Upper Carboniferous and younger rocks. On the modified Bouguer anomaly map (Figure 35ii) these effects have been removed, so that the remaining anomalies indicate differences within the Lower Carboniferous and older rocks. The anomalies shown on the aeromagnetic map of the district (Figure 35iii) are considered to be due to variations within the pre-Carboniferous basement, because rocks of Carboniferous and Mesozoic age do not contain significant amounts of magnetite.

Figure 35ii reveals two major deep-seated structures, one centred north of the north-western corner of the district, the

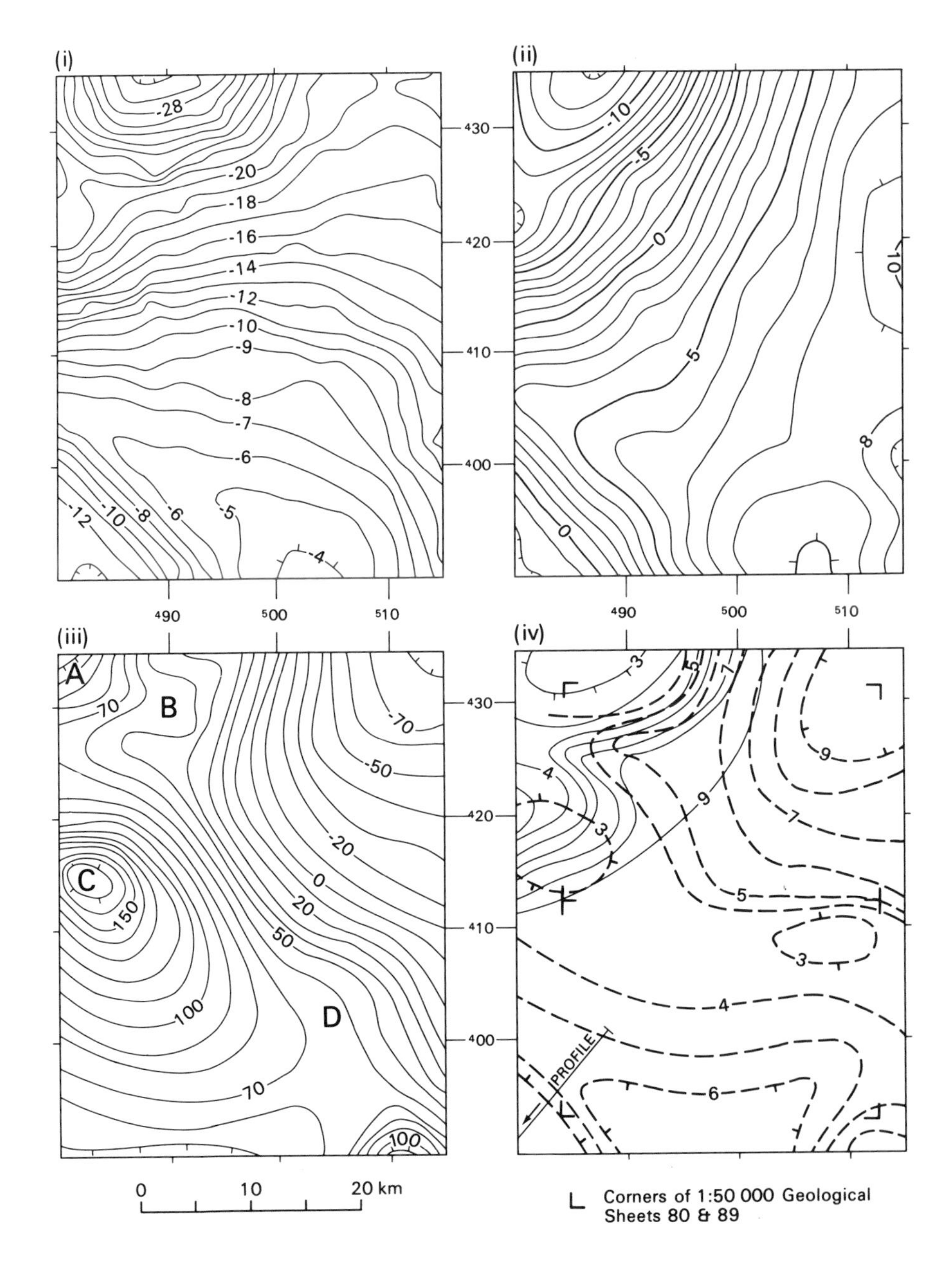

Figure 35 Geophysical maps and basement interpretation of the district

i) Bouguer anomaly map. Contour interval 1 mGal. ii) Bouguer anomaly map after compensation for gravity effect of low-density sediments down to base of Millstone Grit. Contour interval 1 mGal, iii) Aeromagnetic map. Contour interval, 10 nT. Lettered anomalies (A–D) are referred to in the text. iv) Contours, based on geophysical evidence, showing a) depths to surface of postulated Market Weighton granite (from Rollin, 1982) (solid lines), and b) configuration of the magnetic basement surface (broken lines). Contours are at 1 km intervals below OD. Profile shown in Figure 36 other crossing the south-west of the district with a north-westerly trend.

The possibility that the former results from a buried granite has long been recognised. Bott et al. (1978) considered the postulated 'Market Weighton granite' to be of Caledonian age and possibly to form part of the sub-Carboniferous surface. It is calculated to rise to at least 2.5 km below OD near Market Weighton and to descend to 8 km below OD or lower. Such a large, relatively buoyant mass might be responsible for thinning of the Carboniferous rocks in the area and the attenuation of the Jurassic and Lower Cretaceous rocks across the partly coincident Market Weighton Structure. The Bouguer anomaly data suggest that the postulated granite has a north-east to south-west elongation with the major culmination centred [SE 890 380] near Sancton and a secondary one [SE 800 230] near Reedness. Rollin (1982) has interpreted the anomaly in terms of a three-dimensional granite model, and computed contours on the surface of this body have been included in Figure 35iv.

As supporting evidence for the existence of a concealed granite, Bott et al. (1978) suggested that some of the aeromagnetic anomalies (Figure 35iii) represent the effect of magnetic rocks intruded, and partially domed up, by the granite. The anomaly at 'A' is consistent with the presence of magnetic rocks flanking the northern margin of the main granite body outside the area of the figure. The anomalies at 'B' and 'C' may be due to other blocks of magnetic basement rocks at the southern margins of the main and secondary cupolas respectively. The form of the latter anomaly strongly suggests that the magnetic body is shallower to the north, where it would abut against the granite cupola, but that it becomes gradually deeper to the south, on the flanks of the culmination. Bott et al. (1978) have likened the geophysical response of the postulated Market Weighton granite to that of the Wensleydale Granite (Dunham, 1974) underlying the Askrigg Block, where the source of the magnetic anomalies was proved in the Beckermonds Scar Borehole (Wilson and Cornwell, 1982) to be magnetite-bearing Ordovician sediment. There appears to be a deep-seated zone of magnetic rocks running from the Lake District south-eastwards through the Askrigg Block to The Wash, as discussed by Bullerwell (in Bott, 1961). The anomaly at 'C' on Figure 35iii forms part of this zone. The coincidence of the localised bulge in the contours, forming a minor aeromagnetic anomaly 'D' on Figure 35iii, with a poorly defined feature on the Bouguer anomaly map (Figure 35i) suggests the presence of weakly magnetised, but fairly dense rocks, possibly of basic igneous origin and spatially associated with the northern margin of the west-north-westerly-trending anticline in upper Carboniferous rocks across the centre of the district (Figure 37i and ii).

The well-defined Bouguer-anomaly gradient in the south-west of the district (Figures 35i and ii) indicates the northern edge of the Gainsborough Trough. It has been investigated using seismic techniques, and Kent (1966, pp.335–337) proposed a partly faulted and partly monoclinal structure with a vertical displacement of Lower Carboniferous rocks downwards to the south in excess of 300 m. However, the magnitude of the Bouguer anomaly (Figure 36i) would require an increase in thickness of the Millstone Grit in the trough of about 1.2 km, whereas the increase indicated by

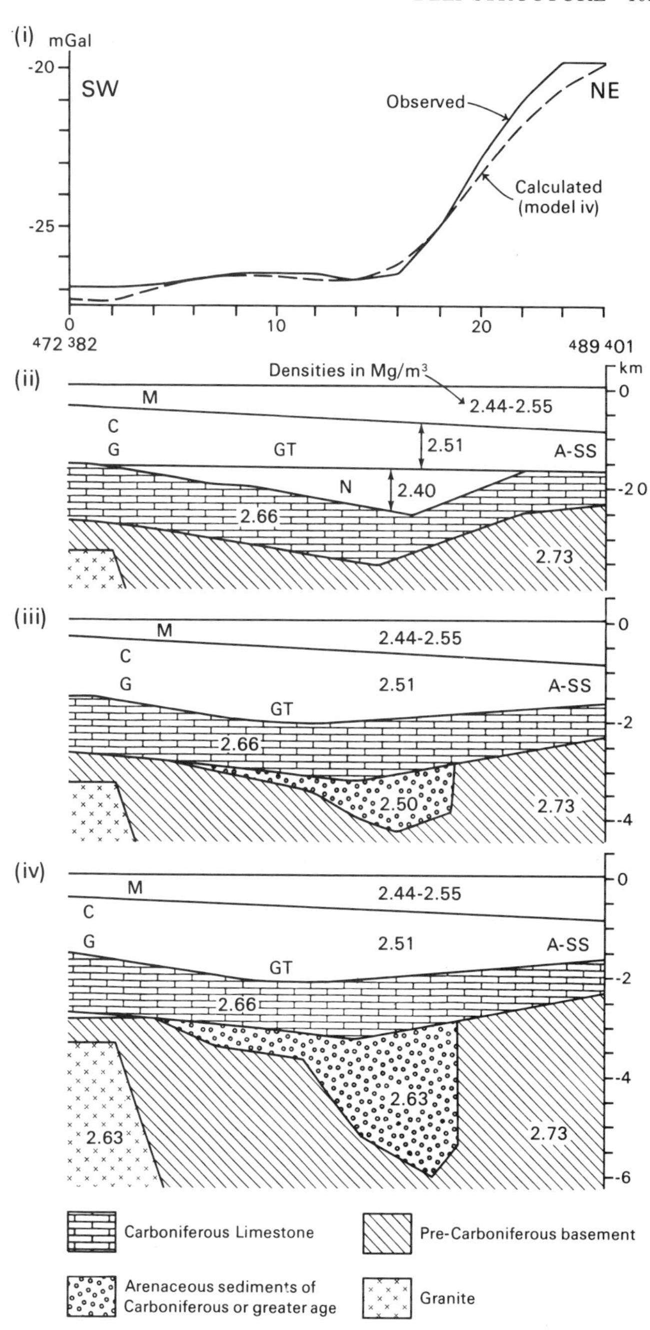

Figure 36 Deep geological interpretation of a Bouguer anomaly profile across the Gainsborough Trough and the Askern-Spital Structure (for location see Figure 35 iv)

i) Observed profile assuming a background value of 14 mGal and calculated profile for model iv. ii) Model in which observed profile is considered to be due mainly to a low-density facies of lower part of Millstone Grit. iii) and iv) Models in which observed profile is considered to be due mainly to the presence of small basins of older, lower density, arenaceous sediments. Model iv) produces the least satisfactory calculated profile compared with the observed profile. The Bouguer anomaly low at the SW end of the profile is assumed to be due to a concealed granite. Abbreviations: N Namurian, G Millstone Grit, C Coal Measures, M Mesozoic rocks; GT Gainsborough Trough, A-SS Askern-Spital Structure seismic results and boreholes is only 300 to 500 m (Kent, 1966, fig. 5; Smith et al., 1973, plate II). It is, therefore, necessary to postulate an alternative source for the anomaly. One possibility is a low-density facies towards the base of the Millstone Grit (Figure 36ii), and others (Figures 36iii and iv) assume changes in thickness of the Lower Carboniferous rocks, possibly combined with earlier development of a restricted basin of arenaceous rocks, as at Eakring (Lees and Taitt, 1946, pp.282–283). It is also possible that the Millstone Grit may thicken more markedly along the northern edge of the Gainsborough Trough than indicated by Kent (1966, fig. 5).

CARBONIFEROUS ROCKS

As yet, it is not possible to contour the pre-Carboniferous surface with any precision. The configuration of the Millstone Grit and overlying rocks can, however, be plotted from seismic reflection surveys, supplemented by boreholes, and structure-contours have been prepared for the base of the Millstone Grit (Figure 37i) and for the Barnsley Coal (Figure 37ii). There is only one major difference between these two maps. The base of the Millstone Grit illustrates, in the extreme south-west of the district, the effect of the Gainsborough Trough, which has an amplitude of between 300 and 500 m. In the Barnsley Coal the influence of this trough is insignificant.

Both maps depict a gentle west-north-westerly-trending anticline crossing the centre of the district and plunging to the east-south-east. On its southern side there is a parallel gentle syncline plunging in the same direction. The area between this syncline and the northern margin of the Gainsborough Trough corresponds to the Askern–Spital Structure (Edwards, 1951, p.114; Kent, 1966, p.347 and plate 21; Kent, 1974, fig. 5), but this is poorly defined within

Figure 37 Structure maps, based largely on seismic evidence, of the Carboniferous and Permian rocks.

i) Contours on base of Millstone Grit. ii) Contours on the Barnsley Coal. iii) modified contours on the Barnsley Coal, post-Permian tilting having been removed. iv) Contours and faults on the base of the Permian rocks. All contours are in metres below OD the district, particularly along its northern edge.

The general regional dip is to the east-north-east, and largely results from post-Permian tilting. Some indication of the pre-Permian structure can, however, be obtained by correcting for this later tilting, and a modified structure-contour map of the Barnsley Coal has been produced by this means (Figure 37iii). This map suggests a line of gentle domes coincident with the anticline crossing the middle of the district, and a line of basins coincident with the flanking syncline to the south. As expected at this stratigraphical level, there is no indication of the margin of the Gainsborough Trough, and little of the Askern–Spital Structure.

PERMIAN AND MESOZOIC ROCKS

Folding

Structure-contours on the base of the Permian rocks (Figure 37iv) illustrate the post-Permian regional tilt to the east-north-east, modified only by subdued expressions of the mid-district anticline and syncline that affect the Upper Carboniferous rocks. The latter effects may be due to post-Permian movements along the fold axes, or may result from the older structures having had some topographical expression when Permian deposition began.

There is little overall structural variation upwards through the Permian, Triassic and Jurassic rocks, other than a slight movement of the regional dip from east-north-east to almost due east, the latter direction being illustrated by contours on the top of the Frodingham Ironstone (Figure 38i). None of the intra-Jurassic unconformities makes any appreciable difference to this pattern. The absence of the Hibaldstow Limestones and the Upper Cornbrash across the Askern–Spital Structure suggests that there might have been minor and periodic intra-Jurassic arching along it. The eastward dip of the Jurassic is, however, largely the result of post-Cretaceous tilting to the north-east. If this tilting is removed from the contours on the top of the Frodingham Ironstone

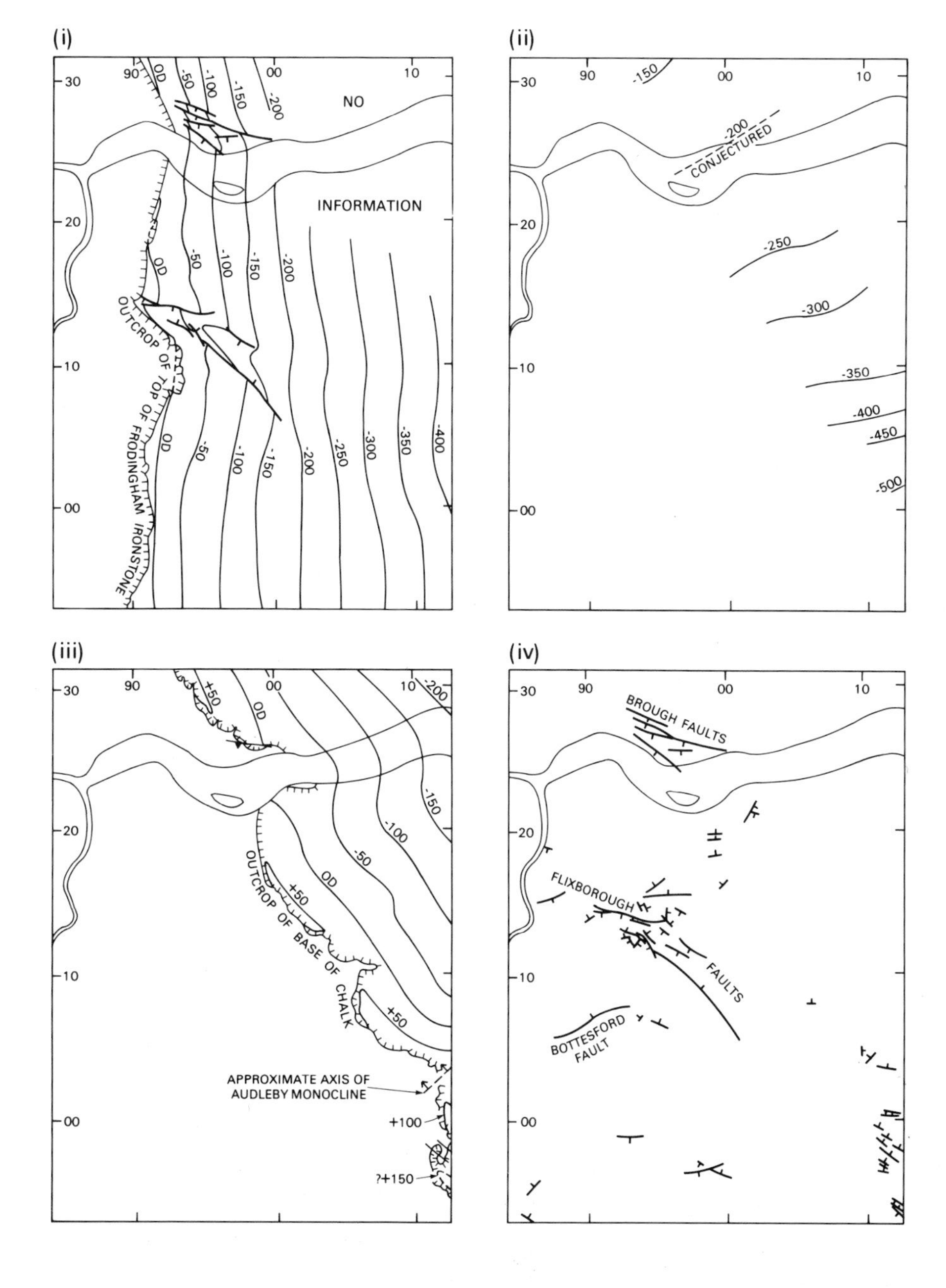

Figure 38 Structure maps of the Mesozoic rocks

i) Contours on the top of the Frodingham Ironstone. ii) Modified contours on the top of the Frodingham Ironstone, post-Cretaceous tilting having been removed. iii) Contours on the base of the Chalk. iv) Faults at outcrop. Contours are in metres above and below OD and are based on borehole evidence where this lies below the Chalk, some indication of the pre-Chalk structure of the Jurassic rocks can be obtained (Figure 38ii). The resulting regional dip is then to the south-south-east, and results from uplift along the Market Weighton Structure which, from late Hettangian to early Cretaceous times, acted as an east–west hinge-line, subject to periodic uplift.

The hinge-line is now manifested by large faults throwing down to the north, which separate the rapidly subsided Cleveland Basin to the north from the more slowly subsided East Midlands Shelf to the south. The structure is thus markedly asymmetrical and can be likened to the northern monoclinal margin of a rigid platform or block. There is no clear southern margin to the Market Weighton Structure. At different times in its history it can be taken variously at: a minor inflexion near Sancton, north of the district (Kent, 1980a); the Brough Faults, active in later Callovian to early Oxfordian times, and possibly earlier (see below); a line marking the northern preserved limits of certain of the Aalenian to early Bajocian strata west of Winterton; the Flixborough Faults, which were active intermittently in early Jurassic, Bajocian and Bathonian times (see below); the Bottesford Fault, which moved intermittently in Sinemurian to Pliensbachian and post-Bajocian times (see below) and which approximately marks the southern limit of the Frodingham Ironstone; and possibly the buried Askern–Spital Structure. However, the above interpretation does not demand any discrete southern margin to the structure, and it is likely that the various lines summarised above mark the transient southern limits of compensating gravitational adjustments at times when the Market Weighton hinge was subject to uplift.

Structure-contours on the base of the Chalk (Figure 38iii) show a fairly uniform regional dip to the east-north-east, the result of post-Cretaceous tilting along the south-western limb of the Wolds Syncline (Donovan, 1968), which plunges south-eastwards along a line running between Driffield and Bridlington. The contours are offset slightly north of Caistor by the north-west facing Audleby Monocline, although the effect of this structure barely extends into the district. An apparently complementary offset in the contours along the Humber may be partly of tectonic origin, particularly as it is aligned with the Brough Faults (see below); but much of this offset, especially along the −50 m OD contour, is attributable to superficial valley-bulging along the estuary (see below). The anomalous trend of the OD contour as it approaches the estuary from both north and south is due to cambering of the Chalk, as also are localised reversals of the contours above OD along the Chalk scarp (see below).

Faulting

Surface faults, except for those believed to be of superficial origin, are shown on Figure 38iv.

Five subparallel faults, all throwing down to the south, have been traced in the Brough–North Ferriby area, and are here termed the Brough Faults. They have the west-north-westerly alignment of many of the faults in the Yorkshire Coalfield, and may be due to renewed movement along earlier structures. The Grantham Formation, the mudstones of the Glentham Formation, the Upper Cornbrash and the Kellaways Clay all die out at outcrop on, or just to the south of these faults, and it is possible that some intermittent movement occurred in Aalenian to early Callovian times. Further movement probably occurred in later Callovian to early Oxfordian times, for a belt of erosion of the Kellaways Beds and overlying mudstones, with reworking of sand from the former into the Brantingham Formation, is aligned with the Brough Faults. One of these faults appears to cut the Chalk, which would imply some post-Cretaceous movement, but the displacement may be due to cambering.

Several faults running from near Flixborough, through Risby Warren and almost to Brigg have been termed the Flixborough Faults. They follow the same trend as the Brough Faults, and are almost in line with faults proved in the Coal Measures around Thorne Colliery, west of the district, which are known to have suffered further movement in post-Triassic times. They are also aligned with, and partly overlie, the mid-district anticline in the Carboniferous rocks (Figures 37i–iii). Between Flixborough and Dragonby the structure is essentially a monoclinal downfold to the south. Eastwards from Dragonby to Broughton Common the faulting has produced a graben enclosing an asymmetric syncline (Figure 38i), and an overall southward downthrow is apparent. Several small domes such as those at [SE 950 154] north of Appleby and at [SE 986 074] near Castlethorpe are present on the flanks of the Flixborough Faults and may be tectonically related to them. The main Mesozoic movement along the faults was clearly post-Callovian and may be post-Cretaceous since they are aligned with faults displacing the Chalk near Nettleton and Claxby. However, localised thickening of the Pecten Ironstone and the Marlstone Rock, major channelling into the Hibaldstow Limestones, and the near coincidence of the fault-belt with the northern limit of the Snitterby Limestone suggest that some movement took place earlier in the Jurassic.

An east-north-easterly trending fault, passing through Bottesford and throwing down to the north, is here termed the Bottesford Fault. It follows another of the trends typical of the Yorkshire Coalfield. Appreciable lithological and thickness changes in the Frodingham Ironstone and Pecten Ironstone, and attenuation of the Grantham Formation, especially in its lower part, take place across this fault and suggest that there was repeated movement along it, although the main phase is clearly post-Bajocian. Nearby, a small fault at New Forest Plantation [SE 950 067] deflects the Redbourne Group outcrops and is probably associated with elongate domes in Scawby Park [SE 974 059]. South of Bottesford small-scale faults cut the Lower Jurassic sequence at Blyton [SK 860 954], Pilham [SK 854 934] and Ings [SK 925 989], whilst an extensional complex affects Middle Jurassic rocks near Waddingham [SK 986 967].

SUPERFICIAL STRUCTURES

The non-diastrophic structures in the district are essentially cambers and valley bulges. These may occur wherever a thick competent rock overlies argillaceous strata along valley slopes or beneath scarps, the weight of the former having squeezed the argillaceous rocks upwards in valley floors and outwards on valley and scarp slopes. In the former, displaced strata form long superficial anticlines or 'valley bulges' along the axes of the valleys and may be overturned and fractured in places; vertical movements of up to 30 m have occurred. On valley and scarp slopes the overlying competent rock has been concomitantly lowered to produce fissures, known as gulls, parallel to the slope; the dip has been reorientated to conform with the slope, thus forming cambers. If the original dip was negligible or in the same direction as the slope, the lowering was accomplished by large blocks of the competent rock becoming detached and sliding down the slope; if its original dip was into the slope it may have been accomplished by rotational movement of blocks between closely spaced faults parallel to the valley (the 'dip-and-fault' structure of Hollingworth et al., 1944) (Plate 4).

Small-scale cambers affect the scarp of the Lincolnshire Limestone in a number of places, and several small anticlines and associated fractures in the Frodingham Ironstone and contiguous mudstones, especially at the bottom of the Winterton Beck Valley, are probably valley bulges. Another possible valley bulge occurs to the west [SE 967 026] of Hibaldstow where a large area has been arched up and eroded to expose the Raventhorpe Beds. The largest superficial structures in the district are, however, associated with the Chalk and the underlying mudstones of the Ancholme Clay Group. The cambering is sufficiently extensive to produce localised reversals of contour trends on the base of the Chalk near the west-facing scarp and adjacent to the Humber (Figure 38iii); the dips of the Chalk towards the estuary shown in horizontal section A′–A″ on the 1:50 000 Kingston upon Hull geological sheet are also caused by cambering. On the south side of the Humber the superficial structures in the Chalk quarries north-east of South Ferriby comprise mainly rotational fractures subparallel to the estuary and accompanied by closely spaced jointing. Because of the rotational effect, the dip of each mass of Chalk is towards the slope (i.e. to the south-east), and is steeper than that of the undisturbed Chalk. In their description of these structures, Higginbottom and Fookes (1971, pp.104–105, plate 17) note that, although the largest observed displacement across any of these fractures is 9 m, many of the 'throws' are only about 1 m and most are less than 0.1 m, while dips of up to 70° occur in the displaced masses. This intense fracturing of the Chalk can be examined in the wave-cut platform at the base of South Ferriby Cliff where, since

Plate 4 Repeated step-faults of superficial origin in the Welton Chalk at Barton upon Humber Quarry. Downthrows are to the right, i.e. towards the Humber Estuary.

the platform lies at the base of the Ipswichian cliff (now concealed behind the present cliff), the superficial movements are presumably pre-Ipswichian in age. On the north side of the Humber, cambering has resulted in the Chalk sliding or flowing towards the estuary along low-angle fractures, with the formation of gulls. The distortions thus produced are well seen in the Melton 'Clay' Pit [SE 970 270].

No obvious bulging or swelling is apparent along the outcrop of the Ancholme Clay, but such structures were noted in boreholes for the Humber Bridge (Higginbottom and Fookes, 1971, p.104), and the western part, at least, of the rockhead ridge running along the middle of the estuary (Figure 40) is presumed to be a major valley bulge.

CHAPTER 7

Quaternary

Quaternary deposits cover nearly 70 per cent of the district, being most extensive and thickest in low-lying areas (Figure 39). They range from clay to gravel, and are the result of a variety of depositional agencies including glacial, periglacial (notably solifluction and aeolian), lacustrine, fluvial, estuarine and marine.

The Quaternary Era is variously estimated to have started between 1.6 and 2.4 million years ago. It has been characterised by repeated major climatic fluctuations, which form the basis for Quaternary chronostratigraphy. In Britain, Mitchell et al. (1973) formulated a sequence of alternating temperate and cold stages, with glacial episodes in some of the more recent cold stages, but the study of oceanic sediments has shown that the succession of climatic changes was much more complex than can be deduced from the incomplete evidence available on land (Bowen, 1978, pp.198–199). However, until means of dating and correlation improve, the land-based stages provide the best available basis for description.

Within the district there are no deposits attributable to the Cromerian or older stages, so that well over half of Quaternary time is unrepresented. The deposits formed in the later part of the era are listed and correlated in Table 4, which also indicates the deduced sequence of events. The correlation is based on the identification of two glacial episodes, one pre-Devensian and the other late Devensian, and of two interglacial episodes distinguished by differences of sea level. Parts of the pre-Devensian correlation are tentative.

The oldest deposits are of glacial origin. Those within the Immingham Channel (see below) are placed in the Anglian because they underlie interglacial deposits of presumed Hoxnian age at Kirmington. Other old glacial deposits in the Ancholme and Trent valleys that extend southwards into the 'chalky' deposits of the East Midlands are also provisionally allocated to the Anglian, but on less certain evidence, and they could conceivably be Wolstonian. The Kirmington interglacial deposits imply a sea level at about 22 m above OD, a level believed to preclude an Ipswichian age, and they are therefore referred to the Hoxnian. No deposits of undoubted Wolstonian age are known in the district, but a cold periglacial episode is suggested by extensive cambering at South Ferriby (p.107). Nor have any Ipswichian deposits been identified with certainty, but an old sea cliff, now concealed beneath Devensian deposits, formed during this interglacial, and some gravels near its foot at about 2 to 3 m above OD may be contemporaneous beach shingle.

Throughout much of the Devensian the district experienced substantial denudation and deep fluvial incision as a consequence of greatly lowered sea level, and solifluction during severe periglacial conditions produced head deposits. Late in the Devensian, ice invaded the north-eastern part of the district and advanced westwards into the Humber Gap. Drainage through the gap was blocked, probably initially by ice and subsequently by till, and a large lake—Lake Humber—formed in the Vale of York and the Ancholme Valley. When the lake filled up with lacustrine sediments, levée deposits formed on the emergent plain and blown sand, indicative of continuing periglacial conditions, accumulated elsewhere. Subsequent fluvial incision, governed by the still low sea level, demolished the blockage in the Humber Gap and cut deep channels across the district. Sea level rose during the Flandrian, and alluvium, estuarine sediments and peat filled these channels, eventually spreading extensively beyond their confines.

Exposures of Quaternary deposits are generally transient, and those existing during the survey were poorer than at several times in the past, particularly with respect to the concealed Ipswichian cliff at Hessle, the interdigitating Devensian glacial and lacustrine deposits at Red Cliff near North Ferriby, and the interglacial deposits at Kirmington. However, many observations of the Quaternary deposits in the district have been published over a long period, and this chapter has drawn heavily upon them.

PRE-DEVENSIAN GLACIAL DEPOSITS

Glacial deposits of pre-Devensian age are mainly confined to three areas—the Immingham Channel north-east of Kirmington, the Ancholme Valley around and south of Brigg, and the Trent Valley south of Scotter.

Immingham Channel

Rockhead contours (Figure 40) show that a channel, deeply incised into Chalk, descends steeply to the north-east from the vicinity of Kirmington. Beyond the district the channel reaches c.73 m below OD under Immingham. The deposits filling the channel range from clay to gravel. They are entirely concealed, at Kirmington by interglacial deposits and elsewhere by Devensian till. The thickest proving of the channel deposits in the district is in a borehole [TA 1218 1307] which records 'boulder clay' (presumably Devensian) to 10.97 m, on gravel (? interglacial) to 12.19 m, on white and red silt to 52.12 m, on chalk gravel to 58.52 m, resting on Chalk at 42.4 m below OD. Several boreholes at Kirmington, including two drilled for the British Association for the Advancement of Science (Lamplugh et al., 1905) and one for the BGS [TA 1031 1163] (Institute of Geological Sciences, 1973), show that rockhead within the channel varies hereabouts between OD and 40 m below OD. They prove the following generalised sequence (below undoubted interglacial deposits) in descending order:

	Thickness m
Sand, yellow, with chalk and flint pebbles and a thin clay layer	1.45 to 4.27
Clay and silt, red and brown near top,	

Figure 39 Map showing approximate limits of principal Quaternary deposits

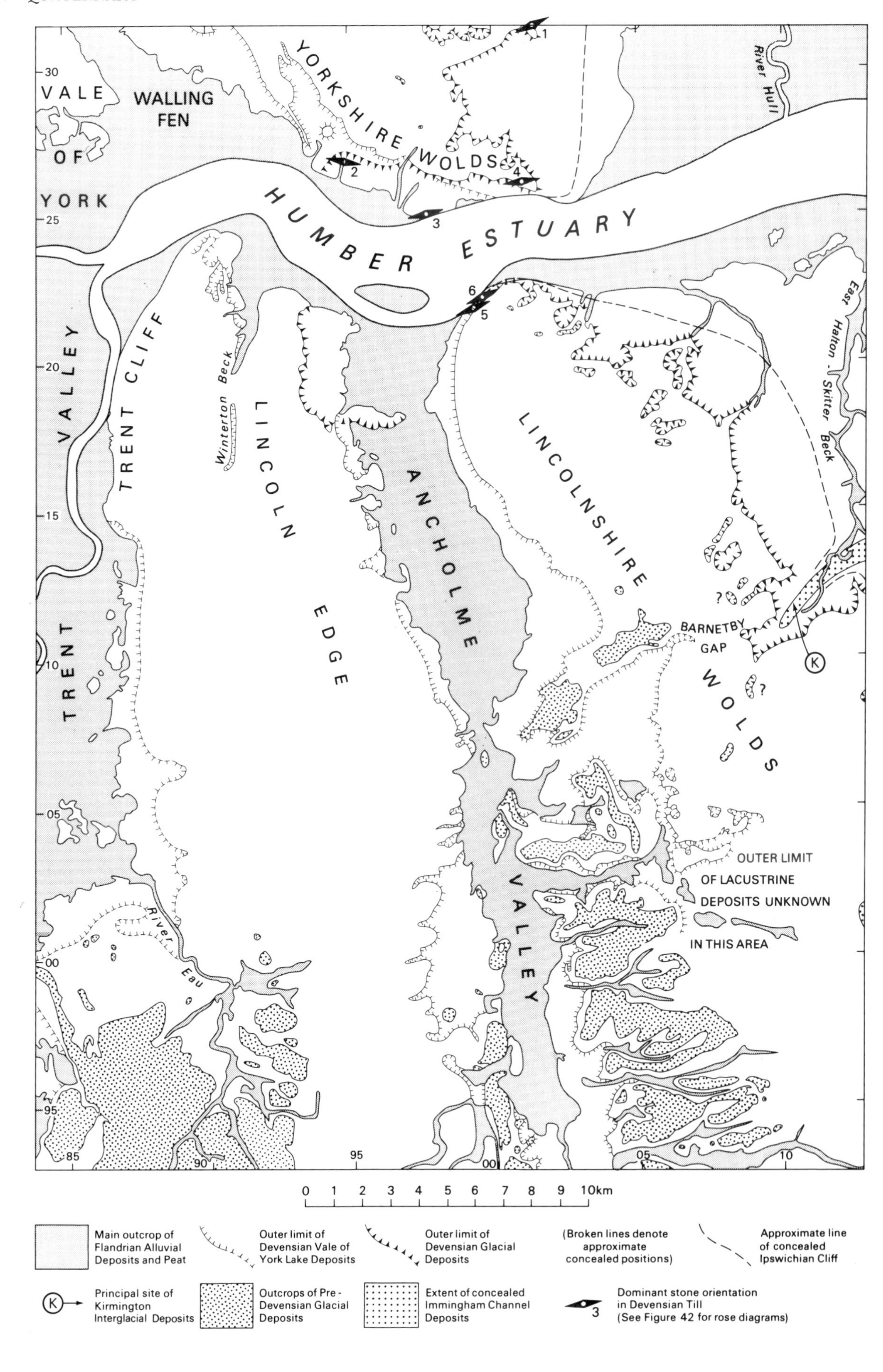

Table 4 Correlation of Quaternary deposits in the district and the deduced sequence of events

Deposits (not necessarily in stratigraphical order within stage divisions)	British Stages (after Mitchell et al., 1973)	Deduced sequence of events
Landslip Tidal Flat Warp Estuarine Alluvium Alluvium Peat Calcareous Tufa Shell Marl	Flandrian (temperate – post-glacial)	Deposition, mainly fluvial and estuarine, consequent on sea level rising to OD
		Fluvial incision due to drop of sea level to below – 20 m OD
Blown Sand Head Dry Valley Deposits First Terrace Terrace, undifferentiated	Devensian (cold-glacial)	Aeolian, solifluction and fluvial depositon
Sand and Gravel, Silt and Clay – Vale of York Glacial Lake Deposits		Deposition in Lake Humber due to glacial blockage of Humber Gap
High-level Laminated Clay and Glacial Lake Deposits (in part) Fluvioglacial Sand and Gravel Glacial Sand and Gravel (in part) Till (in part)		Glacial deposition east of Wolds and in Humber Gap
Non-glacial deposits beneath till (not exposed)		? mainly solifluction deposition
		Prolonged denudation and fluvial incision due to sea level falling to below – 20 m OD
(Possibly some of unexposed deposits at base of Ipswichian cliff)	Ipswichian (temperate – interglacial)	Cutting of now-concealed cliff into Chalk with prolonged sea level at 1 to 3 m OD
(No known deposits)	Wolstonian (cold)	No evidence; possibly mainly denudation
Interglacial Gravel Beach Deposits Interglacial Silt and Clay	Hoxnian (temperate – interglacial) [correlations provisional]	Beach and estuarine deposition related to sea level at c.22 m OD
High-level Laminated Clay and Glacial Lake Deposits (in part) Glacial Sand and Gravel (in part) Till (in part) Immingham Channel Deposits (not exposed)	Anglian (cold-glacial) [correlations provisional]	Glacial deposition
(No known deposits)	Cromerian and older stages	? mainly denudation

Non-correlatable deposits (Clay-with-Flints and Sand and Gravel) of unknown age occur in the district

	Thickness m
grey and purple in middle and yellow near base, containing small stones (of white chalk, red chalk, flint, 'Spilsby Sandstone', 'ferruginous' rock, quartz, basalt and 'porphyrites' in part of the purple clay)	6.55 to 8.84
Gravel, largely or entirely of flint	1.37 to 1.70
Clay, locally sandy and loamy, yellow to reddish brown, with some small stones	1.52 to 3.18
Sand, yellow, 'full of well-rounded quartz grains and specks of chalk, and locally with up to 0.9 m of silt in middle'; contains chalk fragments at base where resting on Chalk, elsewhere with interlaminated clay near base	2.74 to 3.66
Clay, 'lead' coloured to locally red, with small stones (of basalt, 'porphyrite', sandstone, flint, 'grit' and quartz); becomes yellow, with chalk streaks and ? pebbles, near base	0 to 1.90

The Immingham Channel has no drainage hinterland so it cannot have been cut subaerially and, although Shillito (1937) refers to it as a fiord, neither its long profile nor its cross-section suggests ice-gouging. The only feasible explanation is that it was cut by a rapid flow of water, probably carrying abrasive sediment, that was concentrated at the base of an ice-sheet and driven by a substantial hydrostatic head. The infilling deposits, which are well-sorted, are also believed to have formed subglacially. The layers of sand or gravel in the Immingham Channel may result from more than one cut-and-fill cycle. However, the 'yellow sand' at the top of the sequence contains a few degraded dinoflagellate cysts and sparse pollen and spores in its topmost 0.1 m, suggesting some deposition or reworking in the subsequent interglacial. Catt and Penny (1966, p.398) reported foraminifera in the clay below the yellow sand but it is now believed likely that this foraminifera-rich clay overlies the sand and should be interpreted as part of the interglacial sequence (see p.115). The sea level implied by this sequence suggests that it is Hoxnian in age (see p.109); thus the underlying glacial deposits are referred to the Anglian Stage.

Ancholme Valley

Deposits of till, sand and gravel and laminated clay are present around and to the south of Brigg in the Ancholme Valley. They rest directly on Jurassic rocks, for the most part occupying low interfluves of Ancholme Clay to the east of the river.

Till

The till (Ussher, 1890, pp.130–132) is part of the 'Chalky Boulder Clay' of eastern England, and of Straw's (1969) Wragby Till. It is fairly uniform in composition, comprising bluish or brownish grey clay with scattered to abundant erratics, mainly of flint and chalk but also of sandstone (including Elsham Sandstone), quartzite, quartz, igneous and metamorphic rocks and, in the west and north, Lincolnshire Limestone and Lias rocks.

From South Kelsey southwards there appears to be only one till, which in a borehole [TF 0424 9819] at the Bull Inn, South Kelsey, is 5.5 m thick. The suggestion by Ussher et al. (1888, p.135) that two tills at North Owersby are separated by sand and gravel could not be confirmed. Around North Kelsey, however, a lower and an upper till are certainly present, separated by sand, gravel and laminated clay. The lower till is exposed only in degraded ditches, and appears to be not more than 3 m thick and rich in chalk and flint erratics. Slabs of Elsham Sandstone resting on Ancholme Clay at one locality [TA 0330 0082] may have been derived from it. The upper till is exposed in a pit [TA 0430 0133] on Sheepcote Hill, where it comprises 1.9 m of bluish grey clay containing erratics of flint, chalk, sandstone and rolled *Gryphaea*, and rests with a level contact on undisturbed laminated clay. From North Kelsey northwards only one till has been detected. It is locally sandy and even gravelly, with flint and chalk erratics which are especially abundant around and to the north-east of Brigg. Rounded sandstone erratics become more numerous to the west, for example near Cadney, and Ussher (1890, p.131) refers to boulders of 'siliceous grit' in this area. The till is exposed beneath sand and gravel in a pit [TA 043 104] at Melton Gallows Farm. The section is:

	Thickness m
Clay, greenish blue and brown, with erratics of chalk, lignite and several varieties of Lincolnshire Limestone; pockets of bedded sand	1.80
Fine gravel of quartzite pebbles with slabs of Lincolnshire Limestone	c.0.02
Clay, purple-grey, containing abundant, mainly small erratics and a basal pebble layer of flint, chalk, lignite and 'limestone'	1.00
(Ancholme Clay with irregular top surface)	

To the east of these occurrences the only other deposit of till west of the Lincolnshire Wolds forms a small patch of flint and chalk-rich grey clay capping a hill [TA 0953 0388] of Ancholme Clay near Grasby. To the west, small patches of till are present at low elevations near to, and west of, the River Ancholme. These are also flint and chalk-rich, although fragments of Lincolnshire Limestone are more common than farther east. The thickest proving of till in these western locations is 3.1 m in a borehole [SE 9969 0674] on Island Carr.

There is no evidence that these tills represent more than one glaciation, and they are likely to have resulted from the melt of a single ice-sheet, possibly a composite one. On the evidence of the erratic suite Straw (1958; 1963a; 1969; 1983) has advocated that the ice came from the north, either down the eastern side of the Vale of York or across Holderness. The till is taken to be pre-Devensian because of its dissected nature, its perched locations, and the absence of any evidence of Devensian ice in adjacent areas. It is here provisionally assigned, with the Immingham Channel deposits, to the Anglian.

Glacial Sand and Gravel and High-level Laminated Clay and Glacial Lake Deposits

Sand and gravel occur locally below, within, immediately above, and adjacent to but isolated from pre-Devensian till. Deposits of laminated clay intimately associated with the sand and gravel are also described here.

Evidence of sand and gravel beneath pre-Devensian till is confined to the area around and to the south of Howsham and North Kelsey. Small lenses of sand are recorded at or just above the base of the till at Gravel Hill Farm [TF 0598 9713] (Ussher, 1890, p.135). Up to 1.2 m of coarse-grained red sand occurs at the base of the till in the northern part of North Kelsey, and a borehole [TF 0424 9818] at the Bull Inn, South Kelsey, proved 1.83 m of unbottomed fine-grained yellow sand beneath till. No trace was found of the sand beneath the till at Kingerby that was reported by Ussher et al. (1888, p.135), nor of the 1.5 m of gravel beneath till 'between Gadney and Kelsey New Mill' also reported by Ussher (1890, p.131).

The presence of sand and gravel separating two tills at North Owersby (Ussher et al. (1888, p.135) has not been confirmed, and, because one of the pits in question is floored by Ancholme Clay, any sand and gravel present here may lie at the base of the till rather than within it. However, around Gravel Hill Farm [TF 0394 9963], North Kelsey, sand, gravel and laminated clay separate two tills and were formerly exposed in several pits. The 2.4 m of 'South Kelsey

Gravel' with shell fragments (Ussher, 1890, p.135) was, according to Fox-Strangways (BGS ms.), located in one of the pits north of Gravel Hill Farm. At present the only good exposure is in a pit [TA 0430 0133] on Sheepcote Hill, where the sequence below the upper till is:

	Thickness m
Clay, grey, with numerous pink laminae	0.40
Clay, grey, with abundant small chalk fragments	0.05
Clay, red, green and buff, laminated	0.80
Silt, laminated	0.40
Clay, pink, laminated	0.02
Silt, laminated, passing down into cross-bedded sand containing sand-grade chalk fragments	3.50 seen

The contacts between these layers are sharp and undisturbed, and there is no trace of weathering or biogenic activity. This sequence thins to the south-west and locally appears to cut down through the underlying till. It may have originated intraglacially, or as fluvioglacial and lacustrine deposits subsequently covered by flow till.

Sand and gravel deposits resting on till are restricted to the area north of North Kelsey. In some places they extend beyond the till to rest on solid rock. Old pits at Cadney formerly exposed up to 3 m of silt and sand on flint-rich gravels, which also contain pebbles of chalk and sandstone. Farther east, pebbles and sand-grade fragments of chalk are more common, and on Cadney Common [TA 010 044] large, well-rounded flint cobbles have been noted. The deposits compare well with those occurring between the two tills on Sheepcote Hill (p.112), but here there is no trace of an overlying till.

North-east of Brigg there are two substantial deposits of sand and gravel, which rest on a ridge of Ancholme Clay and consist mainly of poorly sorted and stratified coarse gravel with small patches of laminated clay. Ussher (1890, pp.135–136, fig. 3) records 4.6 m of such gravel in a pit south-west of Wrawby, and gravels with laminated clay were formerly worked near [TA 009 075] St Helens. The deposit at Melton Gallows Farm (p.112) consists of about 1 m of unbedded flint gravel with irregular pockets of red sand and laminated clay; it rests with a highly irregular contact on till. To the north-east, near Elsham Top, laminated silts are preserved. The coarseness of the gravel north-east of Brigg implies deposition virtually directly from ice, but the presence of ice-wedge casts and fault-like disruptions at Melton Gallows Farm suggests that some of the lack of sorting and stratification could be due to subsequent cryoturbation.

The isolated sand and gravel deposits occurring near to and west of the River Ancholme range from pebble-free sand to sand with abundant small pebbles, and are apparently no more than 2 m thick. Most of the pebbles are subangular flints, but rounded pebbles of reddened flint, Kellaways Rock and Lincolnshire Limestone are also present. The reddened jasperoid nature of the flint pebbles, as preserved at Redbourne Hayes [SE 999 001], suggests a previous phase of weathering or water-table fluctuations in their history, and in view of the degraded terrace-like form of these deposits it is possible that they had a fluvioglacial or even a fluvial origin.

Trent Valley

Deposits of till and sand and gravel occur south of Scotter, resting on Jurassic and Triassic rocks.

Till

The till, part of the 'Chalky Boulder Clay' (see p.112), is brown to reddish brown and silty, but the upper part in the east is greyer and more clayey. Of the erratics, flint and, below the weathering zone, chalk are ubiquitous, and there are small proportions of far-travelled rocks. Many of the erratics, however, are local, and their distribution reflects the west-to-east variation in the rockhead lithology. 'Skerry' sandstones from the Mercia Mudstone, and various Penarth Group rocks, occur in the west, and limestones, 'ironstones', rolled *Gryphaea* and belemnites, all from the Lias, are present in the east. Layers, lenses and pockets of sand, with and without gravel, are also present in the till.

The brownish clayey silt till around Aisby is locally more than 6 m thick, and at Blyton, where it is at least 3.7 m in places (Gozzard and Price, 1978), reddish brown sandy clay till has been seen beneath sand and gravel along the main street and in a nearby excavation [SK 8523 9446]. In places between Pilham and Laughton till overlies sand and gravel, but it has probably soliflucted into this position from farther east. Two small outcrops of till, apparently emerging from beneath sand and gravel, occur south-east of Laughton. Farther north, the till is largely concealed by blown sand, and is generally thinner and more impersistent, though up to 8 m of greyish brown clayey and silty till have been proved [SE 865 990] north of Park House Farm. A trench cutting one of three small patches of till [SE 8636 0030] near South Hills proved 2 m of brown-weathered bluish grey till containing abundant flints, a few chalk cobbles and some irregular sand masses. The same trench cut one of two small patches of till [SE 8583 0068] near Scotter Wood Farm, exposing 2.4 m of red and purple-green mottled clay till with abundant erratics of flint, chalk and Lias limestone, and intruded by ice-wedge casts filled with coarse-grained red sand. The most north-easterly outcrops, west of Kirton in Lindsey, are of mottled brownish grey clay with abundant flints.

There is no direct evidence of the age of the till which, like similar deposits in the Ancholme Valley (p.112), is provisionally assigned to the Anglian.

Glacial Sand and Gravel

Glacial sand and gravel, closely associated with the Trent Valley till, occur at four localities. At Aisby a lenticular mass of locally clayey sand, up to 5 m thick and containing thin brown clay lenses, occurs within the till (Gozzard and Price, 1978). Only flint pebbles were noted in the sand at the surface, but Ussher et al. (1888, pp.127–128) recorded a suite of both local and far-travelled erratics in pits to the south. The occurrence suggests a denuded kame.

The extensive sand and gravel deposit around Blyton rests partly on till and partly on solid rock. It comprises up to 5 m of sand which is clayey, especially near its base, and contains gravel layers and lenses consisting mainly of pebbles of flint, quartz, quartzite, sandstone and chert. A roadside cutting

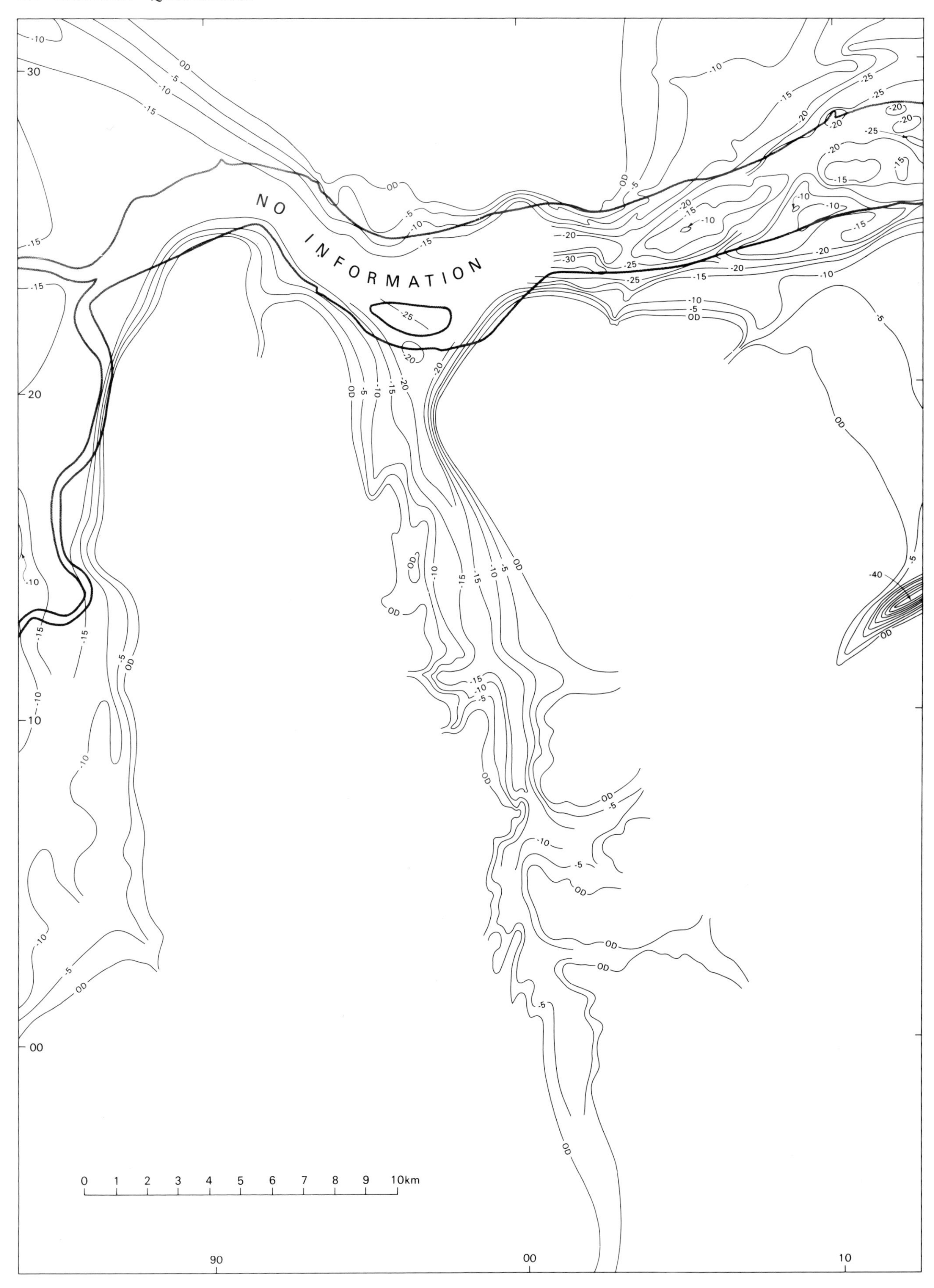

NO INFORMATION
30
20
10
00
90
00
10
0 1 2 3 4 5 6 7 8 9 10km

Figure 40 Map showing rockhead contours at and below OD at 5 m intervals

[SK 8518 9483] in Blyton proved 0.9 m of brownish yellow cross-bedded sand on till, and an old pit [SK 8558 9587] farther north-east exposes 2 m of poorly bedded, locally clayey sand with flint and quartzite gravel. Somewhere nearby, Ussher (1890, p.137) recorded a 2 m section comprising 'two false-bedded masses of gravel, composed of worn fragments of Lias and Rhaetic Beds and angular bits of flint and quartzite pebbles, separated by a 6-inch [0.15 m] seam of fine brown sand'. These deposits are similar in lithology, topographical form and altitude to others that extend discontinuously southwards along the eastern side of the Trent Valley for more than 25 km, and have the appearance of a degraded terrace which may be fluvioglacial or even fluvial in origin.

Locally clayey sand, with gravel in places, caps the ridge of solid rock and till that runs through Laughton to Hardwick Hill. Old pits near Laughton expose up to 3.7 m of sand with gravel consisting mainly of flint, quartzite, sandstone and 'limestone'. The ridge-top location of this deposit invites comparison with deposits in the Doncaster and Snaith areas to the west (Gaunt, 1981, pp.87–88), which are believed to be fluvioglacial in origin.

On Scotton Common a small patch [SE 852 002] of brown clayey sand and sandy clay containing densely packed pebbles of flint, Lias limestone, quartzite, sandstone and coarsely oolitic Lincolnshire Limestone rests on solid rock.

Ussher (1890, pp.136–137) recorded bedded sand and gravel at least 3 m thick in an old pit, not positively identified, about 2 km east of Northorpe railway station.

Other areas

The 'grey boulder clay with flint fragments, and whitish marl overlain by gravel', recorded by Ussher (1890, p.134) from a locality [SE 889 146] south-east of Flixborough has since been removed in quarrying for ironstone. Till recorded north of South Cave by Bisat (1932a) was not seen, but till-like head deposits are present in this area. In Eastfield Quarry [SE 915 324] a displaced mass of Cave Oolite at least 50 m long, 1.2 m thick and horizontally bedded rests, with an intervening 0.8 m-thick greyish brown chalk and flint-bearing clay, on Thorncroft Sands. Stather (1922) recorded a similar section about 150 m to the north, so it appears that the entire intervening hilltop is a giant rafted erratic.

Rockhead contours (Figure 40) show that a narrow channel, incised into Chalk, descends eastwards under part [TA 120 305] of Kingston upon Hull. The clay fill is locally separated from the overlying Devensian till (itself concealed beneath alluvial deposits) by sand and gravel. For example, a borehole [TA 1046 3022] into the channel proved alluvial deposits on 'marl clay' (Devensian till) to 16.15 m, on sand and chalk gravel to 20.42 m, on 'marl clay' to 26. 82 m, on gravel to 27.43 m, on Chalk. The channel, which continues eastwards, appears analogous to the subglacial Immingham Channel (p.109), though a Devensian age cannot be ruled out.

Near Flixborough, two small concentrations [SE 867 153; SE 870 157] of pebbles of sandstone, quartz, flint, Carboniferous limestone, Permian limestone, Lincolnshire Limestone and igneous rocks, with little or no matrix, may be of pre-Devensian glacial origin. An area where Ussher (1890, p.139) recorded 'Old Gravels' at Conesby Bottom [SE 895 144] has since been quarried for ironstone, but the BGS archives contain records dated 1936, from an unknown source, of up to 3.7 m of sand with 'undulating' gravel lenses and also some white silt and tufaceous clay layers. The gravel, 'generally less than pea-size' but including some boulders, contained pebbles of 'ironstone' and flint, some cemented into a tufaceous conglomerate. Sand containing numerous white angular flint fragments is also present at a locality [SE 891 200] north-east of Burton upon Stather. The sand and gravel patches on the Lincolnshire Wolds east of Caistor, north-east of Searby and east of Elsham contain erratic pebbles in addition to flint and chalk; all are sufficiently distant from Devensian glacial deposits to suggest a pre-Devensian glacial origin.

INTERGLACIAL DEPOSITS

Undoubted interglacial deposits are known only at Kirmington and are provisionally referred to the Hoxnian. No Ipswichian deposits have been proved, but a sea cliff, now concealed, was formed at this time and some deposits adjacent to its foot may be Ipswichian in age. Deposits of unknown age, namely Clay-with-flints and Sand and Gravel, which are thought to be of interglacial origin, are described in this section for convenience.

Hoxnian deposits at Kirmington

The interglacial deposits at Kirmington have long been worked in a brick pit [TA 103 115] and a nearby gravel pit [TA 105 117]. The sequence proved by Reid (1885, pp.58–59), the British Association boreholes and excavations (Lamplugh et al., 1905), the excavations by Watts (1959) and by Harland and Downie (1969), and the BGS Borehole [TA 1031 1163] (Institute of Geological Sciences, 1973) is summarised below:

	Thickness m
Till (Devensian)	—
Interglacial Gravel Beach Deposits	
Gravel, mainly of subrounded to rounded flint (hence old name 'Cannonshot gravel') with some quartzite pebbles and a few supposed flint artefacts; upper part cryoturbated	1.37 to 3.35
Interglacial Silt and Clay	
Silt and clay, bluish grey (yellow-brown near top), laminated and calcareous, with thin sand layers and lenses; dinoflagellate cysts, scarce spores and pollen, foraminifera and estuarine molluscs	2.85 to c.5.0

	Thickness m
Peat, mainly comprising *Phragmites*; pollen and spores	0.08 to 0.15
Silt and clay, bluish grey, calcareous, locally sandy; dinoflagellate cysts, scarce spores and pollen, a few wood fragments, foraminifera and molluscs; *Phragmites* rhyzomes extending down from peat	0 to 0.61

Beneath this sequence lies the yellow sand at the top of the Immingham Channel deposits (pp.109–112). No trace has been found of a 'gravel, the lower part occasionally containing shells' which, according to some of the old brick makers, supposedly underlies the sand (Reid, 1885, p.59).

The silt and clay below and above the peat are similar to intertidal deposits in the Humber at present, and most of the contained molluscan species, listed by Reid (1885, p.58), live in the estuary today. The foraminiferal species he noted are described as 'a starved shallow-water fauna', and most would be expected in British estuaries at present. In contrast to this list, the abundant foraminifera found in the silt and clay from the BGS borehole consist almost entirely of *Aubignina perlucida*, identified by Dr K L Knudsen of Aarhus University, who reports (*in litt.*) that this dominance by one species indicates restricted ecological conditions. This species occurs at present off western Britain and Brittany, and its presence at Kirmington suggests slightly warmer water than that of the North Sea today. The dinoflagellate cysts also suggest slightly warmer water and imply estuarine or lagoonal conditions (Harland and Downie, 1969). The pollen and spores are mainly of conifers and ferns respectively which, being particularly robust, may not be fully representative of the contemporary vegetation. No trace has subsequently been found of *Cervus* antlers which, according to Wood and Rome (1868, p.156), were discovered in the 'brick-clay'.

The peat contains, besides *Phragmites*, remains of several other plants, most of them with maritime or brackish water associations; all but one grow around the Humber today (Lamplugh et al., 1905, p.273). The contained tree pollen (Watts, 1959) is dominated by alder, oak and pine, and the non-tree pollen by grasses; the full assemblage suggests an estuarine reed-swamp with adjacent woodland, growing in a climate similar to that prevailing today. The presence of the peat within the silt and clay may be due to a temporary fall of sea level, to the lateral migration of estuarine channels placing the area temporarily beyond tidal reach, or to its growth in a lagoon behind a shingle bar.

The Interglacial Gravel Beach Deposits have been subjected to prolonged energetic abrasive conditions, for most of the flint pebbles are well rounded. They are, therefore, considered to have originated as a marine beach which migrated, probably southwards, over estuarine silt and clay within a tidal sound or creek. They are assumed to be contemporaneous with, or only slightly younger than, the underlying silt and clay, and may initially have formed a bar, as suggested by Straw (*in* Penny et al., 1972, p.31), in the shelter of which flourished the almost monospecific foraminiferal assemblage. The supposed flint implements within them are variably rolled; they include patinated flakes, some simple and others apparently with striking platforms and bulbs of percussion. According to Boylan (1966) 'one may reasonably postulate that a Clactonian and/or Acheulian Industry is represented'. A dozen or so patches of beach deposits, resting on Chalk or protruding though Devensian till, have been mapped within 2 km of Kirmington to the north and east. They occur at elevations similar to that of the gravel pit at Kirmington. Several consist of cannonshot flint gravel, which was also noted by Boylan (1966, p.348) in a well at Ulceby.

That the sequence is interglacial seems obvious from its stratigraphical position and fossil content. Most views on its age are based on its pollen, the supposed artefacts, or its altitude (c.19 to 28 m above OD). According to Watts (1959), the pollen implies a Hoxnian age, but some aspects, such as the high alder content, would not now be taken to preclude an Ipswichian age. The supposed artefacts were referred by Boylan (1966) to the Hoxnian, mainly on their Clactonian affinity, but they may have been reworked into the Interglacial Gravel Beach Deposits, for otherwise their makers were confined by the prevailing high sea level to one or more small islands. A mean sea level of 22 ± 1 m above OD during deposition of the interglacial deposits seems probable, and this agrees with that previously calculated by Straw (*in* Penny et al., 1972). Such a high sea level seems incompatible with an Ipswichian age, when the sea does not appear to have risen above 1 to 3 m above OD in adjacent regions (Gaunt et al., 1974). Moreover, *A. perlucida* was apparently much more common in the North Sea during Holsteinian times (the probable continental equivalent of the Hoxnian) than it has been subsequently (Knudsen, 1986). The interglacial deposits at Kirmington are, therefore, provisionally referred to the Hoxnian, largely on the evidence of the high sea level.

Ipswichian cliff and related deposits

An old sea cliff, cut into Chalk and buried beneath Devensian till, is transected by the present-day cliff at Sewerby, north of Bridlington (Catt and Penny, 1966). It has a beach gravel at its base containing mammalian remains including *Hippopotamus*, which indicate an Ipswichian age (Boylan, 1967). Using borehole evidence, the cliff can be traced southwards (Fletcher, 1985), concealed beneath Devensian till, under Beverley and Cottingham (Newton, 1925), reaching the Humber at Hessle, where it turns west into the estuary (Figure 39). It was formerly exposed at Hessle (Phillips, 1868; Reid, 1885, pp.49–51; Walton, 1895; Sheppard, 1899; 1903; 1908; Crofts, 1906). To the south of the Humber the cliff, again concealed beneath Devensian till, lies a few metres behind the present cliff at South Ferriby, where its basal wave-cut platform is exposed. It is traceable eastwards under Barton-upon-Humber and Barrow upon Humber, where it turns south-eastwards approximately parallel to, and 1 to 2 km west of East Halton-Skitter Beck. The general height of its wave-cut platform at Sewerby and South Ferriby implies a prolonged sea level at 1 to 3 m above OD. The platform at South Ferriby cuts across severely fractured, cambered Chalk (p.107), implying a preceding periglacial episode, probably in the Wolstonian.

Some chalk-rich gravel occurs beneath Devensian till at the foot of the cliff within the district, and may, as at Sewerby, represent contemporaneous beach deposits. The limited borehole evidence available does not, however, allow such gravel to be distinguished from other, more widespread deposits under the till that are thought to be Devensian Head (p.124). On South Ferriby cliff [SE 9952 2212] a mass of calcreted chalk and flint gravel up to 0.6 m thick rests on the wave-cut platform and is overlain by Devensian till. Reid (1885, pp.39, 51) recorded ripple-marked calcreted sand from this locality and also described (1885, pp.56–59) shell-bearing 'Marine Gravels' at various places from Goxhill southwards to beyond Brocklesby. Most of the latter deposits are now identified as Devensian Fluvioglacial Sand and Gravel (p.119). The shells are species of estuarine and marine gastropods and bivalves of temperate to cool aspect, many of which live in the Humber at present, and are presumably reworked. At an unlocated site 'not far from Croxton' they include *Corbicula fluminalis*, a temperate freshwater bivalve now restricted to the North Africa–Kashmir region, but not uncommon in British Ipswichian deposits. This species also occurs in undoubted Devensian fluvioglacial deposits at Kelsey Hill (Catt and Penny, 1966, p.395) and near Paull (Ennis, 1932) in southern Holderness, into which it has clearly been reworked.

In the north-west of the district a borehole [SE 8488 3086] near Newport proved dark brown to black organic silty clay with fibrous plant remains at about 10 m below OD. This depth is comparable to that of Ipswichian deposits at Langham to the west of the sheet area (Gaunt et al., 1974).

Deposits of unknown age

Clay-with-flints

Red to reddish brown chalk-free clay containing numerous flints and a few erratics has been mapped capping a narrow Chalk interfluve at about 43 to 56 m above OD at a locality [TA 035 191] on the northern side of Deepdale, just beyond the Devensian till limit on the Lincolnshire Wolds south of Barton-upon-Humber. Augering suggests that its thickness ranges up to 0.9 m. The deposit has been described by E G Smith and J G O Smart (personal communication in Mathews, 1977, p.236), as were comparable deposits too small to show on the geological map: 'Similar red clay with numerous flints was seen 1.0 km to the north-east in a pipe-trench [TA 045 200] near Field House, just below 30 m above OD, and it also occurs locally in surrounding fields. Some of the exposures show that Clay-with-flints underlies glacial drift. Patches of Clay-with-flints also occur in fields near the cross-roads at Wootton Wold; a pipe-trench [TA 062 150] at 56 m above OD showed up to 1.2 m of red flinty clay containing irregular pockets of broken chalk. This is interpreted as Clay-with-flints frost-heaved into the Chalk rockhead'. These deposits are interpreted as insoluble residues of the Chalk remaining after prolonged chemical weathering, probably in an interglacial episode.

Sand and Gravel of unknown age

Mill Hill [SE 942 278], near Elloughton, which rises to just over 33 m above OD and owes its prominence to the resistant Kellaways Rock near its summit, is capped by sand and gravel, the latter comprising mainly chalk and flint pebbles but with a few examples of 'Red Chalk', Cave Oolite, Carboniferous limestone, and indeterminate sandstone and quartz. The section given by Lamplugh (1887) can be summarised as: soil 0.8 m, on cross-bedded sand and gravel 2.7 m, on yellow sand with pebble layers and mammalian remains 1.5 m, on hard grey clay (probably Ancholme Clay). According to Sheppard (1897) the mammalian remains were found at or near the base of the yellow sand, which was confined to a depression in the underlying clay in the north-eastern part of the hill top. The remains, comprising bones, teeth and tusks, were listed by de Boer et al. (1958, p.198) as *Elephas antiquus*, *E. primigenius*, *Bison priscus*, *Bos primigenius*, *Cervus elephus*, *Cervus* sp. and *Equus caballus*. The list includes both temperate and cold taxa, as typified by the two elephant species, and it is similar to the fauna, also listed by de Boer et al. (1958, pp.197–198), from a much lower-lying site at Bielsbeck, near Market Weighton. The ecologically mixed fauna may imply secondary derivation, although its re-examination might prove enlightening, for Boylan (1981) has shown by re-identification that a fauna from Kirkdale Cave, near Pickering, also formerly thought to be a mixed assemblage, is wholly referable to the Ipswichian.

DEVENSIAN DEPOSITS

Both glacial and nonglacial Devensian deposits occur in the district. The oldest are nonglacial and concealed beneath till. The glacial deposits mainly comprise till and glacial and fluvioglacial sand and gravel. Other nonglacial sediments include various lacustrine, terrace, dry valley, head and blown sand deposits (Table 3), many of which occur widely beyond the limits of the till, and in a few places rest on it. Some of these deposits, notably Head and Blown Sand, range in age into the Flandrian but, for convenience, are described here. Conversely, a small amount of Shell Marl was deposited in the Devensian, but is described together with similar deposits of Flandrian age (p.124).

Deposits beneath till

Boreholes and excavations show that deposits variously described as 'chalk gravel', 'chalk bearings' and 'chalk marl' locally underlie Devensian till east of the Wolds and in the Humber Gap. To the north of the Humber, pale grey silt up to 0.6 m thick and containing angular chalk and flint fragments underlies till in the Kirk Ella railway cutting [TA 016 309 to TA 011 314]. Just beyond the northern boundary of the district in Eppleworth Chalk Pit [TA 0214 3239] a similar deposit, up to 1 m thick, is associated with ice-wedge casts intruded into the Chalk, and is possibly partly loessic in character (Catt et al., 1974).

Banked up against the Ipswichian cliff formerly exposed at Hessle (p.116) is chalk-rich rubble containing mammalian remains which Boylan (1967) regards as Devensian. In the chalk pits west of Hessle, chalk-rich deposits beneath till fill small valleys and gulleys cut into the Chalk, and some [e.g. TA 0220 2569] contain layers and lenses of sand. In another pit [TA 0181 2611] up to 3.6 m of banded reddish brown to yellow, variably sandy 'loam', rich in chalk and flint fragments, rest on Chalk and are overlain by till with an impersistent basal erratic-rich gravel.

South of the Humber, chalk and flint-rich deposits underlie the till east of the Lincolnshire Wolds and thicken eastwards up to 3.6 m near Goxhill. In an old chalk pit [TA 0384 2165] east of Barton-upon-Humber they comprise up to 1 m of pale brownish to pale pink 'loam' containing angular chalk and flint fragments, and rest on an irregular base of fractured Chalk.

These chalk and flint-rich deposits are clearly locally derived, and are thought to be mainly Head formed by solifluction during prolonged Devensian periglacial conditions before the onset of glaciation. It is possible, however, that some of the deposits that lie adjacent to the foot of the concealed Ipswichian cliff may be Ipswichian beach gravel (p.116).

Till

Devensian till is confined to the eastern side of the Yorkshire and Lincolnshire Wolds and to the Humber Gap. The till east of the Yorkshire Wolds—the 'Hessle Clay' of Hall (1867) and Wood and Rome (1868)—has been shown by Madgett and Catt (1978) to be the weathered top of the Skipsea (formerly Drab) Till of Holderness, and its late Devensian age is well established (Penny et al., 1969). The till east of the Lincolnshire Wolds is part of the 'type deposit' of the 'Newer Drift' of Jukes-Browne (1885b) and is also part of the Skipsea Till; according to Straw (1969) it can be subdivided into the Lower and Upper Marsh tills. The matrix of the till is everywhere a grey to purplish brown stiff clay, locally silty and less commonly sandy, which weathers to reddish or yellow-brown shades. The erratics include flint, chalk (except where weathered), Jurassic rocks from north-eastern Yorkshire, Carboniferous (and a few Permian) rocks, quartzite, quartz, and a variety of igneous and metamorphic rocks, including a few from Scandinavia.

To the east of the Yorkshire Wolds the till is up to 9.6 m thick. It descends eastwards beneath alluvial deposits to 18 m below OD, and was exposed in excavations for the Albert and Alexandra docks (Ussher, 1890, p.167) and at Stoneferry (Gaunt and Tooley, 1974). To the west it rises to 100 m above OD within the district [SE 977 313].

North of the Humber up to 2.5 m of silty clay till are exposed in the chalk pits west of Hessle, resting partly on Chalk and partly on chalk-rich deposits (Carruthers, 1948, p.151, fig. 4); erratics from this area are listed by Reid (1885, p.38). Along Red Cliff (Plate 5), south-west of North Ferriby, Stather (1897) and Bisat (1932b) distinguished a lower greyish clay till from an upper reddish brown clay till, separated by a fairly sharp contact and an impersistent thin laminated clay. This contact is now thought to mark the base of the weathering zone, coinciding locally with an impersistent laminated clay within the till similar to one known on South Ferriby Cliff across the Humber. Consequently only a single till is now recognised at Red Cliff (Figure 41). Small lenses of till occur in the overlying lacustrine deposits (p.122) as they do also in East Clough Borehole (Figure 41). An excavation [SE 9738 2526] at the Capper Pass Works exposed 10.5 m of blue-mottled and streaked grey clay till with a weathered zone 2 to 4 m thick, resting on chalk gravel (which may be of lacustrine or solifluction origin). The most westerly exposure is in Elloughton Beck [SE 9476 2684], but Brough No. 1 Borehole farther west proved 5.36 m of grey till resting on rockhead.

South of the Humber up to 2 m of silty and locally sandy clay with a partial cover of sand and gravel occupies gently undulating ground between Winteringham and Winterton to the west of the River Ancholme, and east of Winterton an east-west ridge of clay till, up to 12.5 m thick, protrudes through estuarine alluvium. Straw (1961) refers to these deposits as part of the 'Winteringham – Horkstow Moraine'. Even though they rest on Jurassic rocks, they contain virtually no Jurassic erratics except along their extreme western edge.

Thin layers of till occur within lacustrine deposits that underlie the alluvium at the mouth of the Ancholme, and have been proved, for example, in Low Farm (South Ferriby) Borehole (Figure 41). There are 7 m of sandy clay till exposed on South Ferriby Cliff, much of it weathered reddish brown. It contains a 0.15 m bed of laminated clay (with some silt and sand) about 1 m above its base, and thin sand and gravel layers at higher levels (Figure 41).

The till east of the Lincolnshire Wolds thins westwards from near Goxhill, where it is almost 15 m thick, and rises to a maximum height of 55 m above OD. A small patch [TA 088 090] near Barnetby Wold Farm may, in view of its height well above that of the Barnetby Gap, be pre-Devensian. Among erratics noted near the Humber are 'ironstone' and coal. At Kirmington brownish variably silty and sandy clay till up to 2 m thick contains erratics which include 'Basalt', porphyrites, rhomb-porphyry, and grits (Lamplugh et al., 1905, p.273), and rests locally on the interglacial deposits. A few closed hollows, notably north-east of Thornton Curtis, may be kettle-holes, and Straw (1961) has identified meltwater channels in and south of this area. The till east of the Wolds, both north and south of the Humber, extends farther west along the interfluves than it does in the minor valleys, suggesting that there has been appreciable postdepositional erosion.

The distribution of the till and its erratic content show that the ice entered the district from the north-east. Except possibly between Swanland and Hessle it failed to surmount the Wolds, but intruded westwards into the Humber Gap as far as the Brough and Winterton areas, a conclusion supported by stone-orientation measurements from the till (Figure 42).

Glacial Sand and Gravel

Several deposits of sand and gravel are believed to have formed directly from the Devensian ice because they are closely related to the till, and have a very similar stone content. North of the Humber, clayey sand and gravel capping a low mound of till [TA 0113 3070] north-west of Willerby yielded the following pebble count: flint 17, chalk 11, indeterminate sandstone 22, quartzite 19, quartz 6, igneous rocks 14, metamorphic rocks 2, ?Carboniferous limestone 2, indeterminate rocks 7. Sand and gravel were formerly extracted from a pit [TA 0295 2640] that is now built over.

Plate 5 Quaternary deposits at Red Cliff, North Ferriby. Flandrian peat ('submerged forest') can be seen protruding through beach deposits. The cliff is composed of Devensian deposits, consisting of glaciolacustrine sand and gravel, decalcified at the top, resting on grey till. L 2953

Several mounds of sand that protrude through the lacustrine and alluvial deposits south-east of Brough contain pebbles that are mainly of flint and chalk, but with a few of sandstone, quartz and Lincolnshire Limestone.

South of the Humber, between Winteringham and Winterton, sand and gravel, locally clayey, form extensive ridges and mounds and rest mainly on till, but in places extend on to the Lincolnshire Limestone outcrop. Most of the 7 m exposed in an old pit [SE 949 213] comprise tightly packed, well-sorted, level-bedded gravel consisting mainly of chalk, but with a few pebbles of flint, sandstone and other rocks, in a sand matrix. The number of chalk pebbles decreases markedly upwards in the top 1 m, and they are virtually absent at the surface where a few sandstone ventifacts were found.

To the east of the Lincolnshire Wolds, boreholes show that up to 4 m of sand, locally with some silt and/or gravel, are present locally at the base of the till. Sand and gravel, up to 3 m thick, also occur impersistently within the till and in places protrude through it to the surface as low mounds. At one locality [TA 093 200], sand and gravel have been worked to a depth of 1.2 m below thin till. Patches of sand and gravel in the Burnham area apparently rest on till. Sand resting on Chalk at Yarborough Camp [TA 081 120] may, because of its deep reddish brown colour, be of pre-Devensian age.

Fluvioglacial Sand and Gravel

Deposits of pale brown sand, containing numerous pebbles mainly of chalk and flint but with a few far-travelled erratics, form a terrace-like spread along East Halton-Skitter Beck. At the southern end of their outcrop they occur up to about 18 m above OD, almost reaching the Barnetby Gap. They descend as low as 5 m above OD to the north, and at their northern end, both along East Halton-Skitter Beck and near South End, they appear to pass within till. Trenches and ditches in the north, notably near Thornton Abbey, show that the deposits contain locally numerous pebbles and particles of Westphalian coal. Most of the exposures of shell-bearing 'Marine Gravels' described from south of the Humber by Reid (1885, pp.56–63) are located in or near

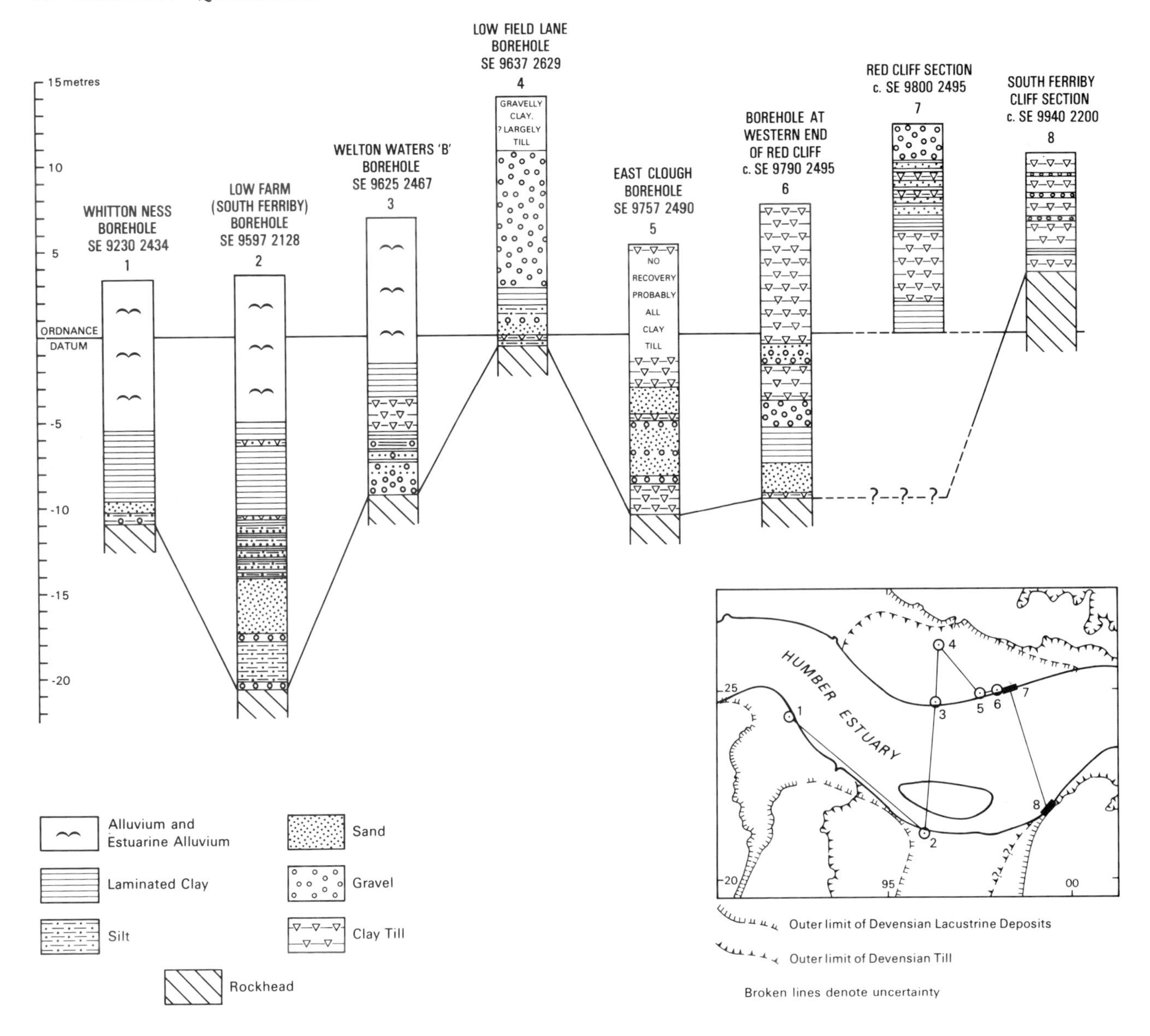

Figure 41 Sections showing Devensian interbedded glacial and lacustrine deposits in the Humber Gap

these deposits, but the shells have probably been reworked from Ipswichian sources (p.117). Ditches and trenches show that the pebbles occur in layers, cross-bedded lenses or irregular pockets, and Reid (1885, p.57) noted in pits near Thornton Station that 'All the Gravels in this neighbourhood show the influence of strong currents, and appear to have been constantly shifting banks'. Up to 3.7 m of sand and gravel were worked in old pits, and up to 4 m have been proved in boreholes.

The terrace-like topography of the sand and gravel, with kettle holes in places, implies subaerial deposition, although this could have taken place along temporary ice-margins or within ice-walled channels. It appears likely that the most northerly deposits were partly overrun by ice or by flow-till from adjacent ice.

High-level Laminated Clay and Glacial Lake Deposits

Kirmington Airport [TA 095 105] is situated on flattish ground consisting largely of brown sandy clay with scattered flint and chalk pebbles, which rests on stoneless clay with gravel layers. An excavation [TA 0925 1082] exposed 0.6 to 1.5 m of brown sandy clay with some pebbles, resting on 0 to 0.15 m of contorted chalk gravel, which overlies at least 1.2 m of reddish brown stoneless laminated clay. These deposits probably formed in a small ice- or moraine-impounded lake.

Vale of York Glacial Lake Deposits

On the geological maps lacustrine Silt and Clay deposits are distinguished from associated Sand and Gravel. For descrip-

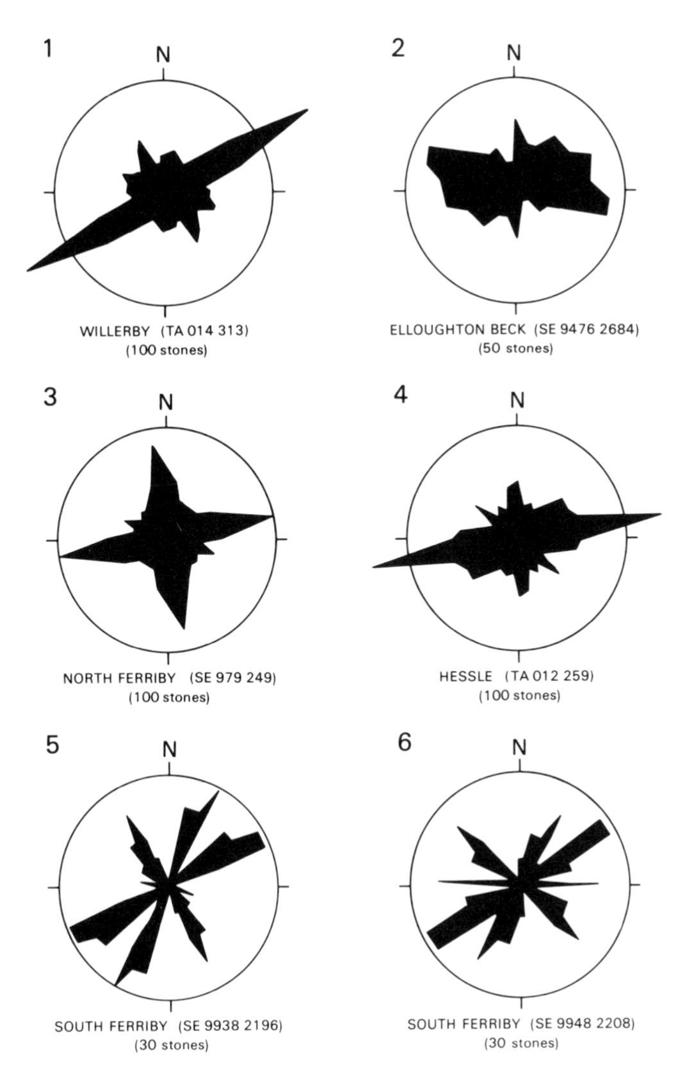

Figure 42 Stone orientations in Devensian till. Sites are shown on Figure 39. Diagrams 1, 3 and 4 are after Penny and Catt, 1967; 2 is after Gaunt, 1976; 5 and 6 are unpublished studies by Mr J G O Smart

tive purposes the latter can be further subdivided into Lower Sand and Gravel, Marginal Sand and Gravel, and Upper Sand, the last including some postlacustrine levée sand. In the Humber area the lacustrine deposits interdigitate with late Devensian till and near Brantingham have yielded a late Devensian radiocarbon date. They are therefore considered to be largely, if not entirely, of late Devensian age. They rest on a buried landscape which contains wide mature valleys incised down to about 20 m below OD and directed into the Humber Gap from the west and south. This landscape is clearly the product of prolonged denudation at a time of low sea level, believed to have been Devensian, before the onset of glaciation in the region. When, in the late Devensian, drainage from this area was blocked by ice in the Humber Gap, an extensive lake—Lake Humber—was formed in the Vale of York and the Ancholme Valley, within which the lacustrine deposits were laid down. After the lake silted up, and following a short interlude when rivers deposited sandy levées on the emergent lacustrine plain, there was deep incision by the main rivers, again down to about 20 m below OD. In many places the lacustrine deposits are now concealed by alluvium, as proved extensively by boreholes in the Trent Valley (Gozzard and Price, 1978; James, 1976; Lovell, 1977).

Lower Sand and Gravel

In the north-west up to 3.3 m of sand and mainly flint and chalk gravel rest on rockhead at depths descending south-eastwards to below 15 m below OD. They are succeeded by up to 5 m of fine- to medium-grained sand which in places passes by interbedding into the overlying silt and clay. In the Trent Valley up to 6.6 m of pebbly and gravelly sand, with pebbles mainly of quartzite, rest on rockhead at depths descending northwards to more than 15 m below OD. They are succeeded by up to 10 m of silty sand which is clayey locally near its top and which, in the more northerly locations, contains abundant coal debris.

Eastwards from Whitton in the Humber Gap, sand, commonly rich in coal debris and locally with gravel, rests close above or directly on rockhead (Figure 41). The gravel consists mainly of flint and chalk, and shelly sandstone probably derived from the Kellaways Rock. In the Ancholme Valley to the south, sand with flint and chalk gravel, apparently not generally more than 2 m thick, rests extensively on rockhead at depths descending northwards to more than 15 m below OD.

The gravel and pebbly sand were probably deposited by the rivers that had earlier incised the buried landscape; the same may also be true for some of the silt in the Humber area. Some of the pebble-free sand may also be fluvial, but its marked variation in thickness over quite short distances suggests that some of it is aeolian, a possibility supported by the presence of ventifacts at this stratigraphical level farther west (Gaunt, 1981, p.90). However, the interbedded and gradational passage into the overlying silt and clay implies some lacustrine deposition or reworking.

Silt and Clay

Most of this deposit comprises grey clays with numerous closely spaced silt and, less commonly, fine-grained sand laminae up to 2 mm thick. In the north-west the deposit is up to 16.5 m thick, and in pits [SE 852 306; SE 861 272] near Newport and North Hall the highest 5 to 6 m are colour-banded in shades of brown, red, purple, blue and grey, and flattened contortions have been noted. A few layers of fine-grained sand up to 0.4 m thick are present near the top and, although the deposit generally is almost stoneless, a few pebbles, cobbles and even small boulders of both local and far-travelled rocks have been found in it. Around Newport the deposit has an almost level surface at about 4 m above OD. Farther south, in the Trent Valley, the deposit is concealed except for a small inlier [SE 874 015] south of Scotterthorpe, where it is up to 4.4 m thick and contains a layer of clayey and silty sand up to 2 m thick. It was formerly worked in old pits south-west of Yaddlethorpe, where there was an overburden of blown sand.

In the Humber area the laminated clay is increasingly interbedded eastwards with sand and gravel and clay till

(Figure 41). In the Winterton Beck valley laminated clay emerges from beneath alluvial deposits to mantle the slope between Whitton and West Halton at up to 8 m above OD. At Red Cliff 1.6 m of laminated clay, silt and sand are exposed beneath till and are probably the laminated clay recorded on the foreshore by Bisat (1932b, p.86). A further 1 m of laminated clay and silt containing thin cross-bedded fine-grained sand lenses with much coal debris overlies the till. It is succeeded locally by over 3 m of interbedded fine-grained sand and laminated clay and silt, which contain lenses of clay till, irregular masses of stoneless clay, scattered erratic pebbles and cobbles, and cross-bedded lenses of coarse-grained sand rich in coal debris. Some of these deposits are severely contorted, and some of the till lenses have been rolled into cylindrical masses. The sequence is capped in places by up to 2 m of unbedded chalk and flint gravel.

In the Ancholme Valley laminated and colour-banded clay emerges locally from beneath alluvial deposits to mantle the lower slopes to east and west. It rises up to 7 m above OD north of Appleby and to 8 m above OD between Bonby and Brigg, where up to 2.8 m of laminated clay are visible in ditches. A trench [TA 0142 0605] across Candley Beck proved 6 m of laminated clay. Between Springfield and Wrawby several thin gravel lenses occur within the clay, and 'cannonshot' flints, presumably derived, are common at the base where seen in a trench [TA 012 085]. The clay around Kettleby Thorpe and Bridge Farm [TA 045 070] is very silty. It thins farther to the east, and near Somerby rises to 11 m above OD, where it passes into micaceous silt and is cryoturbated. It also thins to the south, and rises to 7.5 m above OD.

The regular, uniform laminae and, except in the Humber area, the lack of flowing-water structures indicate the lacustrine origin of the silt and clay.

Marginal sand and gravel

Marginal lacustrine sand and gravel deposits are differentiated into high-level gravelly sands which rise locally up to about 30 m above OD, and low-level virtually pebble-free sands which generally do not rise above about 9 m above OD and which, away from containing slopes, pass laterally into the lacustrine silt and clay described above.

In the Trent Valley the lacustrine silt and clay pass eastwards into fine-grained silty and locally clayey sand that rises eastwards to about 8 m above OD in places. The upper part of this sand has been redistributed by wind and is shown as Blown Sand on the geological maps. Perched higher on the Lias slopes at up to 16 m above OD are small patches of sand and gravel in which, except for some flint, all the pebbles are of locally derived limestones, 'ironstones' and rolled *Gryphaea*. Most of these gravels are the 'Old Gravels' of Ussher (1890, pp.137–141). Two patches [SE 868 144; SE 868 147] south-west of Flixborough form degraded terrace features, and others between Yaddlethorpe and Scotter form small denuded low-angle fans at the mouths of minor valleys such as those of Bottesford Beck [SE 888 064] and the River Eau near Scotter.

To the north of the Humber the silt and clay beneath alluvial deposits on Walling Fen pass eastwards into fine-grained silty and clayey sand up to 3 m thick which, except where redistributed by wind, forms flattish ground up to about 9 m above OD west and south of South Cave. Towards the east this deposit rests on sand and gravel in which the pebbles are almost entirely of Lias rocks, flint and chalk, and which, west of South Cave, locally emerge from below the pebble-free sand to mantle the Lias slopes at up to 23 m above OD. In pits [SE 904 311] west of South Cave a cryoturbated layer topped by a thin pinkish brown silty clay occurs within gently cross-bedded sand and gravel. These deposits were called the 'Gryphite gravel' by Tate and Blake (1876, p.69) because of the abundance in them of rolled *Gryphaea*. To the south-east of South Cave, sand and gravel, in which the pebbles are mainly of Cave Oolite, flint and chalk, cover much of the topographical slack between the Cave Oolite and Chalk scarps and rise locally to about 30 m above OD. Several old pits between Ellerker and Brough, and a number of excavations, showed that the deposits vary from well bedded and well sorted, with fining-upward layers and locally with gentle cross-bedding, to virtually unbedded and unsorted. In a road cutting [SE 9385 2918] near Brantingham a bone fragment found 3.05 m below ground, either at the base or within the sand and gravel, yielded a date of 21 835 ± 1660 radiocarbon years (Gaunt, 1974). To the east of Brough both the pebble-free sand and the sand and gravel maintain the altimetric distinction noted farther north-west. The gravel consists mainly of flint and chalk pebbles. These deposits are 12.50 and 12.35 m thick respectively in Brough No. 2 and Welton Crossing boreholes. In Low Field Lane Borehole (Figure 41) they rest on thin till, are overlain by gravelly clay which is also probably till, and include a thin silt and clay sequence in their lower part. In East Clough Borehole (Figure 41) they are interbedded within three layers of till.

South of the Humber a patch of sand and gravel, in which the pebbles are mainly of Lias rocks, extends up to 21 m above OD west of Whitton. A pit [SE 901 245] exposes up to 2.3 m of interbedded sand and closely packed gravel, the latter varying from well sorted, with fining-upward layers, to virtually unsorted. The top of the deposit is cryoturbated. To the south-east, between Whitton and West Halton, sand and gravel consisting mainly of pebbles of Lias rocks mantle the Lias slope up to 21 m above OD. On the eastern side of the Winterton Beck valley [SE 914 203 to SE 911 193] several small patches of sand and gravel up to 1.3 m thick occur at 12 to 16 m above OD. The contained pebbles are mainly of Lias and Lincolnshire Limestone, but also include flint. In two of these patches laminated silts and clays are interbedded near the base of the deposit. On the floor of this valley farther south, small patches of sand and gravel, locally with interbedded silt and clay, have survived ironstone working [e.g. SE 9070 1687].

On the western side of the Ancholme Valley north of Brigg a few small patches of silty and clayey sand, locally with scattered Lincolnshire Limestone pebbles, rise to 9 m above OD; to the north-east [SE 989 079] of Castlethorpe Hall a substantial limestone gravel is developed. On the eastern side clayey sand, rich in flint and chalk pebbles, mantles the slope from South Ferriby almost to Brigg and rises locally to over 23 m above OD. At South Ferriby the deposit rests on till and its top is cryoturbated. On the western side of the valley to the south of Brigg, silty sand with scattered to locally abundant Lincolnshire Limestone pebbles is at least 2 m thick in places between Hibaldstow and Waddingham. It occupies the lower ground between various minor scarps but rises westwards up several minor valleys to over 16 m above OD in places and is locally calcreted.

On the eastern side of the valley a thick fan-shaped mass of sand and gravel descends south-westwards from the western end of the Barnetby Gap at Melton Ross, where it lies at up to 25 m above OD. The majority of the pebbles are of flint and chalk, the rest being of rocks that are common as erratics in the Devensian till to the north-east. In Kettleby Gravel Pit [TA 038 083], the exposed sequence is: closely packed gravel with some coal fragments 1.22 m, on reddish brown and bluish green mottled laminated clay with silt near its base 1.98 m, on well-bedded, closely packed gravel, coarse near its base, with sand lenses and coal fragments 4 m seen. The clay is recognisable at a few other localities east of Brigg, for example in a railway cutting [TA 036 083]. Farther south, along the eastern side of the Ancholme Valley, there are only a few small patches of pebble-poor silty and clayey sand, some passing westwards into silt and clay, and some isolated on rockhead slopes; none rises above 9 m above OD.

Except for the Barnetby fan, the principal characteristics of the marginal lacustrine deposits are their locally derived pebbles, their occurrence on slopes, their interdigitation in the Humber area with Devensian till, and their differentiation into high-level gravelly deposits and low-level but stratigraphically younger pebble-poor sands. They indicate two phases in the history of the lake. In the earlier phase the water rose to about 30 m above OD and there was sufficient wave or current action to produce gravel. In the later phase, the water did not rise much above 9 m above OD and conditions were quiescent enough to prohibit much gravel deposition; this phase lasted sufficiently long for the lake to fill up with silt and clay virtually to the level of its sandy strandlines (before compaction depressed the top of the silt and clay). Cryoturbation within the marginal deposits near South Cave and localities farther west (Gaunt, 1981, p.91) suggests that between the high- and low-level phases the level of Lake Humber fell transiently to possibly as low as 4 m below OD, presumably due to leakage through the Humber Gap.

The fan of sand and gravel south-west of the Barnetby Gap was probably deposited by meltwater flowing south-westwards into the lake from near Kirmington. Laminated clay within these deposits shows that they were deposited in water at heights compatible with the postulated early high-level lacustrine phase during the maximum extent of the Devensian ice. Stages in the lowering of the lake are marked by terrace features between Barnetby le Wold and Kettleby, and by benches cut into the exposed Ancholme Clay towards Grasby.

Upper Sand

Around Newport, in the extreme north-west, fine-grained silty sand, clayey along its margins, forms low mounds and ridges up to 1.2 m high and rests on the lacustrine silt and clay. In the Trent Valley up to 3 m of silty and clayey sand conceal the silt and clay. North of the River Eau this sand is itself concealed beneath alluvial deposits, but farther south it emerges as the deposits of the First Terrace (see below). In the Ancholme Valley up to 2 m of silty and clayey sand, locally with a few small pebbles at the base, rest on the silt and clay. The sand is concealed north of Brigg, but farther south it protrudes through peat as low mounds and ridges.

Virtually all of this sand is pebble-free. Some may be originally lacustrine and some has obviously been redeposited by wind, but most is believed to have been laid down as levée and floodplain sediments by rivers initiating courses across the emergent lacustrine clay plain. This implies that drainage base-level of areas west and south of the Humber Gap must have remained well above contemporaneous sea level during deposition of the Upper Sand, presumably because moraines blocking the gap remained intact for some time after Lake Humber silted up.

Terrace Deposits

Sand up to 4 m thick, with scattered pebbles and gravelly layers at depth, forms a terrace on the eastern side of the Trent Valley south of the River Eau. It descends northwards from about 7 m above OD and its contained pebbles are small and mainly of flint and quartzite. Its surface has been considerably modified by the action of wind, and the deposit passes northwards beneath blown sand and alluvium into the Upper Sand. Like the last, it is probably a product of river deposition immediately after the disappearance of Lake Humber, and has been classified on the maps as First Terrace.

In various places along the River Eau upstream from Scotter, up to 2 m of sand, with some gravel consisting of locally derived pebbles and flint, forms a wind-modified terrace which descends downstream from about 14 to 7 m above OD. These heights suggest grading either to Lake Humber or to the immediately postlacustrine rivers. It is possible, however, that the thin basal Lias limestones formerly produced a rock-bar across the River Eau at Scotter and, consequently, a perched drainage base-level farther upstream. If so, these deposits might be of Flandrian age, and they are shown on the maps as undifferentiated terrace deposits.

In the Ancholme Valley east of North Kelsey, fine-grained, locally clayey sand forms a discontinuous terrace apparently related to a former stream draining north-westwards from Nettleton.

Dry Valley Deposits

Deposits that are similar to stream alluvium occur in most of the valleys cut into the Chalk of the Wolds and in some of the valleys on the Lincolnshire Limestone. They consist mainly of clayey and sandy silt, commonly with locally derived gravelly layers near and at the base. To the east of the Wolds they transgress locally over Devensian till and, to the west, on the northern side of the Humber, their down-valley extensions, shown on the maps as Head, merge into the Devensian lacustrine deposits. These dry valley sediments originated as alluvium during Devensian periglacial conditions, when frozen groundwater rendered the Chalk and Lincolnshire Limestone impermeable. They have since been added to by soil creep down the sides of the valleys and modified in many places by bourne stream activity.

Head

Deposits of Head accumulated by solifluction processes during Devensian periglacial conditions. Some movement has continued, especially on oversteepened or oversaturated slopes, and the deposits are in many places indistinguishable from those of soil creep, which continue to accumulate at the present day. They normally comprise clay or silt containing locally derived and unworn rock fragments.

Brown sandy clay Head mantles the lower parts of several slopes west of South Cave and around Ellerker, and is locally overlain by Devensian lacustrine deposits and Blown Sand. In Brantingham Dale and adjacent valleys on the western side of the Yorkshire Wolds, grey silty clay Head containing chalk and flint fragments forms down-valley extensions of the Dry Valley Deposits and merges with the Devensian lacustrine deposits.

To the south of the Humber, up to 3 m of brown clay Head, locally sandy, mantles much of the steep eastern slope of the Trent Valley between Whitton and Flixborough. Some of this deposit may pass under Blown Sand, and some thin lenses of it are present within the latter [e.g. SE 8666 1561]. Much of the Head, however, postdates the Blown Sand, having flowed down gullies cut into it. Similar Head mantles the slope southwards to Scotter (where it passes under Blown Sand) and the sides of the Bottesford Beck valley. The Head on the west-facing slope south-west of Winteringham varies from sandy clay to clayey sand. At one locality [SE 9129 1986], now quarried away, it rested on Devensian lacustrine deposits. Downslope it passes under estuarine alluvium.

Reddish brown, silty and clayey sand Head, at least 1.5 m thick locally and containing fragments of Northampton Sand and Grantham Formation rocks, lies on the west-facing slope south of Kirton in Lindsey. Small patches also occur around North Kelsey, Cadney and Howsham in the Ancholme Valley, along the eastern side of which, from Bonby southwards, Head covers much of the slope below the Chalk scarp. North of Caistor it is mainly grey silty clay, but farther south it is sandy in places due to incorporation of detritus from the Spilsby Sandstone and some of the Lower Cretaceous rocks. It underlies, is interdigitated with, and overlies Blown Sand and, locally, is intimately associated with landslip. Around Thornton Abbey, south of Goxhill, Head ranging from silty clay to clayey sand has accumulated in small hollows, some possibly kettle holes, and in shallow channels in the glacial deposits.

Blown Sand

Blown Sand is widespread west of the Wolds. It is characterised by its fine-grained, well-sorted nature, the paucity of pebbles and of interstitial silt and clay, its topographical form (with crescentic dunes in places) and its lack of altimetric constraint. Rounded grains are present but are not ubiquitous, probably because much of the sand has not travelled far; indeed most of it is merely the wind-redistributed top of other underlying sandy deposits. However, patches of sand on more elevated parts of the Lias and Lincolnshire Limestone outcrops, and also east of the River Ancholme between Grasby and Claxby, have probably been blown several kilometres. Thicknesses vary considerably, even over short distances. Up to 6 m of Blown Sand are present locally west of Messingham [SE 8775 0375] and in pits at Fonaby [TF 111 031], north-west of Caistor (Straw, 1963b). Generally, however, the deposit is less than 2 m thick, and there are extensive unmappable spreads less than 0.3 m thick.

The presence of ventifacts at the base of the Devensian deposits farther west (Gaunt, 1981, p.90) and the possibility that some of the sand beneath the lacustrine silt and clay may be aeolian (p.121), suggest that some of the exposed Blown Sand resting on bedrock may predate the late Devensian glacial and lacustrine episodes. Much of the Blown Sand, however, rests on and is clearly derived from sandy lacustrine deposits and is overlain by Flandrian deposits, and is thus referable to the terminal Devensian. This is confirmed by peaty layers within the sand, notably around Messingham, which contain pollen and insects indicating a cold climate and which have yielded dates of 10 055 ± 250 and 10 280 ± 120 radiocarbon years (Buckland, 1982). Mesolithic and a few older artefacts (Jacobi, 1978; Buckland, 1984) occur on and locally within the Blown Sand, notably on Sheffield's Hill [SE 910 158] and Risby Warren, but they are of little dating value in view of subsequent aeolian activity. Remobilisation of Blown Sand probably started with forest clearance in the Bronze Age, if not earlier, and Holland (1975) implied that it was still active in the Iron Age. Even today sand accumulates against hedges and occasionally blocks ditches and minor roads.

Ventifacts up to 0.4 m long are common on drift-free surfaces throughout the district. They are characteristically of sandstone and have dark brown desert varnishes. They probably represent both the aeolian episodes referred to above. Rounded basalt boulders exhibiting onion-skin weathering also occur.

FLANDRIAN DEPOSITS

Most of the widespread Flandrian deposits are of river and estuarine origin, but Peat, some of it concealed, is locally extensive. There are a few small deposits of Shell Marl and Calcareous Tufa, and landslips occur on some unstable slopes. Some of the Terrace Deposits, Head and Blown Sand may also be of Flandrian age.

Shell Marl

The term Shell Marl is generally restricted to calcareous clayey sediments rich in freshwater molluscs, but has been extended here to include some tufaceous deposits containing terrestrial molluscs. Most of the Shell Marl is associated with streams crossing the Lincolnshire Limestone and Snitterby Limestone outcrops, but some lies beneath peat and alluvium farther east, and the 'tufa' underlying peat on the North Ferriby foreshore (Wright and Wright, 1933) is really shell marl.

The main deposits south of the Humber have been described by Preece and Robinson (1984). A few minor patches are present near Appleby, where streams draining the Lincolnshire Limestone reach blown sand or alluvial deposits, but most of them are too small to show on the maps. In the Broughton area (Musham, 1933; 1934; Kennard and Musham, 1937) a small patch occurs near The Follies [SE 958 127], but the most extensive deposits are near Castlethorpe Hall [SE 9859 0772]. Faunal evidence (Preece and Robinson, 1984) shows that they started to accumulate early in the Flandrian in open marshes and pools fed by lime-rich springs, and continued to be deposited after the area became shaded by woodland. A temporary return to open-ground conditions, dated on contained charcoal to 3410 ± 80 radiocarbon years ago, is ascribed to Bronze Age clearance. A similar history was detected elsewhere, for example in Waddingham Beck [SK 963 951]. However, in a stream [SE 979 047] east of Sturton, these authors found that a calcareous silty sand under more extensive tufaceous silt contains a boreal arctic-alpine fauna indicative of open marshes and pools and a climate sufficiently cold to imply a late Devensian age. Shell Marl underlies peat and alluvial deposits west of Brigg (Smith, 1958; Fletcher, 1981) and is occasionally revealed by ploughing farther south [e.g. at SK 992 959 and SK 976 931].

Calcareous Tufa

There are several tufaceous deposits adjacent to calcareous rocks of the Lias and Redbourne groups, most of them associated with lime-rich springs. The 'dragon' at Dragonby is an east–west sinuous ridge [SE 9044 1418], some 23 m long, of hard travertine resting on Blown Sand. A narrow channel running along its crest demonstrates its derivation from a nearby spring at the base of the Northampton Sand. A domed mass of shell-bearing layered tufa exposed in a ditch [SE 9406 1364] on Risby Warren is one of several being precipitated from lime-rich springs rising up along the Flixborough faults. Sandy tufa containing a few freshwater gastropods occurs along Bottesford Beck where lime-rich springs issue from Lias limestones, and north-east of Messingham a spring from a similar source has precipitated a mass of tufa [SE 9070 0548] too small to show on the map. Fragments of tufa ploughed up in a field [SK 926 987] on sandy alluvium west of Kirton in Lindsey may have formed from springs issuing from a nearby fault or the underlying Pecten Ironstone. Tufa coating the bed of, and forming small waterfalls and pools in part of Waddington Beck [SK 9809 9643] may be derived from shell marl farther upstream or from the underlying Snitterby Limestone.

Peat

Peat of Flandrian age occurs in many low-lying areas but, except in the Ancholme Valley south of North Kelsey, it is largely concealed below or within alluvial deposits.

In the Trent Valley the peat in hollows and old channels on the First Terrace deposits is locally at least 1.4 m thick. Peat formerly occurred at the surface over much of the low ground but it has been covered by warp in the last two or three centuries, except for small enclaves such as Butterwick Common [SE 850 062]. This concealed peat, seen in many ditches, is at least 6 m thick locally and extends down to 6 m below OD in places near the present River Trent, where it contains large tree trunks. Thin lenses of peat and peaty clay lie even deeper, possibly as low as 16 m below OD in the alluvial deposits.

On the northern side of the Ouse, Hulme and Beckett (1973) recorded two thin peat layers at c.3.3 m below OD and c.1.4 m below OD within alluvial deposits beneath a Romano-British level near Weighton Lock [SE 875 257]. Concealed clayey peat up to 3.5 m thick, with its base descending to 6.5 m below OD, is extensive under Walling Fen, and an old borehole at Broomfleet proved 2.7 m of peat at about 16.5 m below OD. The lowest of several thin lenses of peat within sandy alluvium at a locality [SE 9476 2684] in Elloughton Beck yielded an age of 3905 ± 105 radiocarbon years. Up to 4.9 m of peat reaches to within 0.3 m of the surface under Brough Airfield, and the nearby Welton Waters 'A' Borehole proved 1.35 m of peat at 3.0 m below OD. On the North Ferriby foreshore, at Red Cliff, up to 1.2 m of peat, commonly referred to as a 'submerged forest' (Plate 5), rests on freshwater calcareous silty shell marl and is overlain by estuarine alluvium (Bisat, 1932b; Wright and Wright, 1933).

South of the Humber, in the lower part of the Winterton Beck valley, up to 1.5 m of peat fills old stream channels at the surface, and more than 2 m of peat are present locally at depth. Much of the peat in the Ancholme Valley is covered by estuarine alluvium and was described by Smith (1958). Two thin layers, one at about 4 m below OD and another at about 2 m above OD and associated with Romano-British remains, are present in alluvial deposits near Ferriby Sluice. Thin peat rests on tufaceous shell marl at depths down to about 4 m below OD and is overlain by estuarine alluvium to the west of Brigg. Probably the same peat is exposed in Coal Dyke End [SE 9922 0781] and has yielded an age of 4050 ± 50 radiocarbon years (Preece and Robinson, 1984). Peat flanking both sides of the estuarine alluvium to the north of Brigg contains large horizontal oak and pine trunks and fragments of yew, birch and hazel. South of Brigg, at a locality [probably c.TF 003 988] near Redbourne Hayes, Smith (1958) proved nearly 1 m of peat at depths down to about 1.3 m below OD, and traces of peat at the surface, now largely destroyed. The extensive peat that emerges from beneath estuarine alluvial deposits from Cadney southwards is generally less than 2 m thick and is being wasted by ploughing and wind erosion; it contains many large fallen oak and pine trunks aligned NE–SW. In this same area thin peat overlies alluvium in several tributary streams.

On both sides of the Humber to the east of the Wolds, thin peat is present at or just above the base of the alluvial deposits. It was seen in dock excavations at Kingston upon Hull (Ussher, 1890, pp.185–187). A recent proving (Gaunt and Tooley, 1974) in the Market Place Borehole [TA 1003 2851] of 2.43 m of peat at 11 m below OD gave two dates of c.6930 radiocarbon years. In excavations [c.TA 102 321] at Stoneferry, up to 0.2 m of peat at about 4 m below OD gave an age of 5240 ± 100 radiocarbon years. Thin layers of peat are present at higher levels within the alluvial deposits. Two thin peats at 1.40 m above OD and 1.78 m above OD within alluvial deposits in a pit [TA 0602 2307] at Barrow Haven gave dates of 2325 ± 60 and 1080 ± 40 radiocarbon years respectively. Peat occurs impersistently within, and filling channels on the surface of the alluvium along The Beck near

Barrow upon Humber and along East Halton–Skitter Beck farther east.

Alluvium and Estuarine Alluvium

The Alluvium and Estuarine Alluvium can rarely be distinguished from each other. At depth these deposits fill narrow channels that are incised to about 20 m below OD under the present rivers and directed towards the Humber Gap (Gaunt, 1981, pp.94–95, fig. 5). Nearer the surface they extend widely beyond the channel confines. The vigorous river incision responsible for the deep channels was unaccompanied by any significant denudation of the soft sediments between the channels, and must have occurred quickly due to a rapid and substantial drop in drainage base-level. This incision, which took place when the morainic blockage across the Humber was finally breached, postdates the post-lacustrine levée Upper Sand and the First Terrace deposits of the Trent Valley. There is evidence farther west that it had ended by about 8500 radiocarbon years ago when shell marl was deposited in a minor channel at Burton Salmon (Norris et al., 1971); it had certainly terminated by 7000 radiocarbon years ago, when sea level was rising (Gaunt and Tooley, 1974). As the sea rose to its present level, alluvium filled the channels and spread thinly over adjacent low ground, commonly covering expanses of peat. Near the bottoms of the deep channels the alluvium ranges from clay to sand, with, in places, a thin basal gravel and thin peats. At higher levels the deposits are mainly silts and clays with impersistent thin peat, and at the surface are largely soft peaty clays, except where silt-rich levées have formed near the rivers.

In the Trent Valley a deep alluvium-filled channel runs northwards from Gunness and is joined north-east of Garthorpe by a similar channel associated with the 'pre-Vermuyden' River Don (Gaunt, 1975). A borehole [SE 8446 1258] near Gunness proved silty clay on peaty clay with sand to 10.7 m, on sand with basal gravel to 19.2 m, on rockhead at 16.1 m below OD. Other boreholes [at c.SE 864 197] farther north show that alluvium within the channel descends to about 20 m below OD. Outside the channel, however, the alluvium is generally less than 6 m thick.

The deep alluvium-filled channel of the Ouse passes south of Blacktoft and under Faxfleet, where, on the evidence of boreholes farther west, it descends to at least 16.1 m below OD. Beyond the channel confines to the north, however, the alluvial deposits are generally less than 3 m thick, and small outliers of Upper Sand protrude through them at Staddlethorpe and Gowthorpe House. A record of peat at about 16.5 m below OD in a borehole at Broomfleet (p.125) probably locates the channel. It is joined nearby by the alluvium-filled channel of the River Foulness, which runs south-eastwards under Walling Fen. Even in the extreme north of the district [SE 8606 3125] the channel of this minor river reaches to at least 8.6 m below OD. The surface alluvium on Walling Fen is soft peaty clay. Broomfleet Island was originally an estuarine silt bank, said to have accreted around an old wreck. By 1880 it had been embanked, and by 1907 joined to the mainland by controlled silting of the intervening channel, Broomfleet Hope. There is evidence of an older accretion to the north, with a northern edge running from Weighton Lock through Broomfleet and past Providence [SE 8895 2563] before curving south-eastwards to Brough. According to Saltmarshe (1920) this area was reclaimed in about 1690. In places the near-surface alluvial clay on Brough Airfield is only 0.3 m thick, but it thickens south-eastwards and extends down to 4.74 m below OD in Welton Waters 'A' Borehole.

On the southern side of the Humber, Whitton Ness Borehole proved soft alluvial silty clay to about 8.3 m below OD, but surface clay farther south in the Winterton Beck valley is commonly only 0.8 to 1.6 m thick. Alluvial deposits, mainly silty clay on sand, reach to about 10.26 m below OD in Winteringham Haven Borehole. Reed's Island off the mouth of the Ancholme valley, originally another estuarine silt bank, was embanked between 1841 and 1886, but has subsequently suffered some erosion.

Low Farm (South Ferriby) Borehole proved alluvial clay to only 3.66 m below OD, so the old incised channel of the Ancholme must reach the Humber farther east, probably near Ferriby Sluice. Its course farther south is uncertain, but at Island Carr, near Brigg, Smith (1958) proved variably peaty clay down to 7 m below OD. Estuarine silty clay was deposited in the Ancholme Valley as far south as Waddington Holmes [TF 012 979], forming heavy carr land. It is locally fossiliferous and contains many salt-tolerant taxa including *Phragmites communis*, *Spartina* sp., pollen, seeds, diatoms, foraminifera, gastropods, ostracods and insects (McGrail, 1981, pp.134–187). Former river courses are recognisable as sinuous depressions, but in the northern part of the valley, on Bonby Carrs [SE 967 148], a former creek network, now reflected by laminated silt ridges with marine shells, is visible. A feature of the estuarine alluvium around Brigg is the occurrence of metallic-like vivianite along vertical joints.

The alluvial deposits at the mouth of the River Hull under Kingston upon Hull comprise a thin impersistent basal sand and gravel succeeded by a thin freshwater or only slightly brackish clay, locally containing thin peat, overlain by shell-bearing estuarine clay and sand, followed by estuarine silt which passes up into clay at the surface. The base of these deposits lies lower than 15 m below OD in the incised channel of the Hull, and in Market Place Borehole they rest on till at 13.72 m below OD. They are much thinner outside the channel, for example reaching only to 4.5 m below OD at Stoneferry, just north of the district (Gaunt and Tooley, 1974). On the opposite side of the Humber, geophysical investigations (Barker, 1982) suggest that a deep channel filled with estuarine alluvium is cut to rockhead and runs parallel to the estuary between Barton-upon-Humber and Goxhill Haven. The alluvium in a pit [TA 0602 2307] at Barrow Haven contained dinoflagellate cysts and ostracods, implying an upward increase in salinity.

Warp

Warp is silt and silty clay deposited by artificially controlled flooding in the last two or three centuries to improve the agricultural quality of land (Gaunt, 1987, pp.23–26). This was done by embanking a tract of ground and allowing sediment-laden river water to enter it through sluice gates at high tide and, having deposited its sediment load, to drain back gently at low tide. Up to 0.3 m of sediment can form in a warping season, mostly between spring and mid-autumn. Warped ground can commonly be identified in the field by its well-drained pale brown silty soil, by levée-like ridges alongside 'warping' drains, by traces of lamination beneath plough level, by the presence of the original soil at depth, and by differences of level at field boundaries. There is also some documentary evidence.

Three areas, all near the western edge of the district, have been warped. North of Blacktoft warping may have begun before 1820 (Sheppard, 1966), continuing in places until 1947. On the western side of the Trent the ground around Amcotts and possibly as far north as Mere Dyke [SE 8500 1660] was warped largely in the 19th century. On the eastern side of the Trent most of the floodplain from Neap House [SE 8610 1324] to the River Eau at Newstead [SE 871 033] has been warped, largely between the 1820s and 1914.

Tidal Flat Deposits

The stretches of tidal flat flying outside the flood defences of the Humber and adjacent parts of the Ouse and Trent consist of soft waterlogged greyish brown silty clay which is continually being eroded and redeposited.

Landslips

There are a number of rotational landslips along the Chalk scarp between South Cave and Brantingham and, on the south side of the Humber, between South Ferriby and Grasby. They have mostly developed at the spring-lubricated contact of the Chalk aquifer with underlying clays, aided by cambering along the scarp. The most extensive landslips in the district, which are virtually continuous along the Chalk scarp, occur south of Nettleton, where the Lower Cretaceous succession is present. There are extensive cambers, with associated slips and mud-flows generated from the squeezed-out Tealby and Roach clays and the uppermost beds of the Ancholme Clay Group, producing very unstable slopes.

Numerous rotational landslips affect the steep slopes of Lias rocks on the eastern side of the Trent between Alkborough and Flixborough. Seepage from the thin limestones may have lubricated these mudstone slips, and undercutting by the river was probably a contributory factor.

A few small landslips below the scarp slope of the Lincolnshire Limestone east of Flixborough may have been initiated by seepages from the base of the limestone; some near Dragonby appear to have been reactivated by ironstone workings downslope.

CHAPTER 8

Economic products and hydrogeology

BUILDING STONE

Virtually all the hard rocks cropping out in the district have been used locally in the past for building stone. The Lincolnshire Limestone was the most important source, particularly its upper part south of the Humber—the Scawby and Hibaldstow limestones. The latter yielded some attractive stone as can be seen, for example, in old buildings at Winterton, and the remains of the Roman Villa [SE 9337 1892] nearby testify to the long history of their use. The Cave Oolite, the lateral equivalent of the Hibaldstow Limestones north of the Humber, has been employed for the construction of many old buildings between South Cave and Brough.

Frodingham Ironstone has been used to advantage on the Sheffield and Elsham estates in and around Scunthorpe. It was commonly incorporated into brick buildings as an edging stone, and often carved. Other readily recognisable rocks in old buildings of the district include Snitterby Limestone, Cornbrash, Kellaways Rock (around South Cave), Tealby Limestone and Chalk (see p.00).

CHALK

Chalk is exploited within the area for cement, whiting, flux, lime burning, agricultural lime and for certain constructional purposes.

Over much of the country it is too weak to have much value as a constructional material, but northern Chalk is significantly stronger and contains less moisture than southern Chalk. Nevertheless, northern Chalk typically contains 8 to 10 per cent moisture and its susceptibility to frost damage limits its application.

There are medium-sized workings in the Ferriby Chalk at Mansgate Hill near Caistor and at Bigby. Both produce ground chalk for agricultural purposes and some flux for use with imported iron ores in sinter at the nearby Scunthorpe Steelworks. A certain amount of chalk from both workings is used for fill or hardcore, but not for any of the more demanding aggregate applications. The rock is normally worked by ripping. Processing consists of a simple screening operation.

At Melton Ross, the Welton Chalk is exploited in a large operation, with a capacity of 3500 to 4000 tonnes per day. Approximately 60 per cent of the output is calcined at the quarry, mainly for use in steel furnaces at Scunthorpe; the uncalcined material is used in sinter at Scunthorpe and a certain amount is sold for constructional purposes. The Chalk here is fairly hard with an approximate ACV of 36, a crushing strength of 35 to 55 MN/m^2 and a moisture content of 8 to 10 per cent. The quarry is worked by explosives.

At South Ferriby the Ferriby and Welton Chalk formations are exploited for cement making. Both chalk and the underlying Ancholme Clay are extracted (Plate 2) and mixed in the ratio 3 parts chalk to one part clay to form the raw material for the plant, which produces 600 000 tonnes of clinker annually and consumes 1.2 m tonnes of chalk and clay. Extraction of the Chalk is by ripping, and the clay is dug directly with a face shovel. The Chalk contains about 8 per cent water and the clay 15 to 20 per cent. The process is based on two coal-fired kilns using the semidry process, in which the raw material is first dried and then treated with sufficient water to form nodules suitable for feeding to the kiln. A typical analysis of run-of-quarry material used for cement kiln feed is:

	Chalk (with minor amounts of flint)	Clay
SiO_2	2.9	57.5
Al_2O_3	0.7	17.7
Fe_2O_3	0.5	6.3
CaO	42.5	1.8
MgO	0.4	1.5
SO_3	nil	3.7
K_2O	0.11	3.1
Loss on ignition	42.4	8.2

At Melton, near Hull, the Welton Chalk was exploited for cement making until 1981, but an associated operation producing whiting continues, the material being selected primarily on the basis of low porosity and colour. About 250 000 tonnes of chalk are extracted annually, although not all of this is converted into whiting. Processing involves crushing to a fine particle size followed by air classification to produce powder in the required size ranges. Whiting manufactured at Melton typically contains about 97.5 per cent $CaCO_3$ with 1.6 per cent silica and other acid insolubles. Powders of several size distributions are produced, the finest being more than 75 per cent finer than 5 microns. The whiting is used mainly as a filler or extender partially to replace the main constituent with a cheaper, inert substitute. A low porosity is of value because it leads to less absorption of the more expensive constituent. The range of application of Melton whiting is extensive. Fine grades are used in paper coating, plastic extrusions, paint and rubber. Other grades are used in putty, mastics, carpet backing, bituminous emulsion, abrasives, vinyl flooring, linoleum, paper loading and polishers. Some of the coarser grades are used for animal feedstuffs, asphalt filler and agricultural lime.

Other chalk workings include quarries at Willerby, near Hull, and at Ulceby; both produce chalk entirely for fill or hardcore. The material is usually extracted directly with a face-shovel, and sold as dug.

CLAY AND MUDSTONE

Argillaceous materials from both Solid and Drift deposits have been exploited throughout the district, but activities are now scattered and small in scale.

Historically, brick and tile making was associated with settlements on or near clay outcrops, and many old workings are now represented by ponds and overgrown pits. Notable among these are old workings in the Mercia Mudstone [SE 873 030] near Scotterthorpe, in Penarth Group mudstones near The Thistles [SE 875 066], in the Lias mudstones below and above the Pecten Ironstone at Santon [SE 929 121] and in Ancholme Clay at Claxby Moor, Moortown Hill, Holton, South Kelsey, North Kelsey and Worlaby (see Ussher, 1890). Lacustrine silt and clay were formerly worked along with Penarth Group mudstone at The Thistles. Estuarine alluvium appears to have provided the best brick-clay; there were three yards on Island Carr on the outskirts of Brigg and others at Brickhills and Thornholme [SE 978 127]; a brickyard at Winterton Holmes produced poor-quality bricks from a mixture of alluvium and till.

There are extensive old brick and tile pits in estuarine alluvium on both banks of the Humber, particularly on the south bank between Barton Cliff and New Holland. This deposit is currently extracted at Broomfleet on the north bank and at Barton-upon-Humber on the south bank, and there is a temporarily idle working at Barrow Haven. Only tiles are made at present. The alluvium is approximately 8 m thick, and pits are pumped dry before the annual supply of clay is extracted during a short summer working season. The alluvium is normally worked by face-shovel or sometimes by scraper, and is crushed and blended before use.

Ancholme Clay workings at South Ferriby for use in cement-making have been described above. The mudstones in the Kirton Cementstones, particularly the so-called Kirton Cement Shale above the Scawby Limestone, have been extensively worked for cement-making at Kirton [SE 951 012] (Plate 1). In the last operations at these works, Coleby Mudstones above and below the Marlstone Rock were dug [SE 933 005] near Mount Pleasant as a substitute for the diminishing reserves of Kirton Cementstones; they were crushed in the mill with limestones from the Kirton sequence.

The long-standing practice of 'doddering' sandy land by spreading soft clay upon the surface continues. Most recently, clay from ironstone workings and other excavations (e.g. irrigation ponds [TA 085 018] and Cadney Reservoir [TA 013 045]) has been transported to the Wolds and used for lining Chalk irrigation ponds.

The laminated clay from the Vale of York Glacial Lake Deposits around Brigg has been used by local potters.

COAL

The Coal Measures are present at depth throughout the district (Chapter 2), but the nearest coal workings, at Thorne Colliery, are some 6 km to the west.

EVAPORITES

Salt is present in the Boulby Halite, the southern limits of which extend at depth under Kingston upon Hull, and in the Fordon Evaporites at a lower level in the Permian. However, salt is abundant in northern England in more easily worked deposits, and it seems unlikely that the resources within the district will attract commercial attention in the forseeable future.

Anhydrite is present at depth in the Permian and Triassic rocks over much of the district. There is little demand for it at present and, like salt, it is more easily worked elsewhere in northern England. Hence formations such as the Hayton Anhydrite and the Fordon Evaporites appear to have little economic value.

Gypsum, a hydrated derivative of anhydrite, forms where anhydrite comes close to the surface and is hydrated by groundwater. The Newark Gypsum near the top of the Mercia Mudstone Group has been worked to the south of the district at Newark and near Gainsborough. The bed is discontinuous and its outcrop lies mainly beneath the floodplain of the River Trent, covered by thick drift. There is no history of working in the district nor knowledge of any worthwhile deposits. However, the bed was intersected in Cockle Pits Borehole where it consisted of 5.02 m of nodular anhydrite underlain by 5.76 m of mudstones cut by thin gypsum veins.

HYDROCARBONS

There are no known occurrences of hydrocarbons in the Permian and Mesozoic rocks of the district, and interest has centred on the Upper Carboniferous rocks, into which boreholes have been drilled at Blyton, Brigg, Broughton, Burton upon Stather, Butterwick and Corringham (see Appendix 1). The only production is from the Corringham field astride the southern boundary of the district, which has yielded modest amounts of oil since exploitation began in 1958 (between 32 000 and 33 000 tonnes up to 1965, when production ceased until 1973, and somewhat more than 20 000 tonnes since).

IRON ORE

Iron has been extracted on a large scale from the Frodingham Ironstone in the Scunthorpe area and northwards almost to West Halton, and on a smaller scale from the Claxby Ironstone at Nettleton.

Frodingham Ironstone

Until recently, the Frodingham Ironstone formed the basis of the steel industry of Scunthorpe. Originally worked by the Romans, it began to be exploited on a large scale in 1859, and by 1888, after the basic steel-making process had been perfected, the annual output exceeded 1 million tonnes of ore. Further expansion occurred in the succeeding century until 1961, when the maximum annual production of 5.6

million tonnes was achieved. In recent years output has fallen drastically in the face of imports of high grade iron ore and the decline in UK manufacturing activity (Figure 43). Production ceased in December 1988.

The Frodingham Ironstone is oolitic and calcareous, a sedimentary iron ore of the minette type. The worked ironstone is typically about 9 m thick in central and southern parts of the orefield, decreasing northwards to 3.7 m at West Halton, which was regarded as the northern limit of practical working. As far south as Ashby the thickness is maintained, but 'dirt' bands increasingly appear and it was not practical to work beyond Brat Hill, 2 km to the south of Scunthorpe. Hence opencast workings (Plate 6) have been confined to a strike distance of about 13 km near Scunthorpe, but mining extended eastwards beneath Risby Warren to Appleby. The deposit continues east of the River Ancholme to the coast at Immingham.

Although nearly 300 million tonnes of ore have been used as a source of iron, the economics of the industry have recently changed to such an extent that the Frodingham Ironstone can be regarded as only a subeconomic resource rather than a reserve. Even when exploitation was at its peak the ore, which on average contained approximately 24 per cent Fe, 22 per cent CaO, 7.5 per cent SiO_2 and 0.35 per cent P, was one of extremely low grade. It was of value

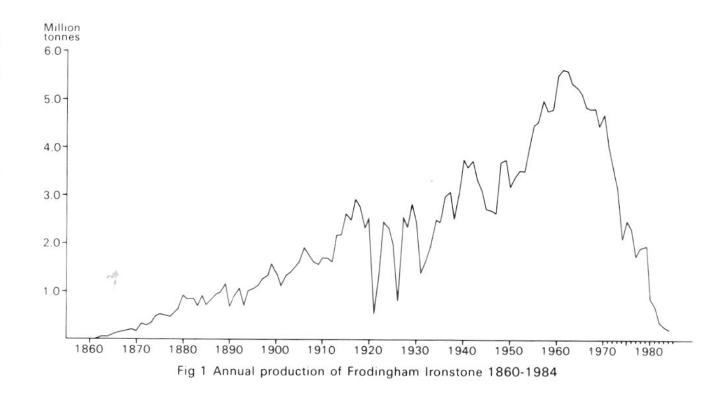

Figure 43 Annual production of Frodingham Ironstone 1860–1984

Plate 6 Abandoned opencast working for Frodingham Ironstone near Bagmoor Farm. The degraded and slipped backwall is in Coleby Mudstones (Lias), but the Pecten Ironstone and Marlstone Rock are visible in the face. A 14378

Table 5 Detailed analysis of Frodingham Ironstone in a borehole [SE 9197 1311] at Santon

Depth to (1 feet)	Total Fe	CaO	SiO_2	MnO	Al_2O_3	MgO	P_2O_5	S	Loss on Ignition
263	17.3	23.4	9.6	1.53	4.51	2.56	0.93	2.05	25.4
264	20.1	22.6	7.9	1.55	4.18	2.67	0.69	2.20	25.1
265	19.0	28.3	4.9	0.92	3.27	1.50	0.70	.555	30.3
266	25.6	23.5	6.0	0.71	3.85	1.51	1.23	.130	26.4
267	30.7	14.0	10.8	0.74	5.69	2.23	0.55	.158	20.9
268	25.2	4.1	27.8	0.50	10.09	2.58	0.42	.257	12.1
269	31.0	4.7	20.5	0.70	8.46	2.61	0.49	.175	15.9
270	25.0	17.5	11.0	1.56	5.34	2.61	0.98	.148	23.7
271	22.9	24.8	6.3	1.06	3.83	1.63	0.82	.592	26.8
272	28.1	21.2	5.3	0.94	3.60	1.67	0.74	.266	26.1
273	27.2	21.3	5.5	0.80	3.61	1.44	0.75	1.12	23.6
274	30.5	18.7	5.3	1.05	3.68	1.62	1.01	.510	23.6
275	33.8	14.7	6.4	1.15	4.22	1.80	1.06	.178	16.0
276	30.7	19.0	5.2	1.12	3.55	1.57	1.18	.352	24.5
277	25.8	21.8	5.8	1.14	3.85	1.84	1.04	.215	27.6
278	20.6	28.4	4.2	0.93	3.02	1.01	0.52	.970	27.7
279	17.2	32.2	2.7	1.53	2.11	1.22	0.79	.070	33.7
280	22.8	26.2	4.8	1.29	3.50	1.37	0.87	.063	28.7
281	18.0	30.7	4.6	1.37	3.33	1.17	0.40	.060	32.1
282	20.3	27.0	4.3	2.34	3.33	1.59	1.31	.290	29.3
283	17.9	29.5	3.9	2.92	2.84	1.63	0.57	.065	32.2
284	20.8	27.2	4.1	2.40	3.11	1.70	0.52	.062	30.8
285	20.7	27.9	4.9	1.40	3.64	1.35	0.55	.038	30.0
286	22.2	27.3	4.1	1.48	3.25	1.31	0.45	.067	29.9
287	28.2	21.8	5.0	1.17	3.92	1.45	0.92	.099	26.7
288	20.2	29.6	3.9	1.06	3.22	1.10	0.68	.050	29.6
289	22.5	24.1	8.3	0.78	5.41	1.52	0.81	.119	24.1
290	26.8	17.4	11.1	0.78	7.00	1.99	0.62	.262	21.9
291	25.6	19.8	8.4	1.40	5.56	1.99	0.56	.228	25.2
292	15.9	30.4	6.3	1.07	4.11	1.35	0.62	.375	30.0
292.5	17.5	28.0	6.5	1.49	3.55	1.90	0.61	.330	30.1
293.5	10.1	11.6	35.5	0.41	9.81	2.15	0.17	.340	18.1
Average	23.3	24.7	5.52	1.33	3.79	1.55	0.70	.277	27.4

The average is taken between 271 and 292.5 ft. The remainder of the bed forms the roof and floor for the underground working.

Samples taken at 1-foot intervals from 262 to 292 feet.

because the high lime content allowed it to be smelted without the addition of limestone. From the 1920s it was usual at Scunthorpe to mix Frodingham ore with siliceous ore from Northamptonshire or Nettleton in the proportion necessary for a self-fluxing blast-furnace charge. More recently, Frodingham ore has been added to high-grade imported ores at Scunthorpe to provide a flux of significant iron value.

The unoxidised ironstone consists essentially of fossil shells and ooliths of goethite and berthierine in a matrix of calcite, berthierine and siderite mudstone. Generally green in colour when fresh, the ironstone becomes yellowish brown and leached of lime on weathering. Details of chemical variation through the ironstone from a representative borehole in the central part of the orefield are shown in Table 5. The lowermost and uppermost parts of the ironstone were left in place during underground mining to form a stable floor and roof for the working. Similarly, in the older opencast operations, ore was left to form a floor for the quarry, although this practise was discontinued in later years.

The position with respect to drilled and proved ore still in place when the British Steel Corporation ceased to extract Frodingham ore is summarised in Figure 44 and Table 6. To the west of the New River Ancholme approximately 81 million tonnes of ore remain in place under an overburden of less than 61 m. Another 164 million tonnes remain under an overburden of 61 to 76 m. Thus 245 million tonnes are deemed to be workable by opencast methods. A further 149 million tonnes of resources exist beneath a cover greater than 76 m west of the New River Ancholme, and are considered to be available for extraction by underground mining. The tonnages refer to the expected yield if the areas were to be worked by conventional methods. In addition, there are reported to be a further 37 million tonnes in the Broughton Common area which have not been drilled.

To the east of the New River Ancholme three companies started drilling programmes during the period 1958–62, when the demand for Frodingham ore was near its maximum, with a view to supplementing their diminishing opencast resources. None of these activities resulted in a mining operation, although one area of 140 km^2 was drilled, at least in outline. In total this area contains a workable resource of about 800 million tonnes, of which some 70 million tonnes near Worlaby can be considered as proved. East of the New River Ancholme the southern economic limit of the orefield has not been defined by drilling nor has the eastward extent been proved beyond Keelby, near Immingham; the full extent of the resource, therefore, remains unquantified.

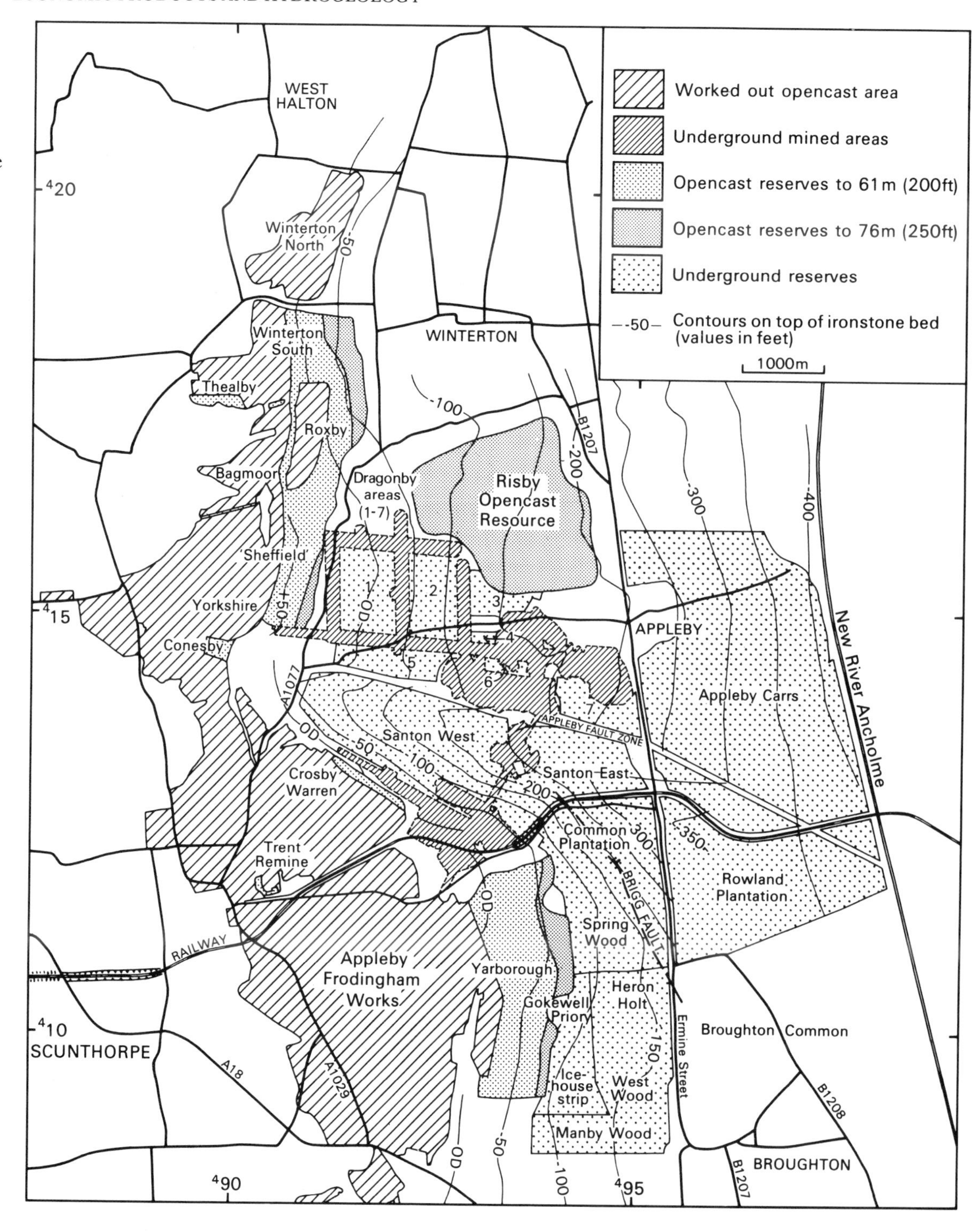

Figure 44 Frodingham Ironstone reserve areas west of New River Ancholme

The large-scale opencast mining operations employed a 25 cu yard walking drag-line, positioned on top of the ironstone, which removed up to 40 m of overburden. Cover in excess of this was usually pre-stripped using scrapers or face shovels. The exposed ironstone was then broken with explosives before being dug with a 4.5 cu yard shovel, loaded into dump trucks and carried directly to the steelworks for crushing and screening. Much of the worked-out area has been returned to agriculture or used for industrial development. Other parts have been used for recreation or waste disposal. The abandoned pits between Crosby Warren and Winterton Cliff farms were incompletely backfilled and are now partly flooded.

Underground working started in 1938 when the original Santon Mine was opened. Dragonby Mine was opened in 1950 and the two were linked under the name of Santon Mine in 1969. Production ceased in 1980. Working was by the room and pillar method with 20 ft (6.1 m) wide headings and 50 ft (15.2 m) square pillars. Of the c.8.5 m of ironstone, 2 m served as a roof to support the overlying clays, and 0.5 m

Table 6 Frodingham Ironstone reserves

a) Open-pit reserves

Reserves to 61 m cover	Area (h_a)	Thickness (m)	Million Tonnes	% Fe
Winterton S	44.99	7.04	7.83	23.0
Roxby	39.54	7.32	7.15	23.6
Bagmoor	19.06	4.63	2.19	23.2
'Sheffield'	62.67	8.05	12.46	23.8
Yorkshire	15.66	8.26	3.20	25.2
Conesby	7.58	8.53	1.60	23.3
Crosby Warren	12.50	8.84	2.73	24.3
Trent Remine	1.99	3.05	0.15	20.6
Yarborough	201.02	8.84	43.91	24.2
Sub total and average	405.01		81.22	24.0
Reserves 61 – 76 m cover				
Winterton S	19.78	6.98	3.41	23.0
Roxby	13.77	7.01	2.39	23.0
'Sheffield'	10.81	7.47	1.99	23.0
Yarborough	47.17	8.84	10.31	24.2
Risby	333.46	7.86	64.80	23.6
Sub total and average	424.99		82.90	23.6
Total	830.00		164.12	23.8

b) Underground reserves west of the River Ancholme

	Area (h_a)	Face thickness (m)	Workable ore (M tonnes)	% Fe
Dragonby areas (1 – 7)	245.48	6.40	18.37	23.8
Santon West	226.58	5.70	15.16	23.7
Santon East	106.35	5.88	7.34	24.6
Common Plantation	91.05	5.46	5.83	22.6
Appleby Carrs	729.08	6.00	51.38	24.6
Rowland Plantation	315.94	5.43	20.12	22.8
Spring Wood	149.57	5.79	10.17	23.5
Gokewell Priory	21.41	5.58	1.40	23.8
Icehouse Strip	67.87	5.46	4.35	23.9
Heron Holt	64.51	5.45	4.20	23.7
West Wood	85.87	5.49	5.53	24.5
Manby Wood	77.94	5.24	4.80	23.1
Total	2181.65		148.65	23.9

of low quality basal material was left to form a floor, leaving a thickness of 6 m to be extracted. Access was by a 46 m shaft at Santon and by an adit near Dragonby through which ore was conveyed by belt to the surface and machinery could enter the workings. It was necessary to secure the roof with roof bolts up to 2 m long on a 2 m grid pattern. In the last year of operation a production rate of 12 000 to 14 000 tonnes per week was achieved. The resources were so large that optimum recovery was not attempted, and pillar dimensions were well in excess of those required for good support. The mine was allowed to flood on abandonment, but since 1984 has been partly dewatered to allow for inspection below roads and services.

The low iron-content has long been recognised as the major inhibiting factor in the use of Frodingham ore, and has directed interest towards beneficiation techniques, particularly chemical leaching and specialised heat treatment because the iron minerals are not susceptible to physical methods of concentration. Research was carried out on controlled sintering of the ore in an attempt to convert the iron to magnetite grains of sufficient size to be recoverable into a concentrate after cooling and crushing. None of these investigations led to a viable process, and Frodingham ore remains unamenable to any form of beneficiation except for the upgrading that occurs during sintering or calcination as carbon dioxide and water are driven off. In addition to the

Figure 45 Workings in Claxby Ironstone

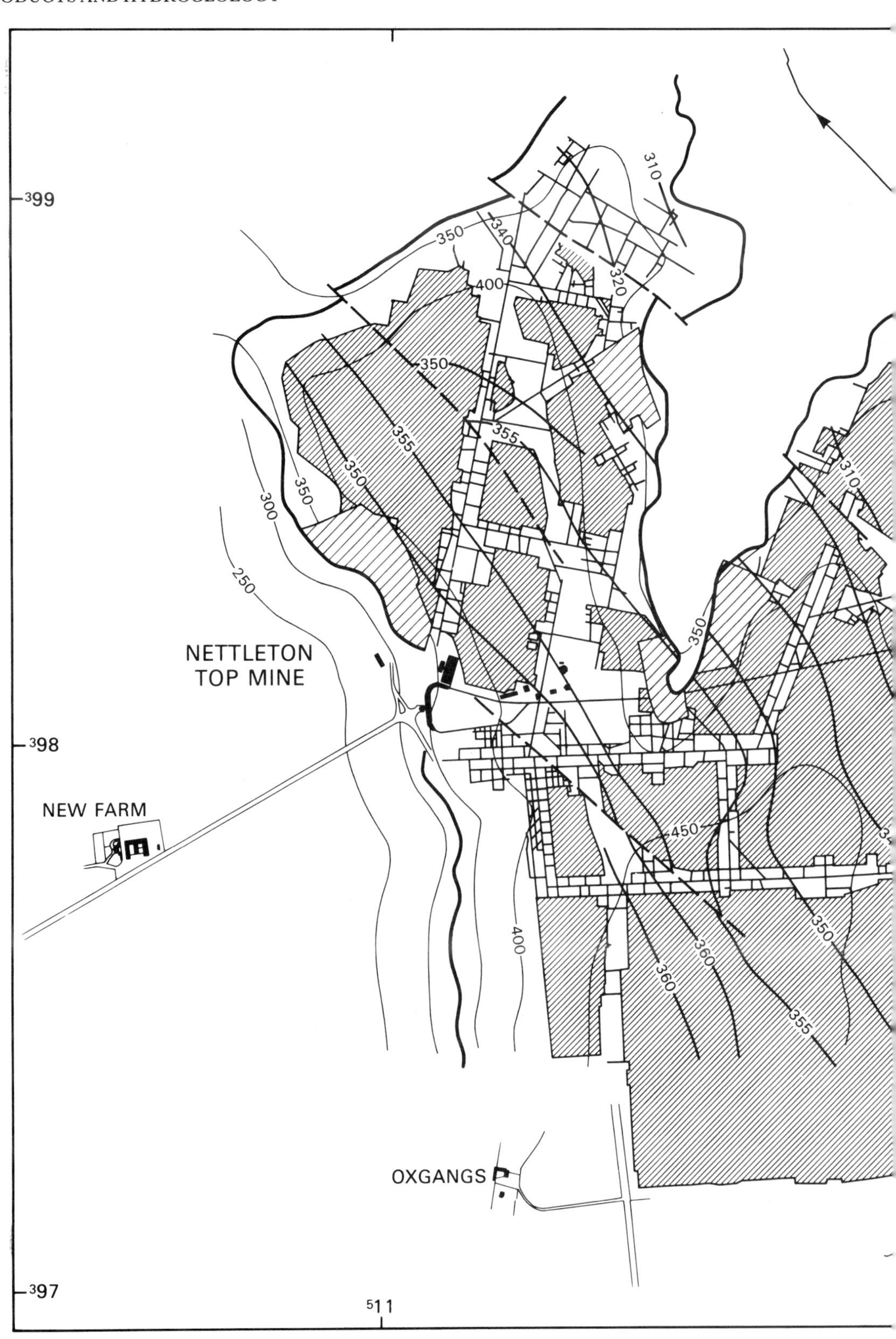

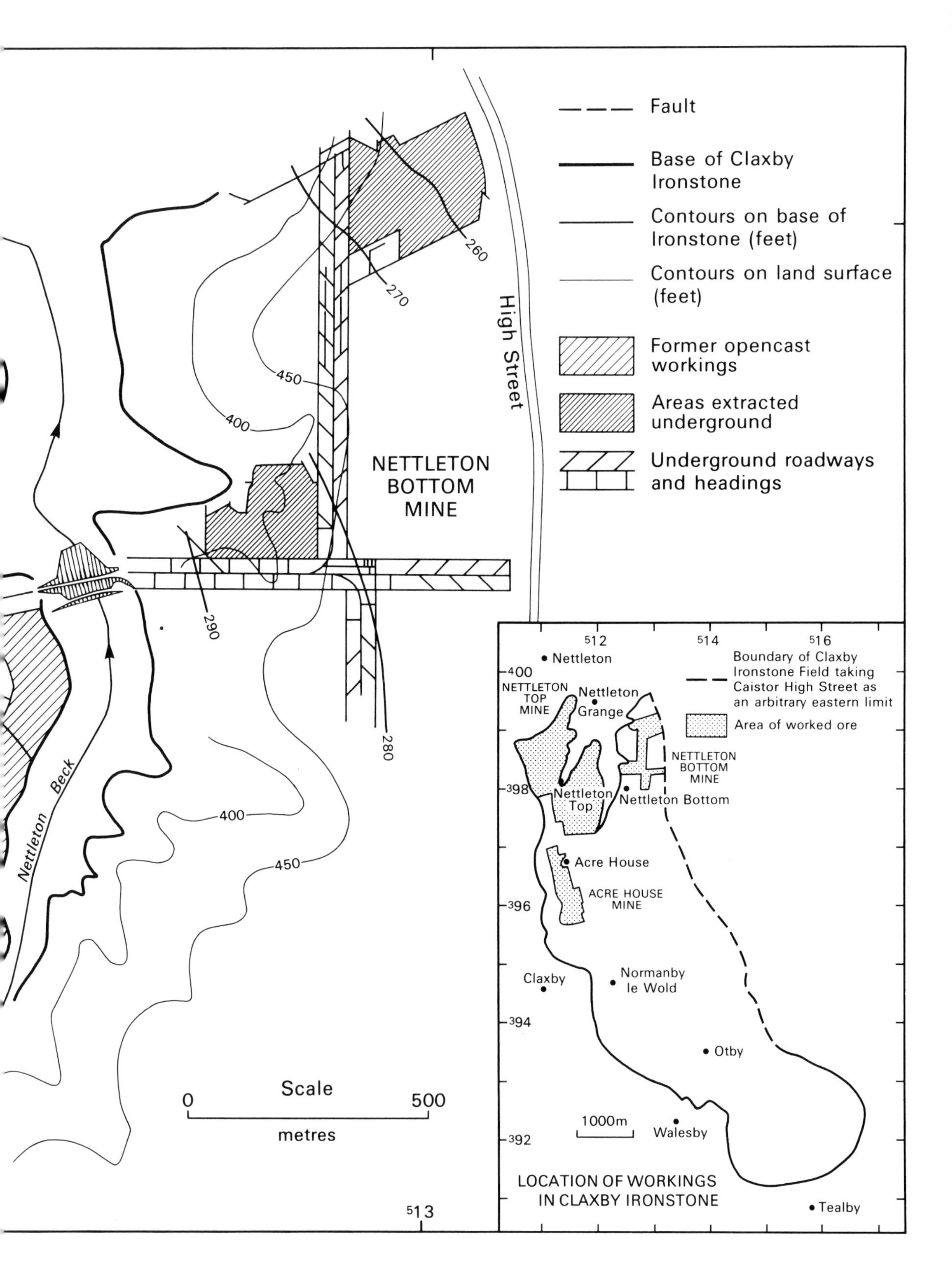
Fault
Base of Claxby Ironstone
Contours on base of Ironstone (feet)
Contours on land surface (feet)
Former opencast workings
Areas extracted underground
Underground roadways and headings
NETTLETON BOTTOM MINE
High Street
Nettleton Beck
260
270
280
290
400
450
Scale
0
500
metres
513
512
514
516
400
398
396
394
392
Nettleton
NETTLETON TOP MINE
Nettleton Grange
Nettleton Top
Nettleton Bottom
NETTLETON BOTTOM MINE
Acre House
ACRE HOUSE MINE
Claxby
Normanby le Wold
Otby
Walesby
Tealby
1000m
Boundary of Claxby Ironstone Field taking Caistor High Street as an arbitrary eastern limit
Area of worked ore
LOCATION OF WORKINGS IN CLAXBY IRONSTONE

low iron content, Frodingham ore suffers from the presence of undesirable constituents, including high phosphorus and alkali contents, which severely limit its application as a ferruginous flux.

Until the mid 1970s the open-hearth steel-making process used at Scunthorpe allowed for the removal of phosphorus from the metal into a basic slag, which was subsequently ground and used on a large scale as a source of phosphate in agriculture. Another important by-product was the large amount of blast-furnace slag, derived from the low-grade Frodingham ore, which was extensively used for roadstone.

Claxby Ironstone

The Claxby Ironstone provided the basis for a medium-sized ore mining operation at two mines at Nettleton near Caistor, which finally ceased production in 1969. The ironstone consists of goethite ooliths set in a fine-grained sideritic and berthierinitic groundmass, more marly in the lower part. The grade of the ore depends to a certain extent on the abundance of the ooliths, and has varied between 20 and 30 per cent Fe. The Claxby ore is siliceous and similar in quality to the poorer types of Northamptonshire ore, the resources of which are so large that there is thought to be little prospect of future Claxby Ironstone workings.

At Nettleton the ironstone worked during the 1960s was typically 4.4 m thick, with the following average composition: Fe 24.5 (weight per cent), SiO_2 16.5, CaO 4.5, P 0.25 and H_2O 18.5. Resources of ore were estimated to be 22 million tonnes when mining ceased, a total of about 6 million tonnes having been extracted. The deposit was worked in Roman times, but extraction did not reach a significant scale until the nineteenth century when some 396 000 tonnes were produced between 1868 and 1881. Later, official output statistics for the working at Nettleton were aggregated with workings from other fields, except for the individual years 1938, 1947 and 1957, which show annual productions of 101 000, 93 000 and 177 000 tonnes respectively.

The earliest large-scale workings were in Acre House Mine (Figure 45), which operated via shaft and adit in the late nineteenth century. In 1919 there were trials in the nearby Walesby–Otby area but, although the ore was of reasonable quality, the cambered strata undulated to such an extent that mining was found to be impractical. Several boreholes put down in 1942–43 in the Nettleton and Walesby areas showed considerable quality variation, and reappraisal of the same data in 1959 concluded that some of them showed no workable ore. Underground drift working in Nettleton Top Mine began in 1929 and continued until 1950. Subsequently there were several opencast operations in the 1950s in the shallower parts of the deposit adjacent to the mine. In 1959 Nettleton Bottom Drift Mine was opened, but by 1969 all extraction had ceased.

No information is available on Claxby ore on the east side of Caistor High Street, and very little on the areas south of Walesby, all of which are excluded from the resource calculation.

Roach Ironstone

No workings are known in the Roach Ironstone, but an analysis of a sample from an exposure 135 m north of Otby House, Walesby, indicates that the Roach is probably of a slightly higher grade than the Claxby Ironstone (c.33 per cent Fe).

LIMESTONE

Apart from the Chalk the only limestone working in the area is around Manton [SE 940 015 and 943 016], where limestones and mudstones of the Lincolnshire Limestone were used in a cement-making operation that closed during the 1970s because of high fuel costs. The limestone quarries now produce crushed and screened material for use as aggregate. The material tends to be susceptible to frost, and is unsuitable for use as concrete aggregate or for coating with bitumen. The Santon Oolite, Raventhorpe Beds and Scawby Limestone are used as hardcore or for foundation material that can be emplaced at a sufficient depth to be insulated from frost damage. The quarries are worked by blasting and face shovel.

OIL SHALE

Bituminous paper shales occur in the Coleby Mudstones above the Marlstone Rock (p.42). Kerogen globules and high yields of gas have been obtained by Mr D Hodgson of Scunthorpe (personal communication) from the sequence at Roxby Mine and Mount Pleasant clay pit.

The Ancholme Clay Group contains several oil shales, grouped into 5 main bands within a vertical distance of about 80 m (Gallois, 1978, 1979). The outcrop of this sequence is largely obscured by Quaternary deposits in the southern part of the district, and the oil shales are cut out by Cretaceous overstep before the Humber is reached. Oil shales were recorded in 1870 immediately beneath the Spilsby Sandstone in Acre House Mine near Nettleton, and kerogen globules have been obtained by Mr D Hodgson (personal communication) from Ancholme Clay samples at 36.2, 50.7 and 256 m depths in Nettleton Bottom Borehole and from an exposure at TF 110 992.

SAND AND GRAVEL

Very little gravel is produced in the district, and the quality is generally inferior to that required for use in concrete. Difficulties arise because of the content of Chalk or Jurassic limestone which, being porous, tends to be susceptible to frost damage. In addition, the dominant flint clasts tend to be flaky, which also limits the usefulness of the deposits.

The sand and gravel resources of the western part of the district (comprising Glacial Sand and Gravel, Older River Gravel (at depth), Vale of York Glacial Lake Deposits, Terrace Deposits, Blown Sand and Alluvium) have been the subject of an assessment survey on behalf of the Department of the Environment (James, 1976; Lovell, 1977; Gozzard and Price, 1978).

Blown Sand is the most important sand resource in the district. It tends to be closely sized, but somewhat lacking in fines for an optimum building sand. Nevertheless, it includes some material adequate for this purpose and is worked at

Fonaby [TA 106 031] near Caistor. It was formerly dug on a smaller scale east-south-east of Wood Farm [SE 8758 0389] and at Sand House Farm [SE 8610 0443] near East Butterwick.

The Vale of York Glacial Lake Deposits are worked only at Kettleby [TA 038 083], where sand and gravel are produced for selected fill.

At Woodshill Quarry, Nettleton [TF 108 996] the Lower Spilsby Sandstone is poorly cemented and composed mainly of closely sized, rounded quartz grains. It is currently worked for sand, mostly for plastering.

SILICA SAND

The Blown Sand deposits of the Messingham area are the principal resource of silica sand in the district. Thicknesses range up to 5 m, with an average of about 2.5 m, and large areas are thus rapidly worked out. At Messingham [SE 911 039] the sands are well sorted, with a very low proportion of fines (5 to 7 per cent $-63\ \mu m$), and consist predominantly of fine- to medium-grained, subrounded to well-rounded quartz grains with some particles of quartzite and chert. They are principally exploited as foundry sands and for the manufacture of coloured glass containers. Processing consists of washing and size classification, principally to remove fines and to adjust the grain-size distribution of the finished product. Physico-chemical properties of the marketable grades of sand are shown in Table 7.

Glacial sand has been excavated in a pit [TA 0430 0133] on Sheepcote Hill, along with laminated clay, as ceramic raw materials.

Kellaways Sand has been worked at Redbourne Hayes [SE 997 002] and near Beck Farm [TF 001 927]. It comprises 5 to 8 m of poorly sorted, fine- to medium-grained sand. Minor quantities of silica sand have also been produced in the past from the Kellaways Beds at South Cave.

Table 7 Typical physico-chemical properties of silica sands, Messingham

	1	2	3	4
SiO_2	95.6	96.0	95.6	95.2
Fe_2O_3	0.25	0.21	0.25	0.26
Al_2O_3	2.12	1.8	2.17	2.41
Cr_2O_3	0.0012	n.a.	n.a.	n.a.
K_2O	1.24	1.0	1.28	1.45
Na_2O	0.15	0.15	0.16	0.15
CaO	0.08	0.1	0.07	0.07
MgO	0.08	0.06	0.08	0.10
Loss on ignition	0.45	0.40	0.44	0.47
Particle size distribution (μm) % retained				
1000	Tr	0.1	Tr	Tr
710	0.3	0.6	0.3	Tr
500	2.2	4.7	1.6	0.2
355	n.a.	22.0	9.3	2.4
250	n.a.	44.7	34.8	28.2
180	n.a.	21.7	35.1	42.7
125	94.3	5.5	14.9	21.7
90	2.3	0.4	2.4	4.1
63	n.a.	Tr	0.1	0.5
-63	n.a.	Tr	0.1	0.2

1 Coloured glass sand
2 Foundry sand (AFS 45)
3 Foundry sand (AFS 55)
4 Foundry sand (AFS 65)

Source: British Industrial Sand Limited

HYDROGEOLOGY

The drainage of the district is into the Humber—on the north bank via the River Hull, on the south bank via the Rivers Trent and Ancholme west of the Wolds, and via minor streams east of the Wolds. To the south of the Humber the responsibility for water supply is shared by the Severn-Trent and Anglian water authorities. For geographical reasons, however, the latter authority supplies all the area to the east of the Trent. Water supply north of the Humber is the responsibility of the Yorkshire Water Authority.

Kingston upon Hull and Scunthorpe are the main demand centres within the area. Kingston upon Hull receives its water supply from an intake on the River Hull at Hempholme Lock and from the River Derwent at Barmby Barrage, respectively north and west of the district, and from boreholes in the Chalk immediately north and west of the city. Scunthorpe is supplied from boreholes in the Lincolnshire Limestone east of Winterton. Boreholes in the Chalk at and around Barrow upon Humber supply the remainder of the district together with potable demands to the east. Surface-water abstraction for public and industrial supply is restricted to the River Ancholme at Cadney. The water is pumped to, and treated at Elsham. It was used for nonpotable supplies in the Immingham area to the east until 1990 when the works were upgraded to supply potable water. In addition to public supply, surface water and groundwater are used by private abstractors for industry and agriculture, the latter mainly for spray irrigation.

Average annual precipitation, affected by topography, ranges from 600 to 650 mm in Lincolnshire and approaches 700 mm on the Yorkshire Wolds. Mean monthly rainfall varies from 40 to 70 mm, but seasonal variation in potential evapotranspiration (annual average is 400 to 500 mm) results in a soil-moisture deficit between May and August and a water-balance surplus between November and March. Recharge to aquifers from precipitation thus mainly takes place during the period of surplus, although rises in water levels after heavy summer storms suggest fissure-flow recharge in the Chalk (Anon, 1978).

Average annual infiltration, and therefore recharge, depends partly on the nature of the aquifer and its cover, if any. At outcrop approximately 260 mm are estimated to infiltrate the Chalk (Anon, 1969, 1971), reducing to 210 mm where covered by alluvial deposits and glacial sands and gravels, and to 25 mm where overlain by till. Approximately 260 mm infiltrate the outcrop of Spilsby Sandstone. Infiltration into the Cave Oolite is estimated to be 125 mm, reducing to 90 mm where covered by alluvial deposits; approximately 195 mm infiltrate the Lincolnshire Limestone at outcrop.

The hydrogeology of the district was originally reviewed by Fox-Strangways (1906) to the north of the Humber, and Woodward (1904) to the south. It is graphically summarised by the 'Hydrogeological Map of North and East Lincolnshire' (Day et al., 1967) and the 'Hydrogeological Map of East Yorkshire' (Taskis et al., 1980).

Quaternary deposits

Till and other impermeable Quaternary deposits confine the underlying Chalk and result in artesian conditions where not affected by abstraction. Where the cover is thin or relatively permeable, Chalk groundwater discharges via springs, locally called blow wells, near Barton-upon-Humber. South of the Humber, glacial sand and gravel directly overlie the Chalk in places and form part of that aquifer: along the confined/unconfined boundary they provide important storage with seasonal movement of the boundary. The Quaternary deposits do not generally contain a groundwater resource, but they may provide limited recharge to the underlying aquifer by induced leakage if water levels fall due to pumping. The bed of the Humber consists predominantly of postglacial deposits that generally form a low-permeability layer between the Chalk and the estuary water. However, small windows of solid rock and localised areas of permeable cover sediment permit hydraulic continuity between the aquifer and the estuary.

Cretaceous

The Chalk is the major aquifer of the district and has been extensively utilised on both sides of the Humber. To the north of the Humber the hydrogeology has been summarised by Foster and Milton (1976) and Foster et al. (1976), and to the south by the University of Birmingham (Anon, 1978). The thickness of the active part of the aquifer is controlled less by stratigraphy than by depth and erosion/solution history. Permeability is enhanced near ground surface by water-table fluctuations, weathering and Pleistocene cryoturbation. The topmost layer of broken chalk ('chalk bearings'), up to 10 m thick locally, increases permeability; conversely, impermeable 'putty chalk' up to 3 m thick also occurs. Rockhead is affected by buried channels and the concealed Ipswichian cliff, which result in very high permeability zones associated with deep weathering and possibly, in the latter case, with beach deposits.

The hydraulic properties of the rock matrix include moderate (0.15 to 0.25) and occasionally high (0.4) porosity, and very low intergranular permeability (10^{-4} to 10^{-3} m/d) and specific yield (<0.003) (Foster and Milton, 1976). Most of the flow and groundwater storage is confined to joints and fissures, enlarged locally by solution. In consequence yields are very variable, and drilling can be speculative; even so, modern large diameter (>0.5 m) boreholes typically yield 50 to 80 litres per second (l/s), and old shafts with adits can produce many times this quantity. Transmissivities are also very variable, but usually lie within the range 100 to 2000 square metres per day (m^2/d). Exceptional values of transmissivity of up to 10 000 m^2/d occur at some sites and these are used for public supply. The storage coefficient ranges from 10^{-3} to 10^{-5}.

Natural discharge from the aquifer is via escarpment springs and dip-slope springs at the contact with confining cover. Springs still flow at Barrow upon Humber and Kirmington, but pumping has virtually dried up all dip-slope springs north of the Humber. Discharge also occurs via blow-wells through the confining cover and into the bed of the Humber. Blow-wells occur along the coast at Barton-upon-Humber, along the River Skitter, and, from hearsay evidence, as perennial springs offshore at Hessle.

The Chalk is, in places, in hydraulic continuity with underlying minor aquifers such as the Carstone and Elsham Sandstone. Water passes in either direction, depending on the relative hydraulic head. These minor aquifers also provide local supplies. Between North Newbald and Market Weighton the Chalk lies directly on the Cave Oolite, and water leaks into the older formation.

Near outcrop the Chalk aquifer contains a 'hard' calcium bicarbonate water with a total dissolved solid (TDS) content of between 250 and 600 milligrams per litre (mg/l) and a chloride ion concentration of 25 mg/l. Water quality deteriorates slightly down-dip due to natural maturation, and saline intrusion occurs along the shore of the Humber. Excessive pumping has induced saline intrusion over an area of approximately 50 km^2 beneath Kingston upon Hull, with chloride values attaining 10 000 mg/l near the mouth of the River Hull. Eastwards from New Holland, along the southern bank of the Humber, saline water is again present in the Chalk. It appears to represent residual water trapped in the aquifer during periods of high sea level (Anon, 1978).

Jurassic

The Spilsby Sandstone is of major importance as an aquifer farther south in Lincolnshire, where it provides good quality water. It has a mean transmissivity of 70 m^2/d (Price, 1957; Anon, 1971) and boreholes typically yield 20 l/s. The formation thins northwards and is overstepped by the Carstone at Grasby. It is, therefore, of minor importance within the district and is used only for a few private supplies.

The Elsham Sandstone, a minor aquifer in the Ancholme Clay Group, is locally in hydraulic contact with the Chalk and Carstone. It is utilised locally in South Humberside where the resources of the Chalk aquifer are considered to be so fully committed that further abstraction licences are not being issued.

The Lincolnshire Limestone is the main aquifer of several in the Middle Jurassic sequence (Foster, 1968; Downing and Williams, 1969; Barker, 1984). The higher aquifers are the Brantingham Formation, the Kellaways Sand, the Cornbrash, the Snitterby Limestone and the Thorncroft Sands, and close below the Lincolnshire Limestone are the Grantham Formation (an aquifer in part) and the Northampton Sand. These aquifers are normally separated from one another by siltstones and mudstones, but hydraulic continuity is common. To the north of Brough the sequence from the Kellaways Rock to the Lincolnshire Limestone may be treated as a single aquifer; so may the Lincolnshire Limestone and the Northampton Sand south of the Humber, in localities where the Grantham Formation is absent or incomplete. Faulting, particularly to the north of the Humber, further complicates the hydraulic relationships of these aquifers, either forming barriers between or connecting aquifers: thus the Brantingham Formation is faulted against, and is likely to be in hydraulic contact with the Kellaways Sand to the north-east of Brough.

The Hibaldstow Limestones and the equivalent Cave Oolite to the north of the Humber form the principal aquifer in the Lincolnshire Limestone. They are predominantly well-cemented, compact, oolitic limestones, in which fissures and fractures provide the main flow-paths and storage, although isotopic evidence suggests that 'inter-granular' and micro-fissure storage is significant (Downing et al., 1977). As is usual in fractured aquifers, yields are variable and drilling is somewhat speculative, but boreholes in the Cave Oolite usually yield about 6 l/s and exceptionally up to 39 l/s. South of the Humber the Lincolnshire Limestone generally yields under 25 l/s with exceptional yields of up to 93 l/s. Groundwater flow in this area is predominantly down-dip to the east, except near outcrop along the estuary, where flow is towards the Humber. To the south of Brigg transmissivity values are generally less than 450 m^2/d, but to the north they range from 450 to 900 m^2/d. The higher values are thought to reflect the proximity of the Humber, which may have encouraged groundwater movement and fissure development during the Quaternary. Where the Lincolnshire Limestone aquifer is confined, the storage coefficient ranges from 10^{-3} to 10^{-5}; where unconfined a specific yield of 0.05 is estimated.

At outcrop the water quality is a 'hard' calcium bicarbonate type containing significant sulphate. Hardness decreases down-dip beneath the confining beds and TDS values of 1000 mg/l are found 4 to 6 km from outcrop, persisting almost to outcrop just to the north of Brigg. The public supply at Winterton Holmes [SE 959 178] (just over 1 km from outcrop) has a high iron content which is reduced by treatment, and the water is softened.

North of the Humber groundwater flows in a south-easterly direction towards the Humber, and is intercepted by boreholes at the Capper Pass Works which tap the Brantingham Formation/Kellaways Sand sequence and the Cave Oolite. Induced saline intrusion has occurred and varies according to abstraction rate and recharge. Chloride levels in the Brantingham Formation/Kellaways Sand have currently steadied at 100 to 200 mg/l, but have been as high as 1700 mg/l. In the two boreholes that tap the Cave Oolite the chloride levels have steadied at approximately 60 mg/l and 300 mg/l and are related to the distance of the holes from the Humber.

The lower Jurassic strata are predominantly mudstones forming confining beds, but the well-jointed Frodingham Ironstone, Pecten Ironstone, Marlstone Rock and Scunthorpe limestones are the source of small springs and a few domestic water supplies, though the water tends to be rich in iron or lime.

Triassic

The Penarth and Mercia Mudstone groups can be regarded as an impermeable confining cover to the Sherwood Sandstone, a major aquifer in adjoining areas. Within the district this aquifer thickens north-eastwards from 300 to 450 m, and its top dips eastwards from 200 to 700 m below OD. At Gainsborough, a little to the south-east of the district, the aquifer is used for public supply and, where confined, transmissivities range from 350 to 850 m^2/d. The water quality is good, but has relatively high sulphate and chloride concentrations, which are likely to increase down-dip.

APPENDIX 1

List of principal boreholes and wells

Confidential logs are marked with an asterisk. Non-confidential logs can be consulted at the National Geosciences Data Centre, BGS, Keyworth.

Borehole and date of drilling	National Grid reference	BGS registration number	Surface level (m) above OD	Stratigraphical range (excluding Quaternary)	Depth (m)
BGS boreholes					
Alandale (1972)	TA 0007 2584	TA 02 NW/68	7.6	Ancholme Clay Group to Glentham Formation	117.5
Blyborough (1972)	SK 9206 9428	SK 99 SW/79	22.3	Coleby Mudstones to Mercia Mudstone Group	167.0
Brough No. 1 (1972)	SE 9443 2677	SE 92 NW/61	9.4	Kellaways Beds to Coleby Mudstones	38.3
Brough No. 2 (1972)	SE 9510 2683	SE 92 NE/41	13.4	Kellaways Beds to Glentham Formation	37.7
Cockle Pits (1972)	SE 9323 2865	SE 92 NW/60	16.4	Lincolnshire Limestone to Mercia Mudstone Group	137.5
Dam Road (Barton) (1972)	TA 0269 2246	TA 02 SW/42	c.4.0	Ferriby Chalk to Ancholme Clay Group	80.1
East Clough (1972)	SE 9757 2490	SE 92 SE/20	5.3	Ancholme Clay Group to Lincolnshire Limestone	75.7
Elsham Quarry (1972)	TA 0371 1323	TA 01 SW/90	c.73.0	Welton Chalk to Elsham Sandstone	46.5
Low Farm (South Ferriby) (1972)	SE 9597 2128	SE 92 SE/19	3.7	Kellaways Beds to Lincolnshire Limestone	46.5
Low Field Lane (1972)	SE 9637 2629	SE 92NE/42	14.0	Brantingham Formation to Lincolnshire Limestone	48.3
Osgodby (1973–4)	TF 0799 9285	TF 09 SE/3	18.0	Ancholme Clay Group	107.7
Welton Crossing (1972)	SE 9552 2613	SE 92 NE/43	8.7	Kellaways Beds to Lincolnshire Limestone	35.0
Welton Waters 'A' (1972)	SE 9544 2476	SE 92 SE/17	5.2	Kellaways Beds to Lincolnshire Limestone	32.7
Welton Waters 'B' (1972)	SE 9625 2467	SE 92 SE/18	7.0	Kellaways Beds to Glentham Formation	24.6
Whitton Ness (1972)	SE 9230 2434	SE 92 SW/52	3.4	Frodingham Ironstone to Scunthorpe Mudstones	37.9
Winteringham Haven (1972)	SE 9353 2303	SE 92 SW/53	5.2	Marlstone Rock to Pecten Ironstone	30.7
Worlaby E (1959)	SE 9975 1596	SE 91 NE/45	c.22.0	Ancholme Clay Group to Frodingham Ironstone	213.5
Worlaby G (1960)	TA 0292 1621	TA 01 NW/15	c.67.0	Welton Chalk to Ancholme Clay Group	142.8
Hydrocarbon boreholes					
Blyton (1960)	SK 8435 9555	SK 89 NW/1	c.4.9	Mercia Mudstone Group to Namurian	1823.3
Brigg No. 1 (1981)	TA 0377 0639	TA 00NW/122	5.6	Ancholme Clay Group to Dinantian	1936.9
Burton upon Stather (1965)	SE 8787 1883	SE 81 NE/2	c.61.0	Frodingham Ironstone to Dinantian	1857.8
Butterwick (1958)	SE 8421 0563	SE 80 NW/1	c.3.1	Mercia Mudstone Group to Dinantian	1694.1
Broughton (1984)*	SE 9463 1076	SE 91 SW/456	63.1	Lincolnshire Limestone to Dinantian	1920.0
Corringham No. 3 (1961)	SK 8905 9352	SK 89 SE/110	c.19.8	Scunthorpe Mudstones to Wesphalian	1626.4

Corringham No. 5 (1959)	SK 8929 9326	SK 89 SE/111	c.19.8	Scunthorpe Mudstones to Namurian	1607.8
Corringham No. 7 (1960)	SK 8963 9300	SK 89 SE/113	c.21.3	Scunthorpe Mudstones to Dinantian	1836.1
Corringham No. 8 (1961)	SK 8967 9362	SK 89 SE/114	c.18.3	Scunthorpe Mudstones to Namurian	1615.7
Corringham No. 9 (1961)	SK 8996 9334	SK 89 SE/115	c.17.7	Coleby Mudstones to Namurian	1642.6
Corringham No. 10 (1975)*	SK 8933 9359	SK 89 SE/121	15.6	Scunthorpe Mudstones to Namurian	1684.0
Nettleton (1972)	TF 1185 9642	TF 19 NW/53	162.6	Ferriby Chalk to Westphalian	1555.7
South Cliffe (1973)	SE 8791 3522	SE 83 NE/8	c.9.5	Mercia Mudstone Group to Westphalian	1070.1
Geothermal borehole					
Nettleton Bottom (1979)	TF 1249 9820	TF 19 NW/54	101.1	Ferriby Chalk to Mercia Mudstone Group	640.2
Water wells					
Barnetby Station (1915)	TA 0527 0983	TA 00 NE/4	18.6	Ancholme Clay Group to Coleby Mudstones	200.3
Burton upon Stather (1925)	SE 8750 1820	SE 81 NE/1	64.6	Scunthorpe Mudstones to Mercia Mudstone Group	81.1
Melton Bottoms 'A' (1920)	SE 9746 2739	SE 92 NE/8A	c.32.0	Welton Chalk to Lincolnshire Limestone	147.8
Melton Bottoms 'B' (1920)	SE 9751 2742	SE 92 NE/8B	31.7	Ancholme Clay Group to Coleby Mudstones	125.6
Coal boreholes					
Crosby (1907–12)	SE 8745 1225	SE 81 SE/8	4.6	Penarth Group to Westphalian	1139.3
North Ewster (1977)*	SE 8363 0389	SE 80 SW/30	4.2	Mercia Mudstone Group to Westphalian	1235.5
Rock Abbey (1979)*	SK 8747 9533	SK 89 NE/117	20.8	Scunthorpe Mudstones to Westphalian	1371.9

APPENDIX 2

List of Geological Survey photographs

Copies of these photographs are deposited for public reference in the Library of the British Geological Survey, Keyworth, Nottinghamshire NG12 5GG. Prints are available on application. The photographs belong to Series A and L as indicated.

SHEET 80 KINGSTON UPON HULL

A 4588	General view of working face in Frodingham Ironstone at Berkeley Pit, Frodingham
A 4589 and 4590	Overburden of Coleby Mudstones and Blown Sand on Frodingham Ironstone at Berkeley Pit No. 2, Frodingham
A 5427–29	Submerged forest and peat on foreshore at North Ferriby
A 8397	Section in thinly bedded Frodingham Ironstone, Thealby Mine
A 8441	A block of Frodingham Ironstone covered with *Cardinia* at Thealby Mine
A 14377 and 14378	Abandoned opencast working for Frodingham Ironstone at Bagmoor Farm, showing Coleby Mudstones with Pecten Ironstone and Marlstone Rock
A 14379	Jurassic–Cretaceous sequence in South Ferriby Quarry
A 14380 and 14381	Burnham Chalk (with tabular flints) on Welton Chalk at Burnham Lodge Quarry
L 1578	Chalk crushing plant at South Ferriby Cement Works Quarry
L 1579	North face of South Ferriby Cement Works Quarry showing multiple oblique jointing
L 1580	East face of South Ferriby Cement Works Quarry
L 1581	Melton Chalk Pit
L 1582–84	Joints and fractures in Welton Chalk, Elsham Quarry
L 1585–88	Repeated step-faults of superficial origin in Welton Chalk, Barton-upon-Humber Quarry
L 1589	Site of South Pier of Humber Bridge at Barton Haven
L 1590	Site of South Anchorage of Humber Bridge at Barton Haven
L 1591	Chalk dug from the Humber Bridge approach-road cutting
L 1592 and 1593	Humber Estuary from North Ferriby Landing
L 1688	Workings in Frodingham Ironstone, Winterton Opencast Mine
L 2951	Ellerker Limestone forming waterfall in garden of Brooklyn House, Ellerker
L 2952	Lower part of Raventhorpe Beds in garden of Brooklyn House, Ellerker
L 2953	Quaternary deposits at Red Cliff, North Ferriby
L 2954	Devensian deposits at Red Cliff, North Ferriby
L 2955	Glaciolacustrine clay at Red Cliff, North Ferriby

SHEET 89 BRIGG

A 4591–3	Frodingham Ironstone, Yarborough Pit, south-east of Scunthorpe
A 6315	Interglacial deposits, Kirmington Brick Pit
A 8396	Section in working face of Frodingham Ironstone, Yarborough Pit, south-east of Scunthorpe
A 8442	Kirton Cementstones in quarry at Cleatham
A 8443	Kirton Cementstones in railway cutting at Kirton in Lindsey
A 8444	View south along Cretaceous escarpment near Nettleton, showing scarp features of the Spilsby Sandstone and Tealby Limestone
A 8445	Scarp features of Spilsby Sandstone and Tealby Limestone at Normanby-le-Wold
A 8448	Section in Welton and Ferriby Chalk in quarry east of Nettleton Church
A 14375 and 14376	Lincolnshire Limestone: uppermost beds of Kirton Cementstones at Kirton Quarries
L 1686	Kirton Cementstones resting on Santon Oolite in quarry near Kirton in Lindsey
L 1687	Marlstone Rock overlying Coleby Mudstones in quarry at Cleatham, near Kirton in Lindsey

APPENDIX 3

Notes on the marker horizons in the Northern Province Chalk

These notes are supplementary to those already documented by Wood and Smith (1978). An asterisk after a name indicates the type locality. NB: in the event of Willerby Quarry becoming back-filled, Little Weighton Quarry can serve as a replacement type locality for the Kirk Ella Marl and Willerby Flints.

Nettleton Pycnodonte Bed Dark grey silty marl 2 to 3 cm thick, with abundant isolated and associated valves of a small *Pycnodonte* (*Phygraea*) sp. (the *Gryphaea vesicularis* bed of Bower and Farmery, 1910, and the correlative of the 'Pycnodonte Event' of northern Germany); rests on a cemented erosion surface and grades up into the Nettleton Stone (q.v.). Named after Nettleton Bottom Quarry* [TA 125 981]. Can be observed in all the larger Cenomanian sections, notably Mansgate Quarry [TA 123 002], Bigby Quarry [TA 059 078], South Ferriby Quarry [SE 991 204]; the bed extends as far south as northern Norfolk.

Nettleton Stone Prominent massive bed of hard, grey, sand-grade chalk just over 1 m thick, overlying and grading up from the Nettleton Pycnodonte Bed (q.v.), and overlain by a thin marl. Corresponds to the 'Nettleton Member' of Jeans (1980) and correlates with the massive chalk bed overlying the 'Pycnodonte Event' of northern Germany. Becomes less well developed towards Norfolk. Named after Nettleton Bottom Quarry* [TA 125 981], but also seen in the other localities listed for the Nettleton Pycnodonte Bed.

Chalk Hill Marls Three marl seams within 25 cm, marking the top of the shell-detrital chalks at the base of the Welton Chalk. The lowest is typically up to 1 cm thick and locally thicker, brown and silty, with much comminuted inoceramid shell, and overlying a bed rich in shell detritus and complete valves of *Mytiloides* sp. The upper two marls are less than 0.5 cm thick, and are black and buttery. Rare, small (1 cm) nodular flints occur between the two black marls, these being the lowest flints seen in the Welton Chalk. Close-set, subvertical to high-angle joints in the immediately overlying bed produce the 'columnar bed' of the Old Series Geological Survey memoirs for the area (e.g. Jukes-Browne and Hill, 1903, pp.478–479, figs. 81, 82). Named after Chalk Hill, which formerly occupied the site of the Melton Ross (working) Quarry*, where fresh, unweathered sections can be examined near the base of the lowest working faces [c.TA 077 110]. Note that the grid reference given for this quarry as [TA 082 112] by Wood and Smith (1978, Appendix, p.277) referred to a very early stage of development of the present enormous working quarry. Extensive weathered sections are found on the back wall of the Elsham lower pit [TA 038 131]; also at Leggat's Quarry, South Ferriby [TA 000 217] and Grasby [TA 089 052] (see Wood and Smith, 1978, fig. 4, where the marls are shown, but not named, immediately beneath the first flints).

Hall Farm Marl A thin, less than 0.5 cm marl found c.2.5 m above the First Main Flint and underlain by a 0.25 m-thick bed with a basal 0.10 m concentration of 'rubbly', locally somewhat iron-stained, indurated chalk with marl envelopes. Conspicuous in quarry faces, locally with an underlying 'columnar bed' comparable with that over the Chalk Hill Marls. Named after the farm adjacent to the Melton Ross (working) Quarry*, where it is observable, but relatively inaccessible, over several hundred metres of section. Best examined at the Elsham upper pit [TA 036 133] and at Leggat's Quarry, South Ferriby.

Yarborough Marl A marl complex, up to 0.25 m thick, immediately underlying the Ferruginous Flint and forming a marked recess in quarry faces. Named after the Yarborough Camp earthworks overlooking the Melton Ross eastern Quarry*, where it is readily accessible over several hundred metres on the north-eastern [e.g. TA 091 120] and south-eastern [e.g. TA 091 117] faces. Observable in all localities given for the Ferruginous Flint by Wood and Smith (1978, Appendix, p.277).

Croxton Marl Thin (up to 1 cm) marl midway between the Grasby Marl and Barton Marl 1; best developed in the now inaccessible underground LPG storage caverns at South Killingholme [TA 175 175], 3 km north of Immingham, where it is underlain by small, closely spaced burrow-form flints. Also seen in a pit at Swallow Vale Farm [TA 175 043] in the Grimsby (Sheet 90) district and in Melton Ross (working) Quarry* [TA 082 112]; in both localities the underlying flints are inconspicuous and widely spaced, locally failing entirely. Marked by an overhang in quarry faces and expressed as a minor low-resistivity spike in resistivity logs. Named after the village and parish of Croxton, north-east of the type locality.

Easthorpe Tabular Flints Three prominent tabular flints in about 1 m of chalk, showing great lateral variation in development, and overlain by a 2 m bed of laminate chalk with small, inconspicuous, flattened burrow-form flints. *Echinocorys* and *Micraster* common in flaser-bedded chalks with *Zoophycos* beneath lowest flint; *Echinocorys* and inoceramids common between lowest and middle flint. Best developed in Enthorpe railway cutting* [SE 9138 4594] in the Beverley (Sheet 72) district where they comprise upper and lower, strongly projecting, non-carious, rectangular flints, respectively 15–20 and 9–15 cm thick, separated by a thinner tabular flint; top tabular double. Seen at Willerby Quarry [TA 014 313], where they are respectively 8, up to 15 and up to 10 cm thick, and contain many inclusions and holes filled with hard chalk; also at Barrow eastern quarry [TA 071 203], where the lowest flint is a continuous tabular up to 6 cm thick, the middle flint is a semitabular occupying a variable position and up to 12 cm thick, and the top flint is double, comprising a lenticular tabular with smaller lenticular tabulars fused to its base. Formerly visible in the Ashby Hill Quarry [TA 240 006] in the Grimsby (Sheet 90) district. Named after the parish of Easthorpe, adjacent to the type locality.

Barrow Flints and nodular chalks Two giant flints, each locally up to 30 cm thick, associated with nodular chalk and separated by laminate chalk with flattened 5 cm-thick lenticular flints. Variable development. At the Barrow eastern quarry* [TA 071 203], the lower flint is complex, comprising fused overlapping lenticular flints totalling 15 cm, while the upper is an overgrown thalassinoid burrow-form flint with a flat rounded top and rounded protuberances below; uniquely at this locality the lower flint is underlain by a complex lenticular flint up to 25 cm thick. At Willerby Quarry [TA 014 313], the lower flint is lenticular and up to 15 cm thick, with many inclusions; the upper is similar and situated beneath yellow-stained indurated nodular chalk approaching the character of a chalkstone. At Ashdale House Pit [TA 1006 1624], the two flints attain their maximum recorded thickness, and the nodular chalk associated with the lower flint yields numerous large *Cremnoceramus deformis*. Formerly visible at the top of the Ashby Hill Quarry [TA 240 006] and tentatively identified in the East Lutton Plantation Pit [SE 9478 6908] in the Beverley (Sheet 72) district.

Kirk Ella Marl Thin, pale brown marl, up to 1 cm thick, underlain by cream and brown, irregular, partially silicified flints typically 10 cm, but locally up to 15 cm thick. Easily recognised marker horizon, but can be confused with the very similar Little Weighton Marl 1 (q.v.) in the absence of the overlying Willerby Flints. Seen in Willerby Quarry* [TA 014 313], at the base of Little Weighton Quarry [SE 981 333] and in the small pit north-east of the Enthorpe railway cutting [SE 9196 4640]. Named after the village of Kirk Ella, south of the type locality.

Willerby Flints Three thick, conspicuous, non-carious flints within some 4 m of chalk. Willerby Flint 1 is lenticular to semitabular, up to 15 cm thick, and overlain by massive chalk with small thalassinoid burrow-form flints and numerous *Echinocorys* and *Micraster bucaillei*. Willerby Flint 2 is a giant (up to 25 cm) tabular with an undulating base and a flat, marl-coated top, succeeded 0.18 m above by a thin marl and lenticular flint; *M. bucaillei* is locally (Little Weighton Quarry) abundant immediately beneath Flint 2. Flint 3 is tabular, up to 10 cm thick and overlain by a bed of yellow indurated chalk with a burrow-form flint and a sharply defined upper limit. Found at Willerby Quarry* [TA 014 313], and in the other localities given for the Kirk Ella Marl.

Riplingham Tabular Flints Three prominent tabular flints within 1.75 m of chalk, recalling the Triple Tabular Flints near the base of the Burnham Chalk but, in contrast to the latter, with the top rather than the bottom flint the thickest and most continuous. Best seen in the lowest face at Little Weighton Quarry* [SE 981 333], where the lowest flint is up to 13 cm thick but may be locally semitabular, discontinuous and very carious; the middle flint is 10 cm thick and overlies yellow indurated chalk as does the top flint, which is 14 cm thick. Also present near the top of Willerby Quarry [TA 014 313], where the middle flint is carious and the top flint is only 8 – 10 cm thick. Formerly visible in the now backfilled railway-cutting [SE 972 337] north-west of Little Weighton Quarry, where *Micraster bucaillei* occurred commonly in the Chalk between the flints. Named after Riplingham Grange, 1 km west of the railway cutting.

Little Weighton Marls Three marls within 3 m of chalk, the lower two each 2 cm thick and forming marked low-resistivity 'spikes' on resistivity logs (the 'Conoco Marls' of Barker and others, 1984); the lowest marl recalls the Kirk Ella Marl in that it rests on an incompletely silicified flint (and is locally overlain by a similar flint). This marl is separated from the middle one by a bed of massive chalk with a continuous (up to 14 cm) tabular flint. The upper marl is overlain by a concentration of valves of *Volviceramus koeneni* in indurated yellow ferruginous chalk. Best seen in Little Weighton Quarry* [SE 981 333], and also present near top of Willerby Quarry [TA 014 313].

Rowley Marls Four marls, each up to 2 cm thick, within 3 m of massive, inoceramid shell fragment-rich chalk, together with several minor marl-coated stylolitic surfaces. Considerable lateral variation in the thickness and persistence of the individual marls. Marl 1 overlies a fossil-bed rich in echinoids and inoceramids. Marls 3 and 4 are replete with inoceramid shell fragments, the former resting locally (Little Weighton Quarry) on vertical 'potstone' type flints; Marl 4 normally rests on a non-carious, lenticular flint, but in the cutting [SE 986 332] south-east of Little Weighton Quarry the flint is large and carious, and the marl is absent. Seen in Little Weighton Quarry* [SE 981 333] (where the lowest marl is at present obscured), the cutting to the south-east and in the Eppleworth Farm cutting [c.TA 021 324], where the fossil bed beneath Marl 1 is particularly well developed. Named after the Rowley Estate, south-west of the type locality, which also gives its name to the parish.

De La Pole Flint Flat-topped, continuous tabular flint 14 cm thick, markedly carious in upper half; overlies incompletely silicified, rounded lenticular flints up to 15 cm thick in the Eppleworth Quarry* [TA 021 324]. Also seen near top of Little Weighton Quarry [SE 981 333], where the flint is completely carious, 12 cm thick, and the underlying flint has not been observed. Named after the hospital and adjacent De-La-Pole Farm [sic] south-east of the type locality.

APPENDIX 4

Ancholme Clay Group Biostratigraphy

The mapping of the dominantly argillaceous Ancholme Clay Group in the Hull - Brigg district has focussed on the recognition of up to three predominantly arenaceous units which have been given formation status. The relatively thick (up to 170 m) intervening and succeeding mudstone sequences have not been named (p.58). Correlation with clay formations outside the district is possible using biostratigraphical evidence from the Nettleton Bottom Borehole (Figures 46, 47) and scattered surface and shallow borehole sections in the district (Figure 49).

The Ancholme Clay Group succession in Nettleton Bottom Borehole has been used as a standard in several applied stratigraphical studies, for example on palynology (Riding, 1987), calcareous micropalaeontology (Wilkinson, 1983) and geophysics (Penn et al., 1986); the details of the section on which these are based have not yet been published.

The sequence in Nettleton Bottom Borehole was logged and collected by Dr T P Fletcher. The stratigraphical interpretation presented here, based upon his log and specimens (BGS specimen numbers BLJ 5001 - 7713; BLK 4482 - 4486), uses the published standard bed-numbered Oxfordian and Kimmeridgian sequences of eastern England (Gallois and Cox, 1976, 1977; Cox and Gallois, 1979, emend. 1981). These standard beds are based on combined lithological and macrofaunal characters. No comparable standard exists for the Callovian sequence, which is well known from the extensive brickpits near Peterborough and Bedford (Callomon, 1968). The Ancholme Clay Group succession in the Nettleton Bottom Borehole is summarised in Figures 46 and 47, by reference to the standard beds of the West Walton Beds, Ampthill Clay and Kimmeridge Clay, indicated by circled numbers, and to the lithostratigraphical divisions mapped within this district. Minor adjustments have been made to the position of the formation and subformation boundaries of Bradshaw and Penney (1982) given in their preliminary report on the borehole; in particular, their Ampthill Clay - Kimmeridge Clay boundary has been raised from 177.00 m to 165.65 m. The equivalents of the upper part of the Ampthill Clay and lower part of the Kimmeridge Clay are much thicker here than in areas to the south. Correlations have been achieved between the

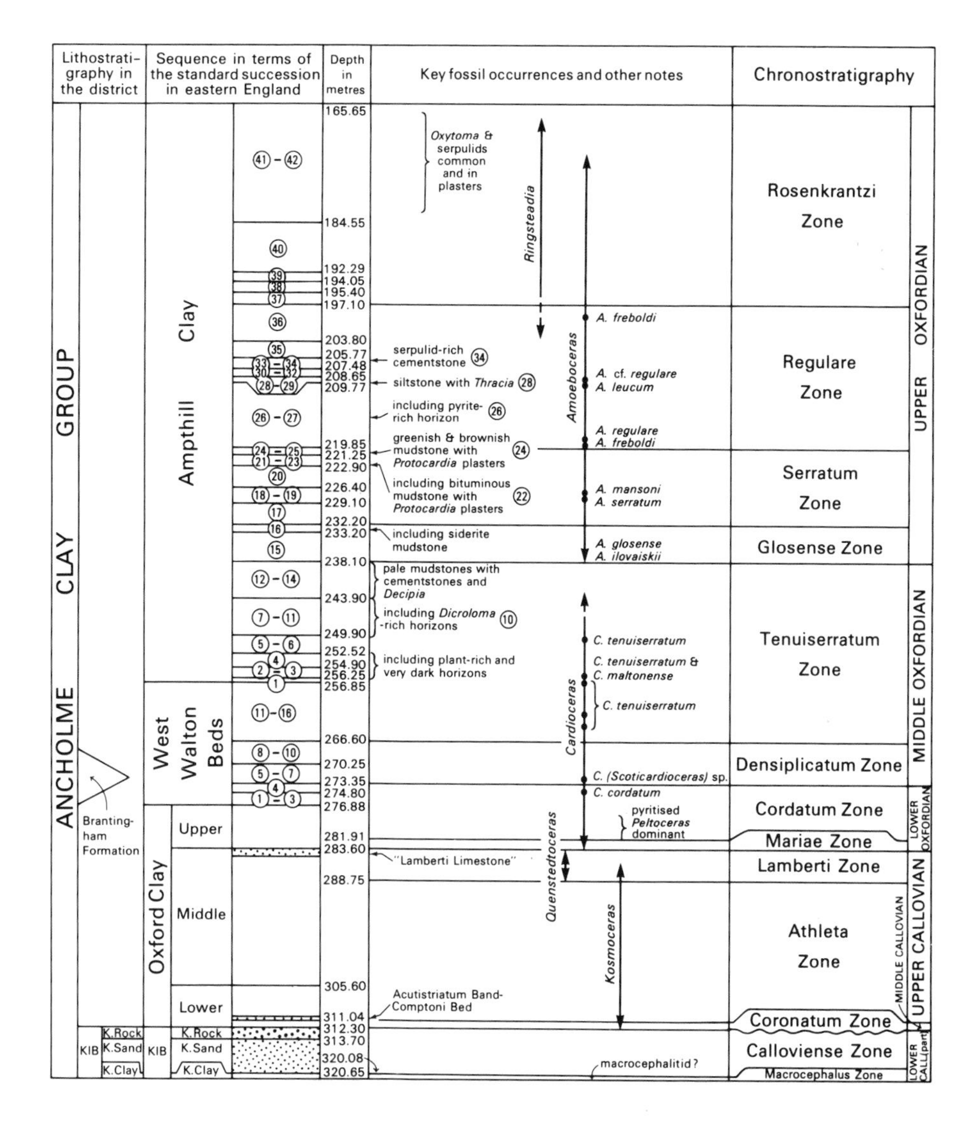

Figure 46 Ancholme Clay Group in Nettleton Bottom Borehole: summary of the Callovian and Oxfordian sequence

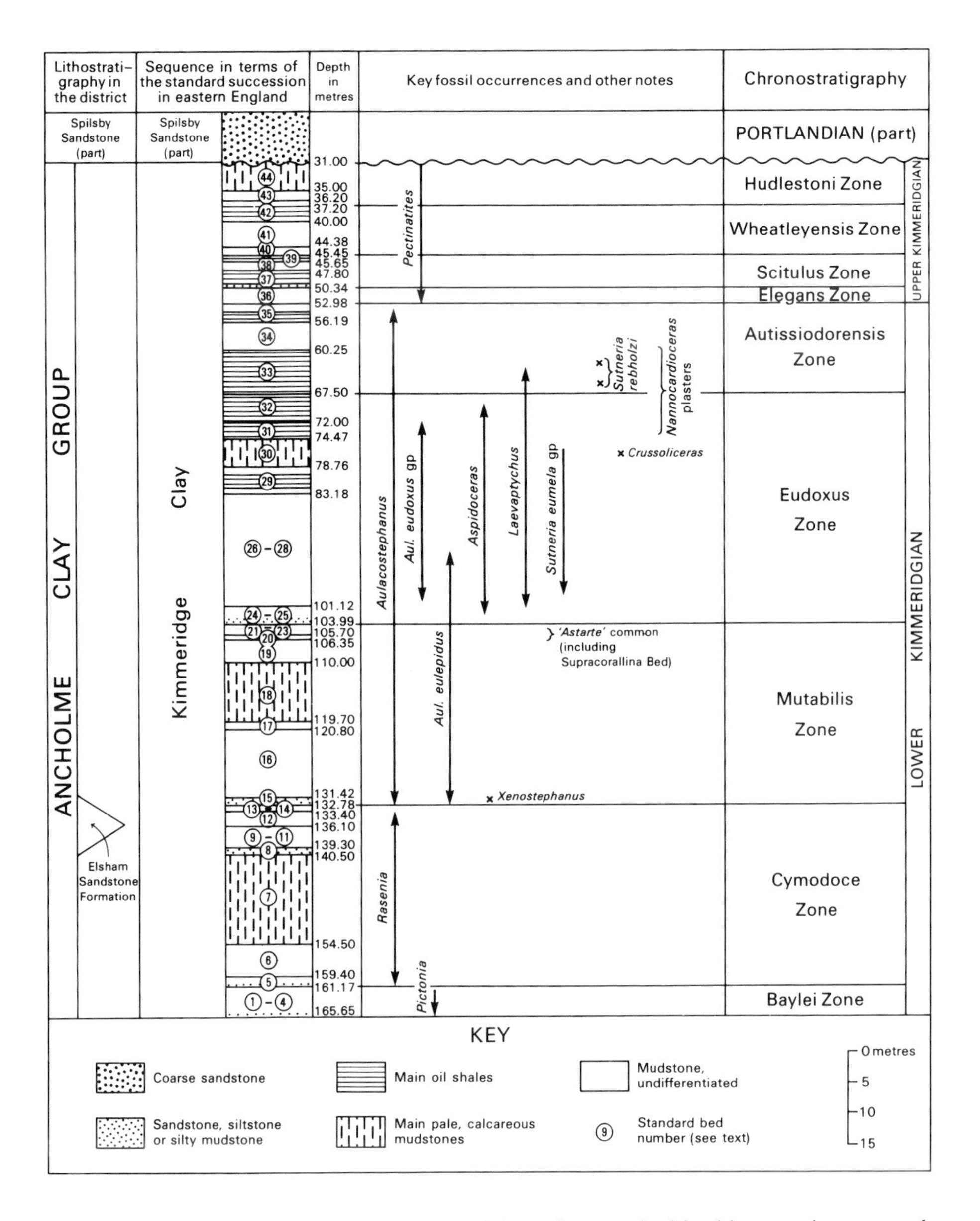

Figure 47 Ancholme Clay Group in Nettleton Bottom Borehole: summary of the Kimmeridgian sequence

Nettleton sequence and sections farther north where the succession is also expanded (see Thomas and Cox, 1988, fig. 2). Correlation with the pit section at South Ferriby, which is the most important of the few exposures in the district, is shown in Figure 48 (Smart and Wood, 1976; Kelly and Rawson, 1983; Birkelund and Callomon, 1985; Ahmed, 1987; Thomas and Cox, 1988). Ammonite ranges and occurrences indicated in Figures 46 and 47 give the basis for the ammonite-based biostratigraphy (zonation).

As well as the cored boreholes in the Humber area (Gaunt et el., 1980), at Worlaby (Richardson, 1979) and those at Nettleton Bottom, Osgodby and Elsham Quarry cited in this memoir, scattered localities in the Ancholme Clay Group in the Hull–Brigg district have yielded stratigraphical information (Figure 49). Many of them were collected by field geologists during the present survey, and additional material was donated by the late Sir Peter Kent. Some of the old brick/clay pits and railway cuttings (e.g. nos. 24, 38, 43, 44, 53, 58) are mentioned by Roberts (1889) and Ussher (1890).

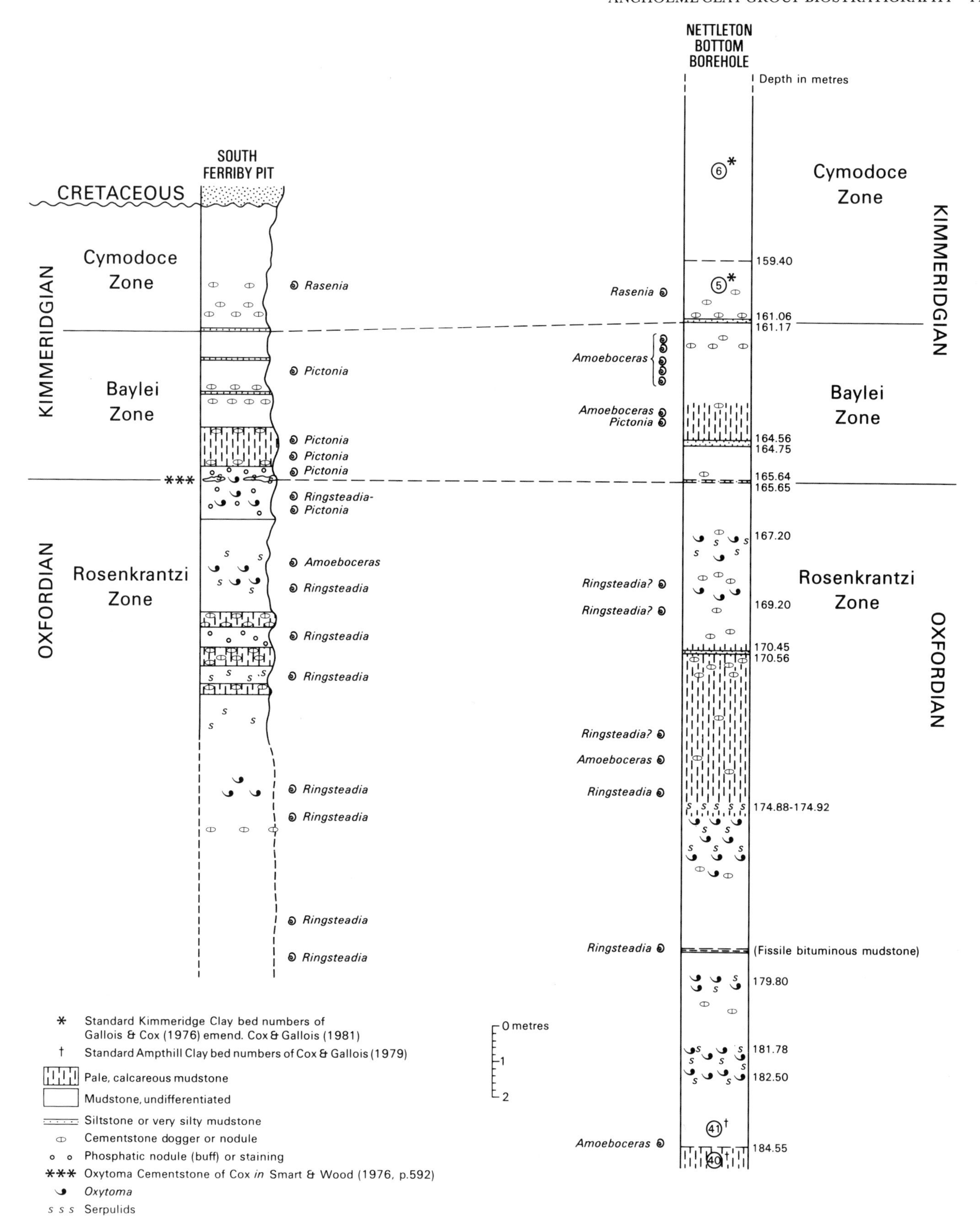

Figure 48 Correlation between Nettleton Bottom Borehole and South Ferriby Pit

NUMBERED FOSSIL LOCALITIES

1. Road cuttings, Elloughton [SE 9415 2889, 9422 2882, 9485 2793]: Densiplicatum Zone
2. Melton Pit [SE 970 270]: Tenuiserratum and Glosense zones (cited Sykes & Callomon, 1979)
3. Melton Bottoms Borehole [SE 9751 2742] 6.7 m – 76.2 m: Glosense Zone to Kellaways Beds (cited Bisat, 1922)
4. Capper Pass C Borehole [SE 9712 2556] c.30 m: Densiplicatum Zone
5. South Ferriby Pit [SE 992 204]: Rosenkrantzi to Cymodoce zones (see text)
6. Elsham LS14 Borehole [TA 0441 1275] c.40 m: ?Lower Kimmeridgian
7. Elsham water pipeline trench (1973) [TA 0239 1157 – 0293 1289]: Rosenkrantzi Zone to Elsham Sandstone, including the Oxytoma Cementstone
8. Elsham water pipeline trench [TA 0256 1183]: Cymodoce Zone
9. Elsham water pipeline trench [TA 0239 1157]: Regulare Zone
10. Old quarry, Elsham [TA 0363 1204]: Elsham Sandstone
11. Catchwater Drain [c.TA 0115 1022]: Tenuiserratum Zone
12. Brigg By-Pass Borehole J [TA 049 110] c.9.6 m: Mutabilis Zone
13. Elsham motorway interchange cutting [TA 0505 1100]: Eudoxus Zone (immediately below Carstone)
14. Southside Plantation, Elsham [TA 0376 1158, 038 115]: Elsham Sandstone
15. Railway cutting, Elsham [c.TA 0250 0991]: Oxfordian – Kimmeridgian boundary (Oxytoma Cementstone)
16. Brigg By-Pass Borehole 72 [TA 018 096] (14.5 m): Glosense Zone
17. Brigg By-Pass Borehole 85A [TA 015 095] (8.5 m): Glosense Zone
18. Brigg By-Pass Borehole 85 [TA 012 095] (7.2 m): Glosense Zone
19. Brigg By-Pass Borehole 48 [TA 0143 0937] 6.5 m: Glosense Zone
20. Brigg By-Pass Borehole 24 [SE 9922 0783] c.9.5 m: Coronatum Zone
21. Brigg By-Pass Borehole 56 [TA 0075 0884] c.3 m: Tenuiserratum Zone
22. Elsham water pipeline trench [TA 011 089, 0119 0895, 0120 0907]: Tenuiserratum Zone
23. Elsham water pipeline trench [TA 0120 0917 – 0920, 0118 0925, 0135 0940]: Tenuiserratum and Glosense zones
24. Railway cutting, Wrawby [c.TA 0305 0940]: Cymodoce Zone
25. Old quarry, Gallows Farm [TA 0433 1050]: Elsham Sandstone
26. Elsham LS12 Borehole [TA 0662 0971] c.25 m: Lower Kimmeridgian
27. Elsham LS11 Borehole [TA 0740 9960] c.36 – 39 m: Eudoxus Zone
28. Elsham water pipeline trench [TA 0147 0761]: Mariae Zone
29. Railway cutting, South Wrawby [TA 0250 0758]: Tenuiserratum and Glosense zones
30. Kettleby Beck [TA 0359 0664]: Oxfordian
31. Faraway Drain [c.TA 0005 0522]: Kellaways Rock
32. Pump house foundations, Cadney Reservoir [TA 0126 0434]: Cordatum Zone, Bukowskii Subzone
33. Owmby Vale Farm Borehole, Grasby [c.TA 068 040] 12 – 28 m: Regulare and Rosenkrantzi zones
34. Loose limestone slabs in ditch, Grasby [TA 0791 0481]: Oxfordian – Kimmeridgian boundary (Oxytoma Cementstone)
35. Sadney Land Drain [TA 0125 0200]: Mariae Zone
36. Sadney Land Drain [TA 0133 0185]: Mariae Zone
37. Roadside ditch, North Kelsey [c.TA 0244 0149]:Middle Oxfordian
38. Brickyard, North Kelsey [c.TA 0558 0263]: Oxfordian – Kimmeridgian boundary (Oxytoma Cementstone)
39. Duke's Drain, Woofham Farm [SK 9996 9852]: Kellaways Rock
40. Spoil from dyke [TF 0028 9860]: Kellaways Rock
41. Ditch [TF 0015 9774]: Kellaways Rock
42. Drain [c.TF 0274 9820]: Oxfordian
43. Brickyard, South Kelsey [c.TF 0420 9900]: Tenuiserratum Zone
44. Clay pit, Moortown Hill [c.TF 0680 9954]: Oxfordian – Kimmeridgian boundary (Oxytoma Cementstone)
45. Irrigation pond, North Kelsey Moor [TA 0830 0185]: Cymodoce Zone
46. Audleby [TA 111 039]: Upper Kimmeridgian
47. Nettleton Beck [TA 0995 0041]: Mutabilis Zone
48. Old River Ancholme [TF 0198 9719]: Callovian

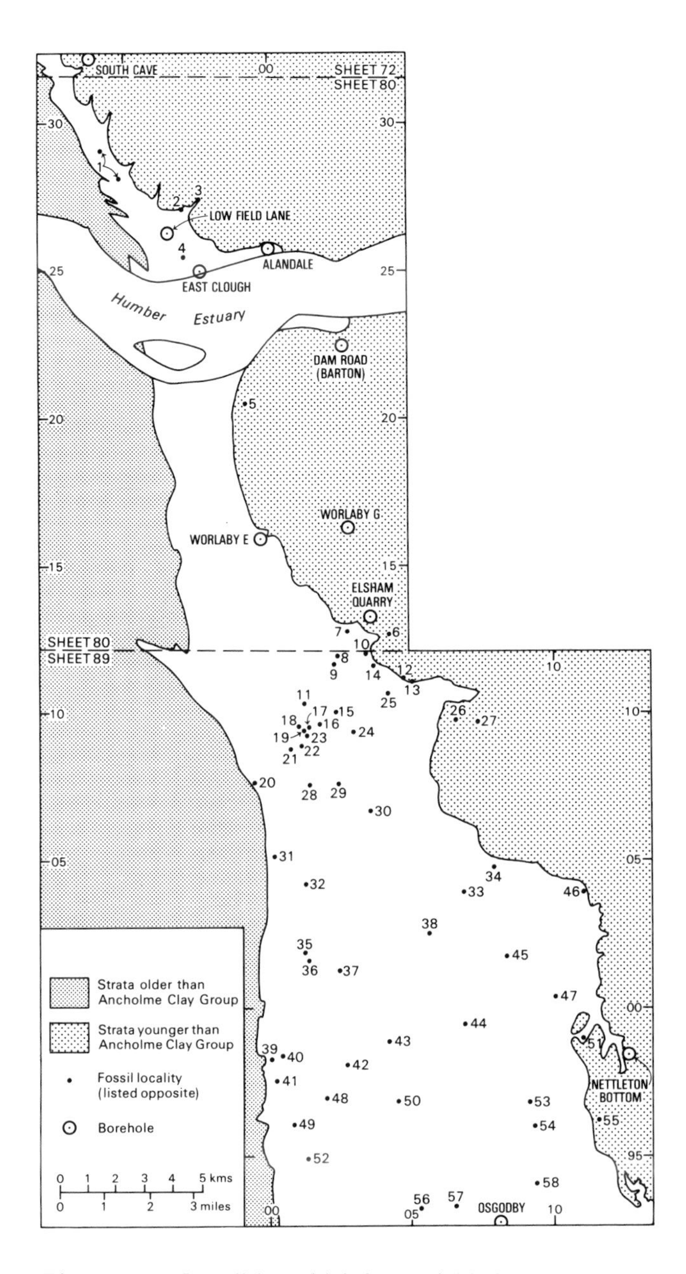

Figure 49 Localities which have yielded biostratigraphical information on the Ancholme Clay Group

49. Black Dyke [c.TF 0077 9629]: Middle Callovian
50. Ditch at south end of South Wood [TF 0446 9700]: Oxfordian
51. Trial sand pit, Nettleton Hill [TF 1085 9889]: Hudlestoni Zone
52. South Ramper Land Drain, Snitterby [TF 012 951]: Middle Callovian
53. Brickyard, Holton [c.TF 0905 9690]: Cymodoce Zone
54. Marling pits, Usselby [TF 0924 9423]: Cymodoce Zone
55. Acre House Mine, Normanby le Wold [TF 1152 9633]: Pectinatus Zone
56. Kingerby Beck, Owersby [TF 0519 9340]: Glosense Zone
57. ?Excavation, Owersby [TF 0637 9341]: Regulare Zone
58. Brickyard, Claxby Moor [c.TF 0925 9617]: Cymodoce Zone

REFERENCES

Most of the references listed below are held in the Library of the British Geological Survey at Keyworth, Nottingham. Copies of the references can be purchased subject to the current copyright legislation.

Ager, D V. 1956. The geographical distribution of brachiopods in the British Middle Lias. *Q. J. Geol. Soc. London*, Vol. 112, 157-188.

Ahmed, S T. 1987. Upper Oxfordian and Lower Kimmeridgian ostracods from the South Ferriby Quarry, South Humberside. *Proc. Yorkshire. Geol. Soc.*, Vol. 46, 267-274.

Andrews, J. 1983. A faunal correlation of the Hunstanton Red Rock with the contemporaneous Gault clay, and its implications for the environment of deposition. *Bull. Geol. Soc. Norfolk*, Vol. 33, 3-26.

Anon. 1969. Survey of water resources. Yorkshire Ouse and Hull River Authorities.

— 1971. Survey of demands, resources and development. Lincolnshire River Authority.

— 1978. South Humberbank Salinity Research Project. Final report to the Anglian Water Authority. (University of Birmingham.)

Arthurton, R S, Booth, S J, Morigi, A N. In preparation. The geology of the country around Great Yarmouth. *Mem. Br. Geol. Surv.*

Ashton, M. 1975. A new section in the Lincolnshire Limestone of South Humberside and its significance. *Proc. Yorkshire Geol. Soc.*, Vol. 40, 419-429.

— 1980. The stratigraphy of the Lincolnshire Limestone Formation (Bajocian) in Lincolnshire and Rutland (Leicestershire). *Proc. Geol. Assoc.*, Vol. 91, 203-223.

Aslin, C J. 1968. Upper Estuarine Series. 223-237 in *Geology of the East Midlands*. Sylvester-Bradley, P C, and Ford, T D (editors). (Leicester University Press).

Bailey, H W, Gale, A S, Mortimore, R N, Swiecki, A, and Wood, C J. 1983. The Coniacian-Maastrichtian Stages of the United Kingdom, with particular reference to southern England. *Newsl. Stratigr.*, Vol. 12, 29-42.

Balchin, D A, and Ridd, M F. 1970. Correlation of the younger Triassic rocks across eastern England. *Q. J. Geol. Soc. London*, Vol. 126, 91-101.

Barker, I C. 1984. The groundwater resources of the Jurassic rocks in the Brough area of North Humberside. Groundwater Section, Rivers Division, Yorkshire Water. [Unpublished].

Barker, R D. 1982. Geophysical surveys near Goxhill, south Humberside. *Proc. Yorkshire Geol. Soc.*, Vol. 44, 119-129.

— Lloyd, J W, and Peach, D W. 1984. The use of resistivity and gamma logging in lithostratigraphical studies of the Chalk in Lincolnshire and South Humberside. *Q. J. Eng. Geol.*, Vol. 17, 71-80.

Barrois, C. 1876. Récherches sur le terrain Crétacé supérieur de l'Angleterre et de l'Irelande. *Mém. Soc. Géol. Nord.*

Bartenstein, H. 1956. Zur Mikrofauna des englischen Hauterive. *Senck. Leth.*, Vol. 37, No. 5/6, 509-533.

Bate, R H. 1967. Stratigraphy and palaeogeography of the Yorkshire Oolites and their relationship with the Lincolnshire Limestone. *Bull. Br. Mus. (Nat. Hist.), Geol. London*, Vol. 14, 111-141.

— 1980. Middle Jurassic ostracods. 34 *in* Mesozoic rocks proved by IGS boreholes in the Humber and Acklam areas. Gaunt, G D, and others. *Rep. Inst. Geol. Sci.*, No. 79/13.

Birkelund, T, and Callomon, J H. 1985. The Kimmeridgian ammonite faunas of Milne Land, central East Greenland. *Geol. Surv. Greenland Bulletin*, No. 153.

Bisat, W S. 1922. New sections near Melton, North Ferriby, Yorks. *Trans. Hull Geol. Soc.*, Vol. 6, 238-243.

— 1932a. On the subdivision of the Holderness boulder clays. *Naturalist*, No. 906, 215-219.

— 1932b. Glacial and post-glacial sections on the Humber shore at North Ferriby. *Trans. Hull Geol. Soc.*, Vol. 7, 83-95.

Bott, M H P. 1961. Geological interpretation of magnetic anomalies over the Askrigg Block. *Q. J. Geol. Soc. London*, Vol. 117, 481-495.

— Robinson, J, and Kohnstamm, M A. 1978. Granite beneath Market Weighton, east Yorkshire. *Q. J. Geol. Soc. London*, Vol. 135, 535-543.

Bowen, D Q. 1978. *Quaternary geology. A stratigraphic framework for multidisciplinary work.* (Pergamon Press.)

Bower, C R, and Farmery, J R. 1910. The zones of the Lower Chalk of Lincolnshire. *Proc. Geol. Assoc.*, Vol. 21, 333-359.

Boylan, P J. 1966. The Pleistocene deposits of Kirmington, Lincolnshire. *Mercian Geol.*, Vol. 1, 339-350.

— 1967. The Pleistocene mammalia of the Sewerby-Hessle buried cliff, east Yorkshire. *Proc. Yorkshire Geol. Soc.*, Vol. 36, 115-125.

— 1981. A new revision of the Pleistocene mammalian fauna of Kirkdale Cave, Yorkshire. *Proc. Yorkshire Geol. Soc.*, Vol. 43, 253-280.

Bradshaw, M J. 1978. A facies analysis of the Bathonian of eastern England. Unpublished thesis, Oxford University.

— and Bate, R H. 1982. Lincolnshire borehole proves greater extent of the Scarborough Formation (Jurassic: Bajocian). *J. Micropalaeontology*, Vol. 1, 141-147.

— and Penney, S R. 1982. A cored Jurassic sequence from north Lincolnshire, England: stratigraphy, facies analysis and regional context. *Geol. Mag.*, Vol. 119, 113-134.

Brandon, A, Sumbler, M G, and Ivimey-Cook, H C. 1990. A revised lithostratigraphy for the Lower and Middle Lias (Lower Jurassic) east of Nottingham, England. *Proc. Yorkshire Geol. Soc.*, Vol. 48, 121-141.

Bralower, T J. 1988. Calcareous nannofossil biostratigraphy and assemblages of the Cenomanian-Turonian boundary interval: implications for the origin and tracing of oceanic anoxia. *Paleoceanography*, Vol. 3, 275-316.

Brasier, M D, and Brasier, C J. 1978. Littoral and fluviatile facies in the 'Kellaways Beds' on the Market Weighton swell. *Proc. Yorkshire Geol. Soc.*, Vol. 42, 1-20.

Bromley, R G, Schulz, M G, and Peake, N B. 1975. Paramoudras: giant flints, long burrows and early diagenesis of chalks. *K. Dansk. Vidensk. Selsk. Geol. Skr.*, Vol. 20/10.

Buckland, P C. 1982. The cover sands of north Lincolnshire and the Vale of York. 148-178 in *Papers in earth studies, Lovatt Lectures*. Adlam, B H, Fenn, C R, and Morris, L (editors). (Norwich.)

— 1984. North-west Lincolnshire 10 000 years ago. 11-17 in *A prospect of Lincolnshire*. Fields, N, and White, A (editors). (Lincoln: Newlands.)

Bujak, J P, and Fisher, M J. 1976. Dinoflagellate cysts from the Upper Triassic of arctic Canada. *Micropalaeontology*, Vol. 22, 44-70.

Bullerwell, W. 1961. Discussion in Geological interpretation of magnetic anomalies over the Askrigg Block. Bott, M P H. *Q. J. Geol. Soc. London*, Vol. 117, 494.

Callomon, J H. 1955. The ammonite succession in the Oxford Clay and Kellaways Beds at Kidlington, Oxfordshire, and the zones of the Callovian Stage. *Philos. Trans. R. Soc.*, Ser. B., Vol. 239, 215-264.

— 1964. Notes on the Callovian and Oxfordian Stages. *C. R. et Mem. Colloq. du Jurassique, Luxembourg, 1962*, 269-291.

— 1968. The Kellaways Beds and Oxford Clay. 264-290 in *Geology of the East Midlands*. Sylvester-Bradley, P C, and Ford, T D (editors). (Leicester University Press.)

Carruthers, R G. 1948. The secret of the glacial drifts Part II. Application to Yorkshire. *Proc. Yorkshire Geol. Soc.*, Vol. 27, 129-172.

Carter, D J, and Hart, M B. 1977. Aspects of mid-Cretaceous stratigraphical micropalaeontology. *Bull. Br. Mus. (Nat. Hist.), Geol.*, No. 29.

Casey, R. 1962. The ammonites of the Spilsby Sandstone, and the Jurassic-Cretaceous boundary. *Proc. Geol. Soc. London*, No. 1598, 95-100.

— 1963. The dawn of the Cretaceous Period in Britain. *Bull. S-East Un. Scien. Socs.*, No. 117, 1-15.

— 1973. The ammonite succession at the Jurassic-Cretaceous boundary in eastern England. 193-266 *in* The Boreal Lower Cretaceous. Casey, R, and Rawson, P F. (editors). *Spec. Issue Geol. J.* No. 5.

— and Gallois, R W. 1973. The Sandringham Sands of Norfolk. *Proc. Yorkshire Geol. Soc.*, Vol. 40, 1-22.

Catt, J A, and Penny, L E. 1966. The Pleistocene deposits of Holderness, east Yorkshire. *Proc. Yorkshire Geol. Soc.*, Vol. 35, 375-420.

— Weir, A H, and Madgett, P A. 1974. The loess of eastern Yorkshire and Lincolnshire. *Proc. Yorkshire Geol. Soc.*, Vol. 40, 23-39.

Christensen, W K. 1982. Late Turonian-early Coniacian belemnites from western and central Europe. *Bull. Geol. Soc. Denmark*, Vol. 31, 63-79.

Clayton, C J. 1984. The chemical environment of flint formation in Upper Cretaceous chalks. 43-54 in *The scientific study of flint and chert*. Sieveking, G D E G, and Hart, M B (editors). (Cambridge University Press.)

Cole, E M. 1886. *Notes on the geology of the Hull, Barnsley and West Riding Railway and Dock*. (Hull: privately printed by M C Peck.)

Colter, V S. and Reed, G E M. 1980. Zechstein 2 Fordon Evaporites of the Atwick No. 1 Borehole, surrounding areas of N E England and the adjacent North Sea. *Contrib. Sedimentol.*, Vol. 9, 115-129.

Cope, J C W, Getty, T A, Howarth, M K, Morton, N, and Torrens, H S. 1980a. A correlation of Jurassic rocks in the British Isles. Part One: Introduction and Lower Jurassic. *Spec. Rep. Geol. Soc. London*, No. 14.

— Duff, K L, Parsons, C F, Torrens, H S, Wimbledon, W A, and Wright, J K. 1980b. A correlation of Jurassic rocks in the British Isles. Part Two: Middle and Upper Jurassic. *Spec. Rep. Geol. Soc. London*, No. 15.

Cox, B M, and Gallois, R W. 1979. Description of the standard stratigraphical sequences of the Upper Kimmeridge Clay, Ampthill Clay and West Walton Beds. *Rep. Inst. Geol. Sci.*, No. 78/19.

— — 1981. The stratigraphy of the Kimmeridge Clay of the Dorset type area and its correlation with some other Kimmeridgian sequences. *Rep. Inst. Geol. Sci.*, No. 80/4.

Crofts, W H. 1906. Notes on the indications of a raised beach at Hessle. *Trans. Hull Geol. Soc.*, Vol. 6, 58-64.

Cross, J E. 1875. The geology of north-west Lincolnshire. *Q. J. Geol. Soc. London*, Vol. 31, 115-130.

Dakyns, J R, Fox-Strangways, C, and Cameron, A G. 1886. The geology of the country between York and Hull. *Mem. Geol. Surv. GB*, Sheets 93 SE, 94 SW and part of 86 (Old Series).

Davies, W, and Dixie, R J M. 1951. Recent work on the Frodingham Ironstone. 85-96 *in* The constitution and origin of sedimentary iron ores: a symposium. Hallimond, A F, Dunham, K C, Hemingway, J E, Taylor, J H, Davies, W, Dixie, R J M, and Bannister, F A. *Proc. Yorkshire Geol. Soc.*, Vol. 28, 61-101.

Day, J B W, Moseley, R, Robertson, A S, and Mercer, I F. 1967. Hydrogeological map of north and east Lincolnshire. (London: Institute of Geological Sciences.)

De Boer, G, Neale, J W, and Penny, L F. 1958. A guide to the geology of the area between Market Weighton and the Humber. *Proc. Yorkshire Geol. Soc.*, Vol. 31, 157-209.

Dikes, W H, and Lee, J W. 1837. Outlines of the geology of Nettleton Hill, Lincolnshire. *The Magazine of Natural History*, Vol. I, No. 11, Art I, 562-566.

Dilley, F C. 1969. The foraminiferal fauna of the Melton Carstone. *Proc. Yorkshire Geol. Soc.*, Vol. 37, 321-322.

Donovan, D T. 1968. Geology of the continental shelf around Britain: a survey of progress. 1-14 in *Geology of shelf seas*. Donovan, D T (editor). (Edinburgh: Oliver and Boyd.)

Douglas, J A, and Arkell, W J. 1928. The stratigraphical distribution of the Cornbrash: I The south-western area. *Q. J. Geol. Soc. London*, Vol. 84, 117-178.

Downing, R A, and Williams, B P J. 1969. The groundwater hydrology of the Lincolnshire Limestone. *Publ. Water Resources Board, Reading, England*, No. 9.

— Smith, D B, Pearson, F J, Monkhouse, R A, and Otlet, R L. 1977. The age of groundwater in the Lincolnshire Limetone, England, and its relevance to the flow mechanism. *J. Hydrol.*, Vol. 33, 201-216.

Drake, H C, and Sheppard, T. 1909. Classified list of organic remains from the rocks of the East Riding of Yorkshire. *Proc. Yorkshire Geol. Soc.*, Vol. 17, 4-71.

Dudley, H E. 1942. The Lower Lias Beds in the Frodingham Railway Cutting, north-west Lincolnshire. *Proc. Geol. Assoc.*, Vol. 53, 152-155.

Dunham, K C. 1948. Petrographical descriptions of two cycles of sedimentation from the Middle Coal Measures, Woodborough Borehole. Appendix VI, 249-253 *in* Marine bands and other faunal marker-horizons in relation to the sedimentary cycles of the Middle Coal Measures of Nottinghamshire and Derbyshire. EDWARDS, W, and STUBBLEFIELD, C J (editors). *Q. J. Geol. Soc. London*, Vol. 103, 209–260.

— 1974. Granite beneath the Pennines in north Yorkshire. *Proc. Yorkshire Geol. Soc.*, Vol. 40, 191–194.

EDEN, R A, STEVENSON, I P, and EDWARDS, W. 1957. Geology of the country around Sheffield. *Mem. Geol. Surv. GB.*

EDWARDS, W. 1951. The concealed coalfield of Yorkshire and Nottinghamshire. *Mem. Geol. Surv. GB.*

— 1967. Geology of the country around Ollerton. (2nd edition, revised). *Mem. Geol. Surv. GB.*

ELLIOTT, R E. 1961. The stratigraphy of the Keuper Series in southern Nottinghamshire. *Proc. Yorkshire Geol. Soc.*, Vol. 33, 197–234.

ENNIS, W C. 1932. Geological work in the Humber area, 1928–1931. *Trans. Hull Geol. Soc.*, Vol. 7, 77–79.

ERNST, G, SCHMID, F, and SEIBERTZ, E. 1983. Event—Stratigraphie im Cenomen und Turon von NW—Deutschland. *Zitteliana.*, Vol. 10, 531–554.

ERNST, G, WOOD, C J, and HILBRECHT, H. 1984. The Cenomanian–Turonian boundary problem in NW Germany with comments on the north–south correlation to the Regensburg Area. *Bull. Geol. Soc. Denmark*, Vol. 33, 103–113.

FENTON, J P G, and FISHER, M J. 1978. Regional distribution of marine microplankton in the Bajocian and Bathonian of Northwest Europe. *Palinologia Num. Extraord.*, Vol. 1, 233–243.

FLETCHER, B N. 1973. The distribution of Lower Cretaceous (Berriasian–Barremian) Foraminifera in the Speeton Clay. 161–168 *in* The Boreal Lower Cretaceous, CASEY, R, and RAWSON, P F (editors). *Geol. J. Spec. Issue* No. 5.

FLETCHER, T P. 1981. Geology of the Ancholme Valley. 189–195 in *The Brigg Raft and her prehistoric environment*, MCGRAIL, S (editor). (National Maritime Museum, Greenwich, Archaeological Series No. 6. BAR British Series 89.)

— 1985. Flamborough and Bridlington Sheets 55 and 65. 1:50 000. (Southampton: Ordnance Survey for British Geological Survey.)

— and KNOX, R W O'B. In preparation. The Frodingham Ironstone. *Rep. Br. Geol. Surv.*

FORSTER, S C, and WARRINGTON, G. 1985. Geochronology of the Carboniferous, Permian and Triassic. 99–113 *in* The chronology of the geological record. SNELLING, N J (editor). *Mem. Geol. Soc. London*, No. 10.

FOSTER, S S D. 1968. Report on the groundwater hydrology and resources of Hydrometric area 26 (Hull River) Part 1. The Jurassic rocks between Market Weighton and the Humber. *Rep. Inst. Geol. Sci. London*, No. WD/68/4. [Unpublished.]

— and MILTON, V A. 1976. Hydrological basis for large-scale development of groundwater storage capacity in the East Yorkshire Chalk. *Rep. Inst. Geol. Sci.*, No. 76/3.

— PARRY, E L, and CHILTON, P J. 1976. Groundwater resources development and saline water intrusion in the Chalk aquifer of North Humberside. *Rep. Inst. Geol. Sci.*, No. 76/4.

FOX-STRANGWAYS, C. 1906. The water supply (from underground sources) of the East Riding of Yorkshire, together with the neighbouring portion of the vales of York and Pickering: with records of sinkings and borings. *Mem. Geol. Surv. GB.*

GALLOIS, R W. 1973. Some detailed correlations in the Upper Kimmeridge Clay in Norfolk and Lincolnshire. *Bull. Geol. Surv. GB*, No. 44, 63–75.

— 1978. A pilot study of oil shale occurrences in the Kimmeridge Clay. *Rep. Inst. Geol. Sci.*, No. 78/13.

— 1979. Geological investigations for the Wash Water Storage Scheme. *Rep. Inst. Geol. Sci.*, No. 78/19.

— and COX, B M. 1976. The stratigraphy of the Lower Kimmeridge Clay of eastern England. *Proc. Yorkshire Geol. Soc.*, Vol. 41, 13–26.

— — 1977. The stratigraphy of the Middle and Upper Oxfordian sediments of Fenland. *Proc. Geol. Assoc.*, Vol. 88, 207–228.

— and MORTER, A A. 1982. The stratigraphy of the Gault of East Anglia. *Proc. Geol. Assoc.*, Vol. 93, 351–368.

GAUNT, G D. 1974. A radiocarbon date relating to Lake Humber. *Proc. Yorkshire Geol. Soc.*, Vol. 40, 195–197.

— 1975. The artificial nature of the River Don north of Thorne, Yorkshire. *Yorkshire Archaeol. J.*, Vol. 47, 15–21.

— 1976. The Quaternary geology of the southern part of the Vale of York. Unpublished PhD thesis, University of Leeds.

— 1981. Quaternary history of the southern part of the Vale of York. 82–97 in *The Quaternary in Britain*. NEALE, J, and FLENLEY, J (editors). (Pergamon Press.)

— 1987. The geology and landscape development of the region around Thorns Moors. 6–30 in *Thorne Moors Papers*. LIMBERT, M (editor). (Doncaster.)

— In preparation. Geology of the country around Goole, Doncaster and the Isle of Axholme. *Mem. Br. Geol. Surv.*

— BARTLEY, D D, and HARLAND, R. 1974. Two interglacial deposits proved in boreholes in the southern part of the Vale of York and their bearing on contemporaneous sea levels. *Bull. Geol. Surv. GB*, No. 48, 1–23.

— IVIMEY-COOK, H C, PENN, I E, and COX, B M. 1980. Mesozoic rocks proved by IGS boreholes in the Humber and Acklam areas. *Rep. Inst. Geol. Sci.*, No. 79/13.

— and TOOLEY, M J. 1974. Evidence for Flandrian sea-level changes in the Humber Estuary and adjacent areas. *Bull. Geol. Surv. GB*, No. 48, 25–41.

GEIGER, M E, and HOPPING, C A. 1968. Triassic stratigraphy of the southern North Sea Basin. *Philos. Trans. R. Soc. London*, Series B, Vol. 254, 1–36.

GETTY, T A. 1973. A revision of the generic classification of the Family Echioceratidae (Cephalopoda, Ammonoidea) (Lower Jurassic). *Pap. Univ. Kansas Pal. Contr.*, No. 63.

GOOSSENS, R F, and SMITH, E G. 1973. The stratigraphy and structure of the Upper Coal Measures in the exposed Yorkshire Coalfield between Pontefract and South Kirby. *Proc. Yorkshire Geol. Soc.*, Vol. 39, 487–514.

GOZZARD, J R, and PRICE, D. 1978. The sand and gravel resources of the country north of Gainsborough, Lincolnshire. Description of 1:25 000 resource sheet SK 89. *Miner. Assess. Rep. Inst. Geol. Sci.*, No. 33.

GRAY, D A. 1955. The occurrence of a Corallian Limestone in East Yorkshire south of Market Weighton. *Proc. Yorkshire. Geol. Soc.*, Vol. 30, 25–34.

HALL, H F. 1867. On the drift sections of the Holderness coast, Yorkshire. *Proc. Liverpool Geol. Soc.*, Vol. 1, 13–38.

HALLAM, A. 1962. A band of extraordinary calcareous concretions in the Upper Lias of Yorkshire, England. *J. Sediment. Petrol.*, Vol. 32, 840–847.

— 1963. Observations on the palaeoecology and ammonite sequence of the Frodingham Ironstone (Lower Jurassic). *Palaeontology*, Vol. 6, 554–574.

— 1981. A revised sea-level curve for the early Jurassic. *J. Geol. Soc. London*, Vol. 138, 735–743.

— Hancock, J M, La Brecque, J L, Lowrie, W, and Channell, J E T. 1985. Jurassic to Paleogene: Part I. Jurassic and Cretaceous geochronology and Jurassic to Paleogene magnetostratigraphy. 118–140 *in* The chronology of the geological record. Snelling, N H (editor). *Mem. Geol. Soc. London*, No. 10.

Hallimond, A F. 1925. Special reports on the mineral resources of Great Britain. Vol. 29. Bedded iron ores of England and Wales: petrography and chemistry. *Mem. Geol. Surv. GB.*

Hancock, J M. 1975. The petrology of the Chalk. *Proc. Geol. Assoc.*, Vol. 86, 499–535.

Harland, R, and Downie, C. 1969. The dinoflagellates of the interglacial deposits at Kirmington, Lincolnshire. *Proc. Yorkshire Geol. Soc.*, Vol. 37, 231–237.

Hart, M B, and Bigg, P J. 1981. Anoxic events in the late Cretaceous chalk seas of North-West Europe. 177–185 in *Microfossils from Recent and fossil shelf seas.* Neale, J W, and Brasier, M D (editors). (Chichester: Ellis Horwood.)

— and Leary, P N. 1989. The stratigraphic and palaeogeographic setting of the late Cenomanian 'anoxic' event. *Q. J. Geol. Soc. London*, Vol. 146, 305–310.

Hemingway, J E. 1974. Jurassic. 161–223 in *The geology and mineral resources of Yorkshire.* Rayner, D H, and Hemingway, J E (editors). (Leeds: Yorkshire Geological Society.)

Higginbottom, I E, and Fookes, P G. 1971. Engineering aspects of periglacial features in Britain. *Q. J. Eng. Geol.*, Vol. 3, 85–117.

Hill, W. 1888. On the lower beds of the Upper Cretaceous Series in Lincolnshire and Yorkshire. *Q. J. Geol. Soc. London*, Vol. 44, 320–367.

— 1902. Note on the Upper Chalk of Lincolnshire. *Geol. Mag.*, Vol. 9, 404–406.

Holland, S. 1975. Pollen analytical investigations at Crosby Warren, Lincolnshire, in the vicinity of the Iron Age and Romano-British settlement of Dragonby. *J. Archaeol. Sci.*, Vol. 2, 353–363.

Hollingworth, S E, and Taylor, J H. 1951. The Mesozoic ironstones of England. The Northampton Sand Ironstone: stratigraphy, structure and reserves. *Mem. Geol. Surv. GB.*

— — and Kellaway, G A. 1944. Large-scale superficial structures in the Northampton Ironstone field. *Q. J. Geol. Soc. London*, Vol. 100, 1–44.

Horton, A. 1977. The age of the Middle Jurassic 'white sands' of north Oxfordshire. *Proc. Geol. Assoc.*, Vol. 88, 147–162.

Howard, A S. 1985. Lithostratigraphy of the Staithes Sandstone and Cleveland Ironstone formations (Lower Jurassic) of north-east Yorkshire. *Proc. Yorkshire Geol. Soc.*, Vol. 45, 261–275.

Howarth, M K. 1962. The Jet Rock Series and Alum Shale Series of the Yorkshire coast. *Proc. Yorkshire Geol. Soc.*, Vol. 33, 381–422.

— 1980. The Toarcian age of the upper part of the Marlstone Rock Bed of England. *Palaeontology*, Vol. 23, 637–656.

— and Rawson, P F. 1965. The Liassic succession in a clay pit at Kirton in Lindsey, North Lincolnshire. *Geol. Mag.*, Vol. 102, 261–266.

Hull, E. 1857. The geology of the country around Cheltenham. *Mem. Geol. Surv. GB*, Sheet 44 (Old Series).

Hulme, P D, and Beckett, S C. 1973. Pollen analysis of the Faxfleet peats. *E. Yorkshire Field Studies*, No. 4, 15–24.

Ingham, F T. 1929. The petrography of the Spilsby Sandstone. *Proc. Geol. Assoc.*, Vol. 40, 1–17.

Institute of Geological Sciences. 1973. *Annual report for 1972*, 132. (London: Institute of Geological Sciences.)

— 1977a. 1:250 000 Series, Bouguer gravity anomaly map (provisional edition), Humber-Trent Sheet, 53°N–02°W. (London: Institute of Geological Sciences.)

— 1977b. 1:250 000 Series, aeromagnetic anomaly map (provisional edition), Humber-Trent Sheet, 53°N–02°W. (London: Institute of Geological Sciences.)

Ivimey-Cook, H C. 1978. Stratigraphical palaeontology of the Lower Jurassic of the Port More Borehole. 80–83 *in* Geology of the Causeway Coast. Vol. 2. Wilson, H E, and Manning, P I. (editors). *Mem. Geol. Surv. Northern Ireland.*

— and Donovan, D T. 1983. The fauna of the Lower Jurassic. Appendix 3, 126–130 *in* Geology of the country around Weston-super-Mare. Whittaker, A, and Green, G W. *Mem. Geol. Surv. GB.*

Jacobi, R M. 1978. Northern England in the Eighth Millenium b.c.: an essay. 295–332 in *The early postglacial settlement of Northern Europe.* Mellors, P (editor).

James, J W C. 1976. The sand and gravel resources of the country north-west of Scunthorpe, Humberside. Description of 1:25 000 resource sheet SE 81. *Miner. Assess. Rep. Inst. Geol. Sci.*, No. 22.

Jarvis, I, Carson, G A, Cooper, M K E, Hart, M B, Leary, P N, Tocher, B A, Horne, D, and Rosenfeld, A. 1988. Microfossil assemblages and the Cenomanian–Turonian (late Cretaceous) Oceanic Anoxic Event. *Cretaceous Res.*, Vol. 9, 3–103.

Jeans, C V. 1973. The Market Weighton Structure: tectonics, sedimentation and diagenesis during the Cretaceous. *Proc. Yorkshire Geol Soc.*, Vol. 39, 409–444.

— 1980. Early submarine lithification in the Red Chalk and Lower Chalk of eastern England: a bacterial control model and its implications. *Proc. Yorkshire Geol. Soc.*, Vol. 43, 81–157.

Jefferies, R P S. 1963. The stratigraphy of the *Actinocamax plenus* Subzone (Turonian) in the Anglo-Paris Basin. *Proc. Geol. Assoc.*, Vol. 74, 1–34.

Judd, J W. 1867. On the strata which form the base of the Lincolnshire Wolds. *Q. J. Geol. Soc. London*, Vol. 23, 227–251.

— 1875. The geology of Rutland. *Mem. Geol. Surv. GB*, Sheet 64 (Old Series).

Jukes-Browne, A J. 1885a. The geology of the south-west part of Lincolnshire, with parts of Leicestershire and Nottinghamshire. *Mem. Geol. Surv. GB*, Sheet 70 (Old Series).

— 1885b. The boulder-clays of Lincolnshire. *Q. J. Geol. Soc. London*, Vol. 41, 114–132.

— and Hill, W. 1903. The Cretaceous rocks of Britain. Vol. 2. The Lower and Middle Chalk of England. *Mem. Geol. Surv. GB.*

— — 1904. The Cretaceous rocks of Britain. Vol. 3. The Upper Chalk of England. *Mem. Geol. Surv. GB.*

Kauffman, E G. 1978. British Middle Cretaceous Inoceramid biostratigraphy. *Ann. Mus. d'Hist. Nat. Nice*, Vol. 4, 1976. Mid-Cretaceous events, reports on the biostratigraphy of key areas, Article 4, 1–12. (Nice.)

KAYE, P. 1964. Some Lower Cretaceous sections in northern England. *Proc. Geol. Assoc.*, Vol. 75, 315-320.

KEEPING, W. 1883. *The fossils and palaeontological affinities of the Neocomian deposits of Upware and Brickhill.* (Cambridge: Cambridge University Press.)

KELLY, S R A. 1977. The bivalves of the Spilsby Sandstone Formation and contiguous deposits. Unpublished PhD thesis, University of London.

— 1980; *Hiatella*—a Jurassic bivalve squatter? *Palaeontology*, Vol. 23, Part 4, 769-781, pl. 96.

— and RAWSON, P F. 1983. Some late Jurassic-mid-Cretaceous sections on the East Midlands Shelf, as demonstrated on a field meeting, 18-20 May 1979. *Proc. Geol. Assoc.*, Vol. 94, 65-73.

KEMPER, E, RAWSON, P F, and THIEULOY, J P. 1981. Ammonites of Tethyan ancestry in the early Cretaceous of north-west Europe. *Palaeontology*, Vol. 24, 251-311.

KENNARD, A S, and MUSHAM, J F. 1937. On the mollusca from a Holocene tufaceous deposit at Broughton-Brigg, Lincolnshire. *Proc. Malac. Soc. London*, Vol. 22, 374-379.

KENNEDY, W J. 1969. The correlation of the Lower Chalk of south-east England. *Proc. Geol. Assoc.*, Vol. 80, 459-560.

KENT, P E. 1937. A lateral change in the Red Chalk. *Trans. Lincs. Nat. Union*, Vol. 9 (for 1935-1938), 166-167.

— 1941. A short outline of the stratigraphy of the Lincolnshire Limestone. *Trans. Lincs. Nat. Union (for 1940)*, Vol. 10, 48-58.

— 1953. The Rhaetic beds of the north-east Midlands. *Proc. Yorkshire Geol. Soc.*, Vol. 29, 117-139.

— 1966. The structure of the concealed Carboniferous rocks of north-eastern England. *Proc. Yorkshire Geol. Soc.*, Vol. 35, 323-352.

— 1970. Problems of the Rhaetic in the East Midlands. *Mercian Geol.*, Vol. 3, 361-373.

— 1974. Structural history. 13-28 in *The geology and mineral resources of Yorkshire.* RAYNER, D H, and HEMINGWAY, J E (editors). (Leeds: Yorkshire Geological Society.)

— 1975. The Grantham Formation of the East Midlands: revision of the Middle Jurassic, Lower Estuarine Beds. *Mercian Geol.*, Vol. 5, 305-327.

— 1980a. Subsidence and uplift in east Yorkshire and Lincolnshire: a double inversion. *Proc. Yorkshire Geol. Soc.*, Vol. 42, 505-524.

— 1980b. *British regional geology: Eastern England from the Tees to the Wash* (2nd edition). (London: HMSO for Institute of Geological Sciences.)

— and BAKER, F T. 1938. Ammonites from the Lincolnshire Limestone. *Trans. Lincs. Nat. Union*, Vol. 9, 169-170.

— and CASEY, R. 1963. A Kimmeridge sandstone in north Lincolnshire. *Proc. Geol. Soc. London*, No. 1606, 57-62.

— and DILLEY, F C. 1968. The Jurassic-Cretaceous junction at Elsham, north Lincolnshire. *Proc. Yorkshire Geol. Soc.*, Vol. 36, 525-530.

KNOX, R W O'B. 1970. Chamosite ooliths from the Winter Gill Ironstone (Jurassic) of Yorkshire, England. *J. Sediment. Petrol.*, Vol. 40, 1216-1225.

— 1984. Lithostratigraphy and depositional history of the late Toarcian sequence at Ravenscar, Yorkshire. *Proc. Yorkshire Geol. Soc.*, Vol. 45, 99-108.

KNUDSON, K L. 1986. Middle and Late Quaternary foraminiferal stratigraphy in the southern and central North Sea. *Striae*, Vol. 24, 201-205.

KRIMGOLTS, G YA, MESEZHNIKOV, M S, SAKS, V N, SHULGINA, N I, and VAKHRAMEEV, V A. 1968. Sur la méthode de l'élaboration des cartes paleobiogeographiques. *Proc. 23rd Int. geol. Congr.*, 239-256.

LAKE, R D, YOUNG, B, WOOD, C J, and MORTIMORE, R N. 1987. Geology of the country around Lewes. *Mem. Br. Geol. Surv.*

LAMPLUGH, G W. 1887. On a mammaliferous gravel at Elloughton in the Humber Valley. *Proc. Yorkshire Geol. Polytech. Soc.*, Vol. 9, 407-411.

— 1896. On the Speeton Series in Yorkshire and Lincolnshire. *Q. J. Geol. Soc. London*, Vol. 52, 179-220.

— STATHER, J W, ANDERSON, T, CARR, J W, CARTER, W L, DWERRYHOUSE, A R, HARMER, F W, HOWARTH, J H, JOHNSON, W, KENDALL, P F, NEWTON, E T, PLATNAUER, H M, REID, C, and SHEPPARD, T. 1905. Investigation of the fossiliferous drift deposits at Kirmington, Lincolnshire, and at various localities in the East Riding of Yorkshire—report of the committee. *Rep. Br. Assoc. Adv. Sci. for 1904*, 272-274.

LANG, W D. 1936. The Green Ammonite Beds of the Dorset Lias. *Q. J. Geol. Soc. London*, Vol. 92, 432-437.

LEES, G M, and TAITT, A H. 1946. The geological results of the search for oilfields in Great Britain. *Q. J. Geol. Soc. London*, Vol. 101, 255-317.

LINDLEY, M, BOWER, F J, and HAWKINS, H S. 1974. The design and promotion of a scheme using mixed water from the Rivers Ancholme, Witham and Trent for industrial supply to Humberside. *J. Inst. Water. Engrs.*, Vol. 28, No. 5, 272-295.

LORD, A R. 1978. The Jurassic. Part 1 (Hettangian-Toarcian). 189-212 *in* A stratigraphical index of British ostracoda. BATE, R H, and ROBINSON, E (editors). *Geol. J. Spec. Issue* No. 8.

LOTT, G K, BALL, K C, and WILKINSON, I P. 1985. Mid-Cretaceous stratigraphy of a cored borehole in the western part of the Central North Sea Basin. *Proc. Yorkshire Geol. Soc.*, Vol. 45, 235-248.

— and WARRINGTON, G. 1988. A review of the latest Triassic succession in the UK sector of the Southern North Sea Basin. *Proc. Yorkshire Geol. Soc.*, Vol. 47, 139-147.

LOVELL, J H. 1977. The sand and gravel resources of the country south-west of Scunthorpe, Humberside. Description of 1:25 000 resource sheet SE 80. *Miner. Assess. Rep. Inst. Geol. Sci.*, No. 29.

MADGETT, P A, and CATT, J A. 1978. Petrography, stratigraphy and weathering of late Pleistocene tills in east Yorkshire, Lincolnshire and north Norfolk. *Proc. Yorkshire Geol. Soc.* Vol. 42, 55-108.

MATTHEWS, B. 1977. Clay-with-flints on the Yorkshire and Lincolnshire Wolds. *Proc. Yorkshire Geol. Soc.*, Vol. 41, 231-239.

MCGRAIL, S. 1981. The Brigg 'Raft' and her prehistoric environment. *National Maritime Museum, Archaeological Series*, No. 6 (British Archaeological Series 89).

MEDD, A W. 1976. A foraminiferal marker horizon in the Ampthill Clay (Upper Jurassic) of eastern England. *Proc. Yorkshire Geol. Soc.*, Vol. 41, 27-34.

MIMRAN, Y. 1977. Chalk deformation and large-scale migration of calcium carbonate. *Sedimentology*, Vol. 24, 333-360.

MITCHELL, G F, PENNY, L F, SHOTTON, F W, and WEST, R G. 1973. A correlation of Quaternary deposits in the British Isles. *Spec. Rep. Geol. Soc. London*, No. 4.

MORBEY, S J. 1975. The palynostratigraphy of the Rhaetian Stage, Upper Triassic in the Kendelbachgraben, Austria. *Palaeontographica*, Series B, Vol. 152, 1–75.

— and DUNAY, R E. 1978. Early Jurassic to late Triassic dinoflagellate cysts and miospores. *Continental Shelf Institute Publication*, No. 100, 47–59.

MORTER, A A, and WOOD, C J. 1983. The biostratigraphy of the Upper Albian–Lower Cenomanian *Aucellina* in Europe. *Zitteliana*, Vol.10, 515–529.

MORTIMORE, R N. 1986. Stratigraphy of the Upper Cretaceous White Chalk of Sussex. *Proc. Geol. Assoc.*, Vol. 97, 97–139.

— and WOOD, C J. 1986. The distribution of flint in the English Chalk, with particular reference to the 'Brandon Flint Series' and the high Turonian flint maximum. 7–20 in *The scientific study of flint and chert*, SIEVEKING, G DE G, and HART, M B (editors). (Cambridge University Press.)

MURRAY, K H. 1986. Correlation of electrical resistivity marker bands in the Cenomanian and Turonian Chalk from the London Basin to Yorkshire. *Rep. Br. Geol. Surv.*, Vol. 17, No. 8.

MUSHAM, J F. 1933; 1934. Land and freshwater shells in the lime deposits around Broughton, near Brigg. In two parts. *Trans. Lincs. Nat. Union*, Vol. 8, 145–150.

MUTTERLOSE, J, PINCKNEY, G, and RAWSON, P F. 1987. The belemnite *Acroteuthis* in the *Hibolites* Beds (Hauterivian–Barremian) of north-west Europe. *Palaeontology*, Vol. 30, 63–645.

NEALE, J W. 1962. Ostracods from the type Speeton Clay (Lower Cretaceous) of Yorkshire. *Micropaleontology*, Vol. 8, 425–484.

— 1974. Chapter 8, Cretaceous. In *The geology and mineral resources of Yorkshire.* RAYNER, D H, and HEMINGWAY, J E (editors). (Leeds: Yorkshire Geological Society.)

NEWTON, C B. 1925. New section across the buried cliff, Holderness. *Trans. Hull Geol. Soc.*, Vol. 6, 290.

NORRIS, A, BARTLEY, D D, and GAUNT, G D. 1971. An account of the deposit of shell marl at Burton Salmon, West Yorkshire. *Naturalist*, No. 917, 57–63.

OAKLEY, K P. 1940. British phosphates. Part II Phosphorites of Lower Cretaceous age in Lincolnshire. *Wartime Pam. D.S.I.R.*, *Geol. Surv. GB*, No. 8.

ORBELL, G. 1973. Palynology of the British Rhaeto–Liassic. *Bull. Geol. Surv. GB*, No. 44, 1–44.

OWEN, E F, RAWSON, P F, and WHITHAM, F. 1968. The Carstone (Lower Cretaceous) of Melton, East Yorkshire, and its brachiopod fauna. *Proc. Yorkshire Geol. Soc.*, Vol. 36, 513–524.

— and THURRELL, R G. 1968. British Neocomian rhynchonelloid brachiopods. *Bull. Br. Mus. (Nat. Hist.), Geol.*, Vol. 16, No. 3.

PACEY, N R. 1984. Bentonites in the Chalk of central eastern England and their relation to the opening of the north east Atlantic. *Earth & Planet. Sci. Lett.*, Vol. 67, 48–60.

PENN, I E, COX, B M, and GALLOIS, R W. 1986. Towards precision in stratigraphy: geophysical log correlation of Upper Jurassic (including Callovian) strata of the Eastern England Shelf. *Q. J. Geol. Soc. London*, Vol. 143, 381–410.

PENNY, L F, and CATT, J A. 1967. Stone orientation and other structural features of tills in east Yorkshire. *Geol. Mag.*, Vol. 104, 344–360.

— COOPE, G R, and CATT, J A. 1969. Age and insect fauna of the Dimlington Silts, east Yorkshire. *Nature, London*, Vol. 224, 65–67.

— and RAWSON, P F. 1969. Field meeting in east Yorkshire and north Lincolnshire. *Proc. Geol. Assoc.*, Vol. 80, 193–218.

— STRAW, A, CATT, J A, FLENLEY, J R, BRIDGER, J F D, MADGETT, P A, and BECKETT, S C. 1972. East Yorkshire. *Quaternary Research Assoc. Field Guide.*

PHILLIPS, J. 1835. *Illustrations of the geology of Yorkshire. Part I. The Yorkshire coast* (2nd edition).

— 1868. Notice of the Hessle drift, as it appeared in sections above forty years since. *Q. J. Geol. Soc. London*, Vol. 24, 250–255.

PINCKNEY, G, and RAWSON, P F. 1974. *Acroteuthis* assemblages in the Upper Jurassic and Lower Cretaceous of north-west Europe. *Newsl. Stratigr.*, Vol. 3, 193–204.

POWELL, J H. 1984. Lithostratigraphical nomenclature of the Lias Group in the Yorkshire Basin. *Proc. Yorkshire Geol. Soc.*, Vol. 45, 51–57.

— and RATHBONE, P A. 1983. The relationship of the Eller Beck Formation and the supposed Blowgill Member (Middle Jurassic) of the Yorkshire Basin. *Proc. Yorkshire Geol. Soc.*, Vol. 44, 365–373.

PREECE, R C, and ROBINSON, J E. 1984. Late Devensian and Flandrian environmental history of the Ancholme Valley, Lincolnshire: molluscan and ostracod evidence. *J. Biogeogr.*, Vol. 11, 319–352.

PRICE, J H. 1957. Hydrogeology of the Spilsby Sandstone. *Rep. Geol. Surv. GB*, No. WD/57/6. [Unpublished.]

PRICE, M. 1987. Fluid flow in the Chalk of England. 141–156 *in* Fluid flow in sedimentary basins and aquifers. GOFF, J C, and WILLIAMS, B P J (editors). *Spec. Publ. Geol. Soc. London*, No. 34.

PRYOR, W A. 1971. Petrology of the Permian Yellow Sands of north eastern England and their North Sea Basin equivalents. *Sediment. Geol.*, Vol. 6, 221–254.

RAMSBOTTOM, W H C, CALVER, M A, EAGAR, R M C, HODSON, F, HOLLIDAY, D W, STUBBLEFIELD, C J, and WILSON, R B. 1978. A correlation of Silesian Rocks in the British Isles. *Spec. Rep. Geol. Soc. London*, No. 10.

RAWSON, P F. 1971. Lower Cretaceous ammonites from north-east England: the Hauterivian genus *Simbirskites*. *Bull Br. Mus. (Nat. Hist.), Geol.*, Vol. 20, No. 2.

— CURRY, D, DILLEY, F C, HANCOCK, J M, KENNEDY, W J, NEALE, J W, WOOD, C J, and WORSSAM, B C. 1978. A correlation of Cretaceous rocks in the British Isles. *Spec. Rep. Geol. Soc. London*, No. 9.

— and MUTTERLOSE, J. 1983. Stratigraphy of the Lower B and basal Cement Beds (Barremian) of the Speeton Clay, Yorkshire, England. *Proc. Geol. Assoc.*, Vol. 94, 133–146.

— and RILEY, L A. 1982. Latest Jurassic–Early Cretaceous events and the 'Late Cimmerian Unconformity' in the North Sea Area. *Bull. Amer. Assoc. Petrol. Geol.*, Vol. 66, No. 12, 2628–2648.

REID, C. 1885. The geology of Holderness and the adjoining parts of Yorkshire and Lincolnshire. *Mem. Geol. Surv. GB.*

RHYS, G H. 1974. A proposed standard lithostratigraphic nomenclature for the southern North Sea and an outline structural nomenclature for the whole of the (UK) North Sea. A report of the joint Oil Industry–Institute of Geological Sciences Committee on North Sea Nomenclature. *Rep. Inst. Geol. Sci.*, No. 74/8.

RICHARDSON, G. 1979. The Mesozoic stratigraphy of two boreholes near Worlaby, Humberside. *Bull. Geol. Surv. GB*, No. 58, 1–24.

RICHARDSON, L. 1940. Field meeting at Lincoln. *Proc. Geol. Assoc.*, Vol. 51, 246–256.

RIDING, J B. 1987. Dinoflagellate cyst stratigraphy of the Nettleton Bottom Borehole (Jurassic: Hettangian to Kimmeridgian), Lincolnshire, England. *Proc. Yorkshire Geol. Soc.*, Vol. 46, 231–266.

ROBERTS, T. 1889. The Upper Jurassic Clays of Lincolnshire. *Q. J. Geol. Soc. London*, Vol. 45, 545–560.

ROBINSON, N D. 1986. Lithostratigraphy of the Chalk Group of the North Downs, southeast England. *Proc. Geol. Assoc.*, Vol. 97, 141–170.

ROLLIN, K E. 1982. *A review of data relating to hot dry rock and selection of targets for detailed study. Investigation of the Geothermal Potential of the UK.* (London: Institute of Geological Sciences.)

ROSE, G N, and KENT, P E. 1955. A *Lingula*-Bed in the Keuper of Nottinghamshire. *Geol. Mag.*, Vol. 92, 476–480.

ROWE, A W. 1904. The zones of the white Chalk of the English coast. IV—Yorkshire. *Proc. Geol. Assoc.*, Vol. 18, 193–296.

— 1929. The zones of the white Chalk of Lincolnshire. *Naturalist*, No. 875, 411–439.

SALTMARSHE, P. 1920. The river banks of Howdenshire, their construction and maintenance in ancient days. *Trans. East Riding Antiquarian Soc.*, Vol. 23, 1–15.

SCHLANGER, S O, ARTHUR, M A, JENKYNS, H C, and SCHOLLE, P A. 1987. The Cenomanian-Turonian Oceanic Anoxic Event, 1 Stratigraphy and distribution of organic carbon-rich beds and the marine $\delta^{13}C$ excursion. 371–399 in *Marine Petroleum Source Rocks.* BROOKS, J, and FLEET, A J (editors).

SELLWOOD, B W. 1972. Regional environmental changes across a Lower Jurassic stage boundary in Britain. *Palaeontology*, Vol. 15, 125–157.

SENIOR, J R, and EARLAND-BENNETT, P M. 1974. The Bajocian stratigraphy of Eastern England: A reply. *Proc. Yorkshire Geol. Soc.*, Vol. 40, 117–118.

SHARP, S. 1870. The Oolites of Northamptonshire, Part I. *Q. J. Geol. Soc. London*, Vol. 26, 354–391.

SHEPPARD, J A. 1966. The draining of the marshlands of south Holderness and the Vale of York. *East Yorkshire Local Hist. Ser.*, No. 20, 3–27.

SHEPPARD, T. 1897. Notes on *Elephas antiquus* and other remains from the gravels at Elloughton, near Brough, east Yorkshire. *Proc. Yorkshire Geol. Polytech. Soc.*, Vol. 13, 221–231.

— 1899. The contents and origin of the gravels around Hull. *Trans. Hull Scient. Fld. Nat. Club*, Vol. 1, 45–51.

— 1900. Notes on some remains of *Cryptocleidus* from the Kellaways Rock of East Yorkshire. *Geol. Mag.*, Vol. 7, 535–538.

— 1901. Geology of the neighbourhood of Brough, east Yorkshire. *Naturalist*, No. 532, 129–144.

— 1903. *Geological rambles in east Yorkshire.* (London.)

— 1908. Bones of reindeer at Hessle. *Naturalist*, No. 622, 424.

SHILLITO, C F B. 1937. The Kirmington fiord. *Trans. Hull. Geol. Soc.*, Vol. 7, 125–129.

SMART, J G O, and WOOD, C J. 1976. South Humberside. 586–593 *in* Field meetings. *Proc. Yorkshire Geol. Soc.*, Vol. 40, 581–600.

SMITH, A G. 1958. Post-glacial deposits in south Yorkshire and north Lincolnshire. *New Phytol.*, Vol. 57, 19–49.

SMITH, A G, HURLEY, A M, and BRIDEN, J C. 1981. *Phanerozoic palaeocontinental world maps.* (Cambridge University Press.)

SMITH, D B. 1968. The Hampole Beds—a significant marker in the Lower Magnesian Limestone of Yorkshire and Nottinghamshire. *Proc. Yorkshire Geol. Soc.*, Vol. 36, 463–477.

— 1970. The palaeogeography of the English Zechstein. 20–23 in *Third symposium on salt*, Vol. 1. RAU, J L, and DELLWIG, L F (editors). (Cleveland, Ohio; Northern Ohio Geological Society.)

— 1974. Permian. 115–144 in *The geology and mineral resources of Yorkshire.* RAYNER, D H, and HEMINGWAY, J E (editors). (Leeds: Yorkshire Geological Society.)

— 1979. Rapid marine transgressions and regressions of the Upper Permian Zechstein Sea. *J. Geol. Soc. London*, Vol. 136, 155–156.

— 1980a. The evolution of the English Zechstein basin. *Contrib. Sedimentol.*, Vol. 9, 7–34.

— 1980b. Permian and Triassic rocks. 36–48 in *The geology of north east England.* ROBSON, D A (editor). *Spec. Publ. Nat. Hist. Soc. Northumbria.* (Newcastle upon Tyne: Hancock Museum.)

— and CROSBY, A. 1979. The regional and stratigraphical context of Zechstein 3 and 4 potash deposits in the British sector of the southern North Sea and adjoining land areas. *Econ. Geol.*, Vol. 74, 397–408.

— HARWOOD, G M, PATTISON, J, and PETTIGREW, T H. 1986. A revised nomenclature for Upper Permian strata in eastern England. 9–17 *in* The English Zechstein and related topics. HARWOOD, G M, and SMITH, D B (editors). *Spec. Publ. Geol. Soc. London*, No. 22.

SMITH, E G, RHYS, G H, and EDEN, R A. 1967. Geology of the country around Chesterfield, Matlock and Mansfield. *Mem. Geol. Surv. GB.*

— RHYS, G H, and GOOSSENS, R F. 1973. Geology of the country around East Retford, Worksop and Gainsborough. *Mem. Geol. Surv. GB.*

— and WARRINGTON, G. 1971. The age and relationships of the Triassic rocks assigned to the lower part of the Keuper in north Nottinghamshire, north-west Lincolnshire and south Yorkshire. *Proc. Yorkshire Geol. Soc.*, Vol. 38, 201–227.

SMITH, W. 1817. *Stratigraphical system of organized fossils.* (London: Williams.)

STATHER, J W. 1897. Notes on the drifts of the Humber Gap. *Proc. Yorkshire Geol. Polytech. Soc.*, Vol. 13, 210–220.

— 1922. On a peculiar displacement in the Millepore Oolite near South Cave. *Proc. Yorkshire Geol. Soc.*, Vol. 19, 395–400.

STOKES, R B. 1975. Royaumes et provinces fauniques de Crétacé établis sur la base d'une étude systématique du genre. *Micraster. Mém. Mus. Nat. Hist. Nat. Paris* (c), Vol. 31.

STRAHAN, A. 1886. Notes on the relations of the Lincolnshire Carstone. *Q. J. Geol. Soc. London*, Vol. 42, 486–493.

STRAW, A. 1958. The glacial sequence in Lincolnshire. *E. Midlands Geogr.*, Vol. 1, 29–40.

— 1961. Drifts, meltwater channels and ice-margins in the Lincolnshire Wolds. *Trans. Inst. Br. Geogr.*, Vol. 29, 115–128.

— 1963a. The Quaternary evolution of the lower and middle Trent. *E. Midlands Geogr.*, Vol. 3, 171–189.

— 1963b. Some observations on the 'cover sands' of north Lincolnshire. *Trans. Lincs. Nat. Union*, Vol. 15, 260–269.

— 1969. Pleistocene events in Lincolnshire: a survey and revised nomenclature. *Trans. Lincs. Nat. Union*, Vol. 17, 85–98.

— 1983. Pre-Devensian glaciation of Lincolnshire (eastern England) and adjacent areas. *Quaternary Sci. Rev.*, Vol. 2, 239–260.

STRONG, T M W. 1960. Lithological and faunal markers useful for well correlation in the Upper Carboniferous of the East Midlands, summarized from unpublished report. Appendix. 50–53 *in* Geological results of petroleum exploration in Britain 1945–1957. FALCON, N L, and KENT, P E (editors). *Mem. Geol. Soc. London*, No. 2.

SWINNERTON, H H. 1935. The rocks below the Red Chalk of Lincolnshire, and their cephalopod faunas. *Q. J. Geol. Soc. London*, Vol. 91, 1–46.

— and KENT, P E. 1976. *The geology of Lincolnshire, from the Humber to the Wash* (2nd edition, revised by Sir Peter Kent). (Lincoln: Lincolnshire Naturalists Union.)

SYKES, R M, and CALLOMON, J H. 1979. The *Amoeboceras* zonation of the Boreal Upper Oxfordian. *Palaeontology*, Vol. 22, Part 4, 839–903, Plates 112–121.

SYLVESTER-BRADLEY, P C. 1947. Yorkshire Naturalist Union excursion to South Cave. *Naturalist*, No. 823, 169–171.

TASKIS, D M, STEEL, M A, CRADOCK-HARTOPP, M A, MOSELEY, R, SHEPHARD-THORN, E R, and DAY, J B W. 1980. Hydrogeological map of East Yorkshire. *Institute of Geological Sciences/Yorkshire Water Authority.*

TATE, R, and BLAKE, J F. 1876. *The Yorkshire Lias*. (London: J. van Voorst.)

TAYLOR, J C M. 1980. Origin of the Werraanhydrit in the UK. Southern North Sea—a reappraisal. *Contrib. Sedimentol.*, Vol. 9, 91–113.

— and COLTER, V S. 1975. Zechstein of the English Sector of the Southern North Sea Basin. 249–263 in *Petroleum and the continental shelf of North-West Europe*. Vol. 1. *Geology*. WOODLAND, A W (editor). (Barking: Applied Science Publishers Ltd.)

THOMAS, J E, and COX, B M. 1988. The Oxfordian–Kimmeridgian Stage Boundary (Upper Jurassic): Dinoflagellate cyst assemblages from the Harome Borehole, North Yorkshire, England. 313–326 in *Review of Palaeobotany and Palynology*, Vol. 56.

THOMPSON, B. 1921. Excursion to Northamptonshire. *Proc. Geol. Assoc.*, Vol. 32, 219–226.

— 1930. The Upper Estuarine Series of Northamptonshire and northern Oxfordshire. *Q. J. Geol. Soc. London*, Vol. 86, 430–462.

TRUEMAN, A E. 1918. The Lias of south Lincolnshire. *Geol. Mag.*, Vol. 5, 64–73, 101–111.

USSHER, W A E. 1890. The geology of parts of north Lincolnshire and south Yorkshire. *Mem. Geol. Surv. GB*, Sheet 86 (Old Series).

— JUKES-BROWNE, A J, and STRAHAN, A. 1888. The geology of the country around Lincoln. *Mem. Geol. Surv. GB*, Sheet 83 (Old Series).

VERSEY, H C. 1925. The beds underlying the Magnesian Limestone in Yorkshire. *Proc. Yorkshire Geol Soc.*, Vol. 20, 200–214.

— 1931. Saxonian movements in East Yorkshire and Lincolnshire. *Proc. Yorkshire Geol. Soc.*, Vol. 22, 52–68.

— and CARTER, C. 1926. The petrography of the Carstone and associated beds in Yorkshire and Lincolnshire. *Proc. Yorkshire Geol. Soc.*, Vol. 20, 349–365.

WALTON, F F. 1895. Some new sections in the Hessle gravels. *Proc. Yorkshire Geol. Polytech. Soc.*, Vol. 12, 396–406.

WARRINGTON, G. 1974. Trias. 145–160 in *The geology and mineral resources of Yorkshire.* RAYNER, D H, and HEMINGWAY, J E (editors). (Leeds: Yorkshire Geological Society.)

— 1977. Palynology of the White Lias, Cotham and Westbury Beds and the Keuper Marl of the Steeple Aston Borehole. 40–43 *in* Stratigraphy of the Steeple Aston Borehole, Oxfordshire. POOLE, E G (editor). *Bull. Geol. Surv. GB*, No. 57.

— 1978. Palynology of the Keuper, Westbury and Cotham Beds and the White Lias of the Withycombe Farm Borehole. 22–28 *in* Stratigraphy of the Withycombe Farm Borehole, near Banbury, Oxfordshire. POOLE, E G (editor). *Bull. Geol. Surv. GB*, No. 68.

— 1982. Palynology of cores from the basal Lias and the Permian (?)—Triassic sequence of the Winterborne Kingston borehole, Dorset. 122–126 *in* The Winterborne Kingston borehole, Dorset, England. RHYS, G H, LOTT, G K, and CALVER, M A (editors). *Rep. Inst. Geol. Sci.*, No. 81/3.

— 1987. Triassic palynology. 131–135 *in* Geology of the country around Chipping Norton. HORTON, A, and others. *Mem. Br. Geol. Surv.*

— AUDLEY-CHARLES, M G, ELLIOTT, R E, EVANS, W B, IVIMEY-COOK, H C, KENT, P E, ROBINSON, P L, SHOTTON, F W, and TAYLOR, F M. 1980. A correlation of Triassic rocks in the British Isles. *Spec. Rep. Geol. Soc. London*, No. 13.

WATTS, W A. 1959. Pollen spectra from the interglacial deposits at Kirmington, Lincolnshire. *Proc. Yorkshire Geol. Soc.*, Vol. 32, 145–152.

WEDD, C B. 1920. Frodingham (Lincs) District (Lower Lias). 71–105 in *Spec. Rep. Miner. Resour. GB, Mem. Geol. Surv. GB.* Vol. 12. Iron ores: bedded ores of the Lias, Oolites and later formations in England. LAMPLUGH, G W, WEDD, C B, and PRINGLE, J.

WHITEHEAD, T H, ANDERSON, W, WILSON, V, and WRAY, D A. 1952. The Mesozoic Ironstones of England. The Liassic Ironstones. *Mem. Geol. Surv. GB.*

WILKINSON, I P. 1983. Biostratigraphical and environmental aspects of ostracoda from the Upper Kimmeridgian of eastern England. 165–218 *in* Applications of Ostracoda. MADDOCKS, R F (editor). *Proc. Eighth International Symposium on ostracoda.*

— and MORTER, A A. 1981. The biostratigraphical zonation of the East Anglian Gault by Ostracoda. 163–176 in *Microfossils from recent and fossil shelf seas.* NEALE, J W, and BRASIER, M D (editors). (Chichester: Ellis Horwood.)

WILSON, A A, and CORNWELL, J D. 1982. The Institute of Geological Sciences borehole at Beckermonds Scar, north Yorkshire. *Proc. Yorkshire Geol. Soc.*, Vol. 44, 59–82.

WILSON, V. 1948. *British regional geology: east Yorkshire and Lincolnshire.* (London: HMSO for Geological Survey of Great Britain.)

WOOD, C J, and SMITH, E G. 1978. Lithostratigraphical classification of the Chalk in North Yorkshire, Humberside and Lincolnshire. *Proc. Yorkshire Geol. Soc.*, Vol. 42, 263–87.

— ERNST, G, and RASEMANN, G. 1984. The Turonian–Coniacian stage boundary in Lower Saxony (Germany) and adjacent areas: the Salzgitter–Solder Quarry as a proposed international standard section. *Bull. Geol. Soc. Denmark*, Vol. 33, 225–238.

WOOD, S V, and ROME, J L. 1868. On the glacial and postglacial structure of Lincolnshire and south-east Yorkshire. *Q. J. Geol. Soc. London*, Vol. 24, 146–184.

WOODWARD, H B. 1895. The Jurassic rocks of Britain. Vol. V, the middle and upper oolitic rocks of England (Yorkshire excepted). *Mem. Geol. Surv. GB.*

— 1904. The water supply of Lincolnshire from underground sources: with records of sinkings and borings. *Mem. Geol. Surv. GB.*

WRIGHT, C W. 1941. Brachiopods from Nettleton, Lincs. *Naturalist*, No. 796, 269–270.

— and WRIGHT, E V. 1933. Some notes on the Holocene deposits at North Ferriby. *Naturalist*, No. 920, 210–212.

WRIGHT, J K. 1977. The Cornbrash Formation (Callovian) in North Yorkshire and Cleveland. *Proc. Yorkshire Geol. Soc.*, Vol. 41, 325–346.

— 1978. The Callovian succession (excluding Cornbrash) in the western and northern parts of the Yorkshire Basin. *Proc. Geol. Assoc.*, Vol. 89, 239–261.

ZEIGLER, P A. 1982. *Geological atlas of western and central Europe.* (The Hague: Shell International Petroleum Maatschappij BV.)

FOSSIL INDEX

Taxa are indexed without the qualifications that are used in the text to indicate instances or degrees of uncertainty in determination. Fossils identified or listed at generic level only are listed *before* name species of that genus.

Italic page numbers refer to figures.

Acanthoceras jukesbrownei (Spath) *85*
A. rhotomagense confusum (Guéranger) 86
Acanthodiscus 73
Acanthopleuroceras sp. *42*
Acanthothiris broughensis Muir-Wood 52
A. crossi (J F Walker) 50, 51, 52
Acanthotriletes spp. *28*
Acesta sp. 75
Acesta longa Ro 74, 75
Acroteuthis lindseyensis Swinnerton 69
A. paracmonoides (Swinnerton) 73
A. partneyi Swinnerton 69
A. subquadratus (Roemer) 72
A. (Acroteuthis) explanatoides (Pavlow) 70, 72, 73
A. (A.) paracmonoides arctica Blüthgen 70
A. (Boreioteuthis) rawsoni (Pickney) 74
A. (Microbelus) 69
Actinocamax bohemicus Stolley 89, 100
A. plenus (Blainville) 87, 88, *89*
Aegasteroceras 36
A. sagittarium (Blake) 36, 42
Aegoceras (Androgynoceras) lataecosta (J de C Sowerby) 37, 40, 42
A. (A.) maculatum (Young & Bird) 40, 42
A. (Oistoceras) figulinum (Simpson) 42
Aegocrioceras bicarinatum (Young & Bird) 74
Aetostreon latissimum (Lamarck) 77
A. subsinuata (Leymerie) 75
A. subsinuata carinatoplicata (Renngarten) 72
Agassiceras 34
A. scipionianum (d'Orbigny) 42
Alisporites 25, 26, *28*
Allocrioceras 97
A. angustum (J de C Sowerby) *89*
Amaltheus 40
A. bifurcus Howarth *42*
A. margaritatus de Montfort *42*
A. stokesi (J Sowerby) *37*, 40, *42*
A. subnodosus (Young & Bird) *37*, 40, *42*
A. wertheri (Lange) *42*
Ammodiscus 55
'*Ammonites serpentinus*' Reinecke 44
Amoeboceras 66, *145*, *147*
A. bauhini (Oppel) *59*
A. cricki (Salfeld) *59*
A. freboldi Spath *59, 145*
A. glosense (Bigot & Brasil) *59, 145*
A. ilovaiskii (M Sokolov) *59, 145*
A. leucum Spath *59, 145*
A. mansoni Pringle *59, 145*
A. newbridgense Sykes & Callomon *59*
A. nunningtonense Wright *59*
A. regulare Spath *59, 145*
A. schulginae Mesezhnikov *59*
A. serratum (J Sowerby) *59*
A. transitorium Spath *59*
A. (Amoebites) spp. *59*
A. (Nannocardioceras) spp. *59*
Amphicytherura bartensteini Kaye & Barker 74, 76
A. roemeri (Bartenstein) 74
Amphidonte 87
Androgynoceras *37*
Angulaticeras (Sulciferites) sp. *42*
Annulispora cicatricosa (Rogalska) Morbey, 1975 *28*
Anthracoceratites sp. 7
Anthracomya phillipsii (Williamson) 14
Anthraconaia 13
A. modiolaris (J de C Sowerby) 14
A. pruvosti (Tchernyshev) 14
Anthracosia 13, 14
Anthraconauta phillipsii (Williamson) 14
Anthracosphaerium dawsoni (Brown) 12
Antiquicyprina lincolnshirensis Kelly 69
Apatocythere simulans Triebel 74
Apoderoceras 39
A. aculeatum (Simpson) *42*
Aratrisporites fimbriatus (Klaus) Mädler emend. Morby, 1975 *28*
Arenobulimina macfadyeni Cushman 77
Arnioceras 34, 36
A. bodleyi (J Buckman) 36, *42*
A. miserabile (Quenstedt) *42*
A. semicostatum (Young & Bird) 36, *42*
A. (Metarnioceras) sp. *42*
Aspidoceras *146*
Astarte *146*
Asteroceras 36
A. stellare (J Sowerby) 36
A. suevicum (Quenstedt) 36, *42*
Atreta sp. 82
Aubignina perlucida (Heron-Allen & Earland) 116
Aucellina 82, 83, 84
A. gryphaeoides (J de C Sowerby) *81*, 82–84, 86
A. uerpmanni Poluroff *81*, 82, 84, 86
Aulacostephanoides mutabilis 66
A. (Aulacostephanites) eulepidus 66
Aulacostephanus 66, *146*
A. eudoxus (d'Orbigny) *146*
A. eulepidus (Schneid) *59, 146*
A. mutabilis (J de C Sowerby) *59*
Aulacoteuthis 75
A. speetonensis (Pavlow) 75
Auritulinasporites triclavis Nilsson, 1958 *28*
Austiniceras 86–88

Balanocrinus 40
Bathichnus paramoudra Bromley, Scholz & Peake 79
Bathrotomaria 99
B. speetonensis Cox 74
B. swinnertoni Cox 70
Batioladinium longicornutum (Alberti) Brideaux, 1975 74
Beaniceras 40
B. luridum (Simpson) 40, *42*
Belemnocamax boweri Crick 86
Beyrichia 14
Bifericeras sp. *42*
Bilinguites gracilis (Bisat) 7
Biplicatoria 81, 82–84
B. ferruginea Cooper *81*, 82, 83
B. hunstantonensis Cooper *81*, 82, 83
Birostrina 83
B. concentrica (Parkinson) *81*, 82, 83
B. concentrica concentrica 83
B. sulcata (Parkinson) *81*, 82, 83, 84
Bison priscus (Bojanus) 117
Bojarkia (Bojarkia) bodylevskii (Shulgina) 70
Bos primigenius Bojanus 117
Botryococcus 46, 54, 55
Bourgueticrinus sp. 99
Brachythyris sp. 5
Burrirhynchia leightonensis (Lamplugh & Walker) 77

Caddasphaera halosa (Filatoff, 1975) Fenton et al., 1980 46
Caenisites sp. 36, *42*
Calamospora mesozoica Couper, 1958 *28*
Caloceras *33*, 34
C. bloomfieldense Donovan *42*
C. intermedium (Portlock) *42*
C. johnstoni (J de C Sowerby) *42*
Camarozonosporites rudis (Leschik) Klaus, 1960 *28*
Camerogalerus cylindricus (Lamarck) 86, 87
Camptonectes (Camptonectes) sp. 70
C. (Mclearnia) cinctus (J Sowerby) 70, 72, 73, 74, 75, 76
C. (M.) sp. 70
Caneyella 7
Canningia reticulata Cookson & Eisenack, 1960 75
Capillarina diversa rubicunda Cox & Middlemiss *81*, 82
Capillithyris 86
Carbonicola 12, 14
C. bipennis (Brown) *9*, 12
C. cristagalli Wright *9*, 12
C. obtusa Hind 12
C. oslancis Wright *9*, 12
C. proxima Eagar 8, *9*
C. rhomboidalis Hind 12
Carbonita 14
C. humilis (Jones & Kirby) 12
C. salteriana (Jones & Kirby) 12
Cardinia 32, 36, 46
C. regularis Terquem 26, 32
Cardioceras 65, 66, *145*
C. blakei Spath *59*
C. cawtonense (Blake & Hudleston) *59*
C. cordatum (J Sowerby) *59, 145*
C. costicardia S S Buckman *59*
C. densiplicatum Boden *59*
C. kokeni Boden *59*
C. praecordatum Douvillé *59*
C. scarburgense (Young & Bird) *59*

C. schellwieni Boden *59*
C. tenuiserratum (Oppel) *59*, *145*
C. zenaidae Ilovaisky *59*
C. (Cardioceras) ashtonense Arkell *59*, 65
C. (C.) persecans (S S Buckman) *59*, 65
C. (Maltoniceras) maltonense (Young & Bird) *59*, 65, *145*
C. (Plasmatoceras) spp. *59*, *64*
C. (Scoticardioceras) spp. *59*, *64*, *145*
C. (Subvertebriceras) zenaidae *64*
C. (S.) densiplicatum *64*
C. (S.) costellatum S S Buckman *64*
Carnisporites spp. *28*
Catacoeloceras sp. *42*
Ceratomya 50
Cervus 116, 117
C. elaphus Linnaeus 117
Chasmatosporites magnolioides (Erdtman) Nilsson, 1958 *28*
Chlamydophorella trabeculosa (Gocht) Davey, 1978 74
Chlamys valoniensis (Defrance) 26
Chomatoseris 57
Chondrites 32, 38, 48, 50, 88
Cingulizonates rhaeticus (Reinhardt) Schulz, 1967 *28*
Citharina discors (Koch) 74
Classopollis torosus (Reissinger) Balme, 1967 25, 26, *28*
Clydoniceras discus (J Sowerby) 57
Coccolithophoridae 78
Coeloceras pettos Hyatt *42*
Concinnithyris 83, 88
C. albensis (Leymerie) 93
C. burhamensis Sahni 86
C. protobesa Sahni 88
C. subundata (J Sowerby) *81*, 82, 86
Contignisporites problematicus (Couper) Döring, 1965 *28*
Conulus 99
C. subrotundus Maurell 88, 93
Converrucosisporites luebbenensis Schulz, 1967 *28*
Convolutispora microrugulata Schulz, 1967 27, *28*
Corbicellopsis claxbiensis (Woods) 70
Corbicula fluminalis (Müller) 117
Corbula 48, 50, 54
C. hebridica Tate 56
C. isocardaeformis Harborr 74
Coroniceras *42*
C. alcinoe (Reynès) 36
C. crossi (Wright) 36
C. gmuendense (Oppel) 36
C. scunthorpensis (Spath) 36
Cremnoceramus? 98
C. deformis (Meek) 99, 143
C. erectus (Meek) 98
C.? rotundatus (Fuege) 98
C. schloenbachi (Böhm) *89*, 99
C? waltersdorfensis (Ander) *89*, 98
C. waltersdorfensis hannovrensis (Heinz) *98*
Crendonites sp. 69
Cretirhynchia 98
C. cuneiformis Pettitt 97
C. minor Pettitt 97
Cribroperidinium sepimentum Neale & Sarjeant, 1962 74
Crioceratites 74
C. hildesiensis (Koenen) 74
C. woekeneri (Koenen) 74
Crucilobiceras sp. *37*, 38, *42*
Crussoliceras *146*
Ctenidodinium elegantulum Millioud, 1969 75
Curvirimula 7, 12
C. subovata (Dewar) 12
C. trapeziforma (Dewar) 12
Cuspidaria ibbetsoni (Morris) 54
Cyclonephelium compactum Deflandre & Cookson, 1955 88
Cyclothyris mirabilis (Walker) 77
Cylindroteuthis 61
Cymatiosphaera spp. 26, *28*, 28
Cymbites sp. *42*
Cyrtothyris cyrta arminiae Middlemiss 75
Cystispongia bursa Schrammen 97
Cythereis (Rehacythereis) senckenbergi Triebel 73
Cytherella fragilis Neale 74
Cytherelloidea pulchra Neale 74
Cytheropteron (Eocytheropteron) nova Kaye 76
Cytherurinae 74

Dactylioceras *37*, 41, 42, 45
D. anguiforme (S S Buckman) *42*
D. pseudocommune *42*
D. toxophorum (S S Buckman) *42*
D. (Dactylioceras) pseudocommune Fucini 42, 43
D. (Orthodactylites) clevelandicum Howarth 41, *42*
D. (O.) semicelatum (Simpson) 37, *42*
D. (O.) tenuicostatum (Young & Bird) 37, 41, *42*
Dapcodinium priscum Evitt, 1961 27, *28*, 28
Deflandrea 88
Deltoidospora spp. 27, *28*
Dichadogonyaulax sp. 54
Dicranodonta 70
D. benniworthensis (Kelly) 72
Dicroloma *145*
Didymotis *89*, 98
Diploceras cristatum (Brongniart) *81*
Diplocraterion 32, 45
Discoloripes fischerianus (d'Orbigny) 69
Ditrupa (Tetraditrupa) sp. 76
Dorothia kummi (Zedler) 72, 73, 74
?Dunbarella 14

Echinocorys 97–100, 143, 144
E. sphaerica (Schlüter) 86, 87
Echioceras 38
E. quenstedti (Schafhautl) *42*
E. raricostatoides (Vadasz) *42*
E. raricostatum (Zieten) *42*
?Edmondia 14
Ektyphocythere betzi (Klingler & Neuweiler) 34
Eleganticeras sp. *37*, *42*, 43
Elephas [= *Palaeoloxodon*] *antiquus* Falconer & Cautley 117
E. primigenius Blumenbach 117
Elonichthys 7, 12
Endemoceras 73
E. regale (Pavlow) 73
Entolium 56
E. demissum (Phillips) 70
E. laminosum (Maurell) 86
E. nummulare (Fischer de Waldheim) 70
E. orbiculare (J Sowerby) 69, 74, 75, 76, 77
Eoderoceras armatum (J Sowerby) *42*
Eolepas 51
Eopecten studeri (Pictet & Roux) 82
Eotrapezium concentricum (Moore) 26
Eparietites 36
E. denotatus (Simpson) 36, *42*
Epiaster michelini (Agassiz) 89, 97
Epilaugeites sp. 69
Epistomina ornata (Roemer) 73
Epophioceras 36
E. landriotii (d'Orbigny) 36
Eprolithus floralis (Stradner, 1962), Stover 1966 88
Equus caballus Linnaeus 117
Erymnoceras spp. *59*, *60*, 62
Estheria 14
Euagassiceras *33*, 34, 36
E. resupinatum (Simpson) *42*
E. terquemi (Reynès) *42*
Euaspidoceras sp. *59*
Eucoroniceras sp. *42*
Euestheria sp. 12, 15, 26
Euphemites sp. 7
Euritycythere dorsicristata Wilkinson 74
E. parisiorum Oertli 73
Exogyra conica (J Sowerby) 77
Exogyra sinuata (J Sowerby) 74
Exophthalmocythere anterospinosa Neale 73

Fabanella 72
F. boloniensis (Jones) 72
Fastigatocythere juglandica (Jones) 55

Gagaticeras *37*, 38
G. finitimum (Blake) 42
G. gagateum (Young & Bird) 42
G. neglectum (Simpson) 42
Geisina 13, 14
G. arcuata (Bean) 12
Gemmellaroceras tubellum (Simpson) *42*
Genicularia (Glandifera) rustica (J de C Sowerby) 86
Gervillella 48, 52
Gervillia praecursor (Quenstedt) 32
Gibbirhynchia 41, 97, 100
Gibbithyris subrotunda (J Sowerby) 97
Gigantoproductus 5
Gleviceras sp. *42*
Gliscopollis meyeriana (Klaus) Venkatachala, 1966 *28*
Glomerula gordialis Schlotheim 75
Glomospira gordialis (Jones & Parker) *64*, 66
Glyptocythere 55

G. guembeliana (Jones) 55
Goniomya sp. 74
Grammatodon compressiusculum (Rouillier & Vossinsky) 69
G. schourovskii (Rouillier & Vossinsky) 69
Granuloperculatipollis rudis Venkatachala & Góczán emend. Morbey, 1975 *28*
Gryphaea *33*, 34, 36, 38, 61, 62, 122
Gryphaeostrea canaliculata (J Sowerby) 82
Gyrolepis alberti L. Agassiz 26
Gyrolithes 88
Gyrostrea osmana (Wollemann) 74, 75, 77

Haplocytheridea parallela (Kaye) 76
Harmatosia crassa (d'Archiac) 86
Harpoceras elegans (J Sowerby) *37*, *42*, 43
H. exaratum (Young & Bird) *37*, *42*, 43
H. falciferum (J Sowerby) *37*, *42*, 43, 44
Hartwellia hartwellensis (J de C Sowerby) 69
Hechticythere frankei frankei (Triebel) 72
H. hechti (Triebel) 72
Hecticoceras sp. *59*
Hedbergella 88
Hemiaster 98
H. morrisii Woodward 86
Hemicrinus canon (Seeley) 82
Hemimicroceras spp. 38, *42*
Hiatella foetida (Cox) 69
Hibolites jaculoides Swinnerton 74
Hildaites murleyi (Moxon) 42
Hildoceras bifrons (Brugnière) 44
Hippopotamus 116
Holaster 82, 86
H. subglobosus (Leske) 86, 87
H. trecensis (Leymerie) 87
Hyalostelia sp. 7, 8
Hyperlioceras 49, 52
H. rudidiscites S S Buckman 52
Hyphantoceras reussianum (d'Orbigny) *89*, 97
Hyposalenia sp. nov. 86
Hypoturrulites spp. 86
Hystrichodinium furcatum Alberti, 1961 74
H. voigtii (Alberti, 1961), Davey, 1974 74

Idiognathoides sinuatus Harris & Hollingsworth 7
Infulaster *89*, 93, 98, 99
I. excentricus (Woodward) 93
I. hagenowi (d'Orbigny) 97
Inoceramus 83, 84, 93, 97
I. anglicus Woods *81*, 82, 83
I. annulatus Goldfuss *89*, 99
I. atlanticus Heinz 87
I. crippsi Mantell 86
I. cuvierii J Sowerby 93
I. gibbosus Schlüter 100
I. lezennensis Barrois 100
I. glatziae Flegel 97, 98
I. inaequivalvis Schlüter 93, 97
I. lamarcki Parkinson 93, 99
I. lamarcki geinitzi Tröger 93
I. lamarcki stuemckei Heinz 97
I. lissa (Seeley) *81*, 82, 83
I. lusatiae Andert 93
I. modestus Heinz 97
I. pictus J de C Sowerby 87, 88
I. websteri Mantell 99
Iotrigonia scapha (Agassiz) 74
Isastraea 55, 57
Ischyosporites variegatus (Couper) Schulz, 1967 *28*
Isthomocystis distincta Duxbury, 1979 74

?Jamesonites 39

Kallirhynchia sharpi Muir-Wood 55
K. yaxleyensis (Davidson) 57
Kepplerites sp. *59*, *60*, 61
Kerberites kerberus S S Buckman 69
Kingena concinna Owen 86
K. elegans Owen *89*, 97
K. spinulosa (Davidson & Morris) *81*, 82
Kosmoceras spp. *59*, *60*, 62, *145*
K. (Kosmoceras) sp. 62
K. (Lobokosmokeras) duncani (J Sowerby) 62
K. (Spinikosmokeras) sp. 62
Kosmoceras (Zugokosmokeras) medea Callomon 60
Kraeuselisporites reissingeri (Harris) Morbey, 1975 *28*
Kyrtomisporis laevigatus Mädler, 1964 *28*

Laevapytchus *146*
Lamellaerhynchia rawsoni Owen 75
L. rostriformis (Roemer) 72, 73
L. walkeri claxbiensis Owen & Thurrell 73
Leioceras opalinum (Reinecke) 45
Lenticulina 66
L. crepidularis (Roemer) 73
L. guttata (Ten Dam) 72
L. ouachensis wisselmanni Bettenstaedt 74
L. schreiteri (Eichenberg) 72
Leptechioceras *37*, 38
L. macdonnelli (Portlock) *42*
Lewesiceras 88, 97
L. mantelli Wright & Wright *89*
Leymeriella tardefurcata (d'Orbigny) *81*
Limbosporites lundbladii Nilsson, 1958 *28*
Limea granulatissima Wollemann 74
Lingula 7, 12, 13, 14, 25, *33*, 34
L. kestevenensis Muir-Wood 55
L. mytilloides J Sowerby 7, 8, 12, 14
L. 'mytilus' 14
Linotrigonia (Oistotrigonia) ornata (d'Orbigny) 74
Liostrea 32
L. hebridica (Forbes) 55
Liparoceras *42*
L. bronni Spath 39
Lithostrotion junceum (Fleming) 5
Lobothyris edwardsi (Davidson) 41
L. punctata (J Sowerby) 41
Lophocythere propinqua Malz 55
L. transversiplicata Bate 55
Lucina hauchecornei Wollemann 74
Ludwigia 45
Lunatisporites rhaeticus (Schulz) Warrington, 1974 *28*
Lyapinella laevis (Phillips) 70, 72
Lyriomorphoria postera (Quenstedt) 26
Lytoceras *37*, 40
L. crenatum (S S Buckman) *42*
L. fimbriatum (J Sowerby) *42*

Macrocephalites (Dolikephalites) typicus Blake 57
Mammites nodosoides (Schlüter) 88
Mandelstamia sexti Neale 72
Mantelliceras dixoni Spath *85*
Marginulinopsis cephalotes (Reuss) 77
Matronella cottigenda (Kaye) 83
Meleagrinella? 56
Menjaites 72
Metoicoceras geslinianum (d'Orbigny) *28*, *85*, *89*
Metopaster sp. 100
Meyeria rapax (Harbort) 74
Micraster 97, 98, 143
M. bucaillei Parent *89*, 97, 99, 100, 144
M. coranguinum (Leske) 100
M. corbovis Forbes *89*, 97, 98
M. gibbus (Lamarck) 99
M. leskei (Des Moulins) *89*, 97
M. normanniae Bucaille *89*, 98, 99
Micrhystridium spp. 26, *28*, 28, 46
Microbiplices sp. *59*
Microderoceras birchi (J Sowerby) 36
Micropneumatocythere quadrata Bate 55
Microreticulatisporites fuscus (Nilsson) Morbey, 1975 *28*
Modestella festiva Owen 77
Modiolus 70
M. imbricatus J Sowerby 55
Monticlarella 88
M. carteri (Davidson) 86
M. jefferiesi Owen 87
Mortoniceras (Mortoniceras) inflatum (J Sowerby) 81, 82
M. rostratum (J de C Sowerby) *81*
Moutonithyris dutempleana (d'Orbigny) 83
Mucroserpula sp. 75
Muderongia simplex Alberti, 1961 74
M. tetracantha (Gocht) Alberti, 1961 74, 75
Myalina compressa Hind 14
Myophorella claxbiensis (Kelly) 70
M. intermedia (Fahrenkohl) 69
M. rawsoni 72
M. (Pseudomyophorella?) rawsoni Kelly 72
Mytiloides 87, 88, 93, 143
M.? dresdenensis (Tröger) 98
M. labiatus (Schlotheim) *28*, 88, *89*
M. mytiloides (Mantell) 88, *89*
M. striatoconcentricus Gümbel 97

Naiadites 12, 13, 14
N. flexuosus Dix & Trueman 12
N. subtruncatus (Brown) 12
Najdinothyris 97
Nannocardioceras *146*
Nannoceratopsis 48, 50
N. gracilis Alberti, 1961 54
Nanogyra thurmanni (Etallon) 70
Natica oppelii Moore 26
Neithea 77
N. quinquecostata (J Sowerby) 77
Neocomites? trezanensis (de Loriol) 72

Neocrassina nummus (Sauvage) 69
Neocythere (Neocythere) protovanveenae Kaye 74
Neohibolites 82, 84
N. ernsti Spaeth *81*, 82, 83, 84
N. minimus Miller 77, 81, 83
N. minimus pinguis Stolley 77
N. oxycaudatus Spaeth *81*, 83
N. praeultimus Spaeth 81, 82, 83
N. ultimus (d'Orbigny) 83, 84, 86
Neomicroceras sp. *37*, 38, *42*
Neomiodon 48, 50, 54, 56
Nerineopsis aculeatum (Shomon & Newron) 74
Nevesisporites bigranulatus (Levet-Carette) Morbey, 1975 *28*
Nicaniella (Trautscholdia) claxbiensis (Woods) 70
Nielsenicrinus cretaceus (Leymerie) 82

Obovothyris stiltonensis (Walker) 57
Odontochitina operculata (O. Wetzel) Deflandre & Cookson, 1955 77
Oistoceras *37*, 40
Orbiculoidea 7, 14
O. cincta (Portlock) 12
O. nitida (Phillips) 7, 8
Orbirhynchia 93, 97, 98, 99
O. compta Pettitt 88, 93
O. dispansa Pettitt 97
O. herberti Pettitt 93
O. mantelliana (J de C Sowerby) 86
O. multicostata Pettitt 87, 83
O. wiesti (Quenstedt) 87, 88
Ornatothyris (Actactosia) 83, 86, 87
O. (A.) obtusus (J de C Sowerby) 86
O. (A.) pentagonalis Sahni *81*, 82
O. (Ornatothyris?) spp. 86
Ornithella subcalloviensis (Douglas & Arkell) 57
Osangularia schloenbachi (Reuss) 77
Ovalipollis pseudoalatus (Thiergart) Schuurman, 1976 25, 26, 27, *28*
Oxynoticeras 38
O. oxynotum (Quenstedt) 38, *42*
O. simpsoni (Simpson) 36, *42*
Oxyteuthis 76
O. germanica Stolley 76
Oxytoma *145*, 166, *147*
O. pectinatum (J de Sowerby) 77
O. seminudum (Dames) 86, 87, 88

Palaeoechioceras sp. 37, 38, *42*
Paltechioceras 38
P. boehmi (Hug) *42*
Panopea 74
P. neocomiensis (Leymerie) 70
Paraconularia sp. 7
Paracraspedites sp. 77
Paracymbites dennyi (Simpson) 42
Parallelodon 48, 52
Paranodiceras sp. *42*
Paranotacythere (Paranotacythere) anglica (Neale) 74
P. (P.) blanda (Kaye) 74, 76
Parexophthalmocythere rodewaldensis Bartenstein & Brand 74
Parsimonia antiquata (J de C Sowerby) 86
Pavlovia sp. 69
Pectinatites *59*, 66, 67, 69, *146*
P. (Arkellites) hudlestoni Cope 67
P. (Pectinatites) eastlecottensis? (Salfeld) 67
Pectinatites spp. *59*, 66, 69, *146*
Peltoceras *59*, *60*, 62, *145*
Peregrinoceras 72
P. albidum Casey 70
P. bellum Sasonova 70
P. rosei Casey 70
P. subpressulum (Bogoslovsky) 70
P. wrighti (Neale) 70
Perinosporites thuringiacus Schulz, 1962 *28*
Perisphinctes 65
Phlebopteris woodwardii (Leckenby) 46
Phoberocysta neocomica (Gocht) Millioud, 1969 74
Pholadomya 48, 50, 69
Phragmites 116
P. communis (Trinius) 126
Phricodoceras taylori (J de C Sowerby) *42*
Pictonia spp. *59*, 66, *146*, *147*
Pinna 38, 39, 70
P. subcuneata Eichwald 69
Pityosporites scaurus (Nilsson) Schulz, 1967 *28*
Plagiostoma 32
P. planum (Roemer) 72
P. ferdinandi (Weerth) 76
P. subcardiiformis (Greppin) 70
Planolites 13, 14, 88
P. opthalmoides (Jessen) 7, 8, 14
Platyceramus 100, 101
P. mantelli (Barrois) 100
Platypleuroceras *37*, 39, *42*
P. bituberculatum Tutcher & Trueman 39
Platysomus sp. 8, 12
Pleuroceras *37*
P. hawskerense (Young & Bird) 41
P. spinatum (Bruguière) *37,* 41, *42*
Pleuromya 69, 72
P. peregina (d'Orbigny) 70
P. uniformis (J Sowerby) 69
Plicatula inflata (J de C Sowerby) 86
P. minuta Seeley 82
P. producta (Rouillier & Vossinsky) 69
Polycingulatisporites bicollateralis (Rogalska) Morbey, 1975 *28*
Polymorphites *37*, 39
P. lineatus (Quenstedt) *42*
P. mixtus (Quenstedt) *42*
P. quadratus (Quenstedt) *42*
P. trivialis (Simpson) *42*
Polyptychites gravesiformis (Pavlow) 72
Praeglobotruncana helvetica (Bolli) 90
P. praehelvetica (Trujillo) 90
Praeoxyteuthis jasikofiana (Lahusen) 75
P. pugio (Stolley) 75
Praeoxyteuthis spp. 75
Procyprina 75
P. centralis Casey 70
?Productus 14
Prognocythere triquetra Bate 55
Promicroceras 36, 38, *42*
P. planicosta (J Sowerby) 36, *42*
Propectinatites sp. *59*
Proplanulites sp. *59*, 62
Protocardia *34*, *145*, 48
P. rhaetica (Merian) 26, 32
Protocythere strigosa Grosdidier 74
P. triplicata (Roemer) 72, 76
Protogrammoceras paltum (S S Buckman) 41, 43
Protohaploxypinus microcorpus (Schaarschmidt) Clarke, 1965 *28*
Protopinus scanicus Nilsson, 1958 *28*
Protosacculina macrosacca Schulz, 1967 *28*
Pseudoceratium pelliferum Gocht, 1957 74, 75
Pseudogalathea? 7
Pseudojacobites farmeryi (Crick) *89*, 97
Pseudolimea echinata Etheridge 86
Pseudomelania 52
Pseudomytiloides sp. 43
Pseudonodosaria vulgata (Bornemann) 73
Pseudopecten 39
P. equivalvis (J Sowerby) 39
Pseudotrapezium 48
Psiloceras 16, 28, 30, 32, *33*
P. planorbis (J de C Sowerby) 32, *42*
Psilophyllites *33*, 34
P. hagenowi (Dunker) *42*
Pycnodonte (Gryphaea) 87
Phrygaea 82, 143
P. (Phrygaea) vesicularis (Lamarck) 82

Quadraeculina anellaeformis Malijavkina, 1949 *28*
Quasihermanites bicarinata Gründel 74
Quenstedtoceras *145*
Q. leachi (J Sowerby) *59*
Q. mariae (d'Orbigny) *59*
Q. woodhamense Arkell *59*
Q. (Lamberticeras) spp. *59*

Rasenia *59*, 66, *146*, *147*
Rastellum colubrinum (Lamarck) 77
R. macropterum (J Sowerby) 74, 75, 76, 77
Rectithyris? 86
Resatrix neocomiensis (Weerth) 74
Reticuloceras reticulatum (Phillips) 7
Rhabdoderma 7, 12
Rhadinichthys 7, 8, 12
Rhaetavicula contorta (Portlock) 26
Rhaetipollis germanicus Schulz, 1967 *28*
Rhaetogonyaulax rhaetica (Sarjeant) Loeblich & Loeblich emend. Harland, Morbey & Sarjeant, 1975 26, 27, *28*
Rhazella 63
Rhizocorallium 32
Rhizodopsis 12
R. sauroides (Williamson) 12
Rhynchonelloidea inflata Douglas & Arkell 57
Ricciisporites tuberculatus Lundblad, 1954 26, *28*
Ringsteadia spp. *59*, 66, *145*, 147
Rotalipora 86, 87
Rotularia umbonata (J Sowerby) 82
Rouillieria ovoides (J Sowerby) 69, 70
R. tilbeyensis (Davidson) 72, 73

R. walkeri (Davidson) 73
Rugitela hippopus (Roemer) 73
R. rugosa Owen 73, 75

Sanguinolites sp. 7
Saracenaria bononiensis (Berthelin) 77
Scapanothynchus raphiodon (Agassiz) 100
Scaphitodites scaphitoides (Cocquand) *59*
Schizophoria 7
Schlotheimia *33*, 34, *42*
S. similis Spath 34
Schuleridea hammi (Triebel) 76
S. rhomboidalis Neale 76
S. trigonalis (Jones) 55
Sciponoceras anterius Wright & Kennedy *89*
S. bohemicum anterius Wright & Kennedy 88
Schloenbachia sp. 86
Sellithyris sella lindensis Middlemiss 73
S. 48
Serpuloides 7, 8, 13
S. stubblefieldi (Schmidt & Teichmuller) 7, 8
Sigaloceras *59*, *60*, 61, 62
S. enodatum *59*
S. (Catasigaloceras) enodatum *60*, 61
Simbirskites toensbergensis (Weerth) 74
S. (Craspedodiscus) 74
S. (C.) discofalcatus (Lahusen) 74
S. (C.) juddii Rawson 74
S. (Milanowskia) concinnus (Phillips) 74
S. (M.) speetonensis (Young & Bird) 74
S. (Simbirskites) virgifer (Neumayr & Uhlig) 74
Sonninia (Fissilobiceras) sp. 52
Sowerbya longior Blake 69
Spartina 126
Spirorbis 12, 13, 14
Spondylus dutempleanus d'Orbigny 99
Stegoconcha plotii (Lycett) 70
Steinmannia 43
Sterisporites *28*
Stereocidaris sceptifera (Mantell) 99
Sternotaxis 93
S. placenta (Agassiz) *89*, 97, 98
S. plana (Mantell) 89, 93, 97
?Strepsodus 14
Subastieria sulcosa (Pavlow) 73, 74
Subcraspedites 69
S. (Subcraspedites) claxbiensis Spath 77
S. (Swinnertonia) 77
S. (S.) cristatus Swinnerton 69
S. (S.) primitivus Swinnerton 69
S. (Volgidiscus) lamplughi Spath 77
Subprionocyclus neptuni (Geinitz) *89*, 93
Sulcatisporites pinoides Nilsson, 1958 *28*
Surites 77
Sutneria eumela (d'Orbigny) *146*
S. rebholzi (Berckhemer) *59*, *146*
Sverdrupiella mutabilis Bujak & Fisher, 1976 27, *28*
Synolinthia? 100

Tamarella oweni Peybemes & Calzadra *81*, 82
Taramelliceras *59*
Tasmanites *28*, 28
Terebratulina 86
T. lata R Etheridge 78, *89*, 93, 97
T. martiniana (d'Orbigny) 82
T. striatula (Mantell) 97, 99
Tetraditrupa 74
Tetrarhynchia tetrahedra (J Sowerby) 41
Thalassinoides 48, 79, *81*, *85*, 86
T. paradoxica (Woodward) 82, 86
Tiltoniceras 43
T. antiquum (Wright) *37*, *42*, 43
Todisporites minor Couper, 1958 *28*
?Tomaculum 14
Toxaster retusus (Lamarck) 74
Tragophylloceras sp. *37*, 39, *42*
Triancoraesporites spp. *28*
Trigerastrea 55
Trigonia hemisphaerica Lycett 50
Trochocyathus conulus (Phillips) 74
Tsugaepollenites? pseudomassulae (Mädler) Morbey, 1975 *28*
Turnus amphisboena (Goldfuss) 88
Tutcheria cloacina (Quenstedt) 26

Uptonia 39, *42*

Vaginulopsis humilis praecursoria Bartenstein & Brand 75
Vermiceras *33*, 34
V. conybeari (J Sowerby) 34, *42*
V. scylla (Reynès) *33*, 34, *42*
Veryhachium 28
Vesicaspora fuscus (Pautsch) Morbey, 1975 25, *28*
Vitreisporites pallidus (Reissinger) Nilsson, 1958 *28*
Volgidiscus 69
Volviceramus 100, 101
V. involutus (J de C Sowerby) *89*, *95*, 100
V. koeneni (Müller) 89, *95*, 100, 144

Waehneroceras *33*, 34
W. iapetus Spath 42
Watznaueria barnesae (Black *in* Black & Barnes, 1959) Perch Nielsen, 1968 88
Witchellia 50, 52

Xenostephanus *59*, 66, *146*
Xipheroceras 36, *42*

Zebrasporites 28
Zoophycos 48, 98, 143

GENERAL INDEX

Italic page numbers refer to figures.
Bold page numbers refer to tables.
Page numbers followed by P refer to plates.

Aalenian Stage *29*, *44*, 44, 45, 46, 106
Abdy Coal 13
Acheulian Industry 116
Ackworth Division 14
Ackworth Rock *11*, 14, 15
Acre House 75, *135*
Acre House Mine 67, *135*, 136
ambygonium Zone 68
arctica 70
Acroteuthis Assemblage 4 72
Acroteuthis Assemblage 5 73
Acroteuthis Assemblage 7 74
Acton Rock 14
Acutistriatum Band-Comptoni Bed 62, *145*
Aegiranum (Mansfield) Marine Band *11*, 13-14
aeromagnetic anomalies *102*, 102, 103
Aisby 113
Aislaby Group **16**, 17, *18*, 20-21
Alandale Borehole *53*, 57, *60*, 61, 62, 63, *64*, 140, *148*
albae/bosquetiana ostracod Zone 83
albae/vinculum ostracod zonal assemblage 83
Albert and Alexandra Docks 118
Albian Stage *68*, 71, 77, **78**, 82, 83, 84, 86
albidum Zone *68*, 70, 72
Alkborough 127
Alluvium **111**, 126, 136
Alportian Stage **5**
?Alton Coal *9*
alumina 36
Amaliae (Norton) Marine Band 8, *9*
amblygonium Zone *68*
Amcotts 127
Ammonite Beds *85*, 86
Ampthill Clay 57, 66, *145*, 145
Ancaster Beds 51
Ancholme Clay Group 3, *29*, 29, 30, 57-70, 107, 108, 128, 129, 136
  biostratigraphy 145-148
  Callovian and Oxfordian sequences *145*
  chronostratigraphical correlation *58*
  Kimmeridgian sequence *147*
  sections *63*, *64*
Ancholme, New River *132*
Ancholme, River *2*, 2, 126, 137
Ancholme Valley 109, *110*, 112-113, 121, 122, 123, 124, 126
Anglian Stage 109, **111**, 112, 113
Anglian Water Authority 137
Anglo-Paris Province 77
anguiformis Zone *68*, 69
'Angulata' Clays 30
angulata Zone *31*, *33*, 34
anhydrite 17, 19, 20, 21, 22, 23, 25, 129
Anisian Stage 24
ankerite 38
apatite 54
aplanatum Subzone *37*, 38, *42*
Appleby 48, 50, 106, 122, 124, 130
Appleby Carrs ironstone reserves 57, *132*, **133**
Appleby Fault Zone *132*
Appleby Frodingham Works *132*
Appleby Station Borehole 54
Aptian-Albian boundary 76
apyrenum Subzone 41, *42*
aquifers
  Cretaceous 138
  Jurassic 138-139
  Triassic 139
Arctic Boreal Ocean 67
Arnsbergian Stage **5**
arsenic 36
artefacts, Mesolithic 124
Asbian Stage **5**, 5
Ashby 34, 35, 130
Ashby Hill Quarry 79, 99, 143
Ashdale House 99
Ashdale House pit *91*, 143
Ashover Grit 7
Askern-Spital Structure 57, 60, 104-105, 106
Aspen Farm 35
aspidoides Zone *44*, 55
Athleta Zone *58*, *59*, 62, *63*, *145*
Atterby 50, 57
Audleby 71, 73, 76
Audleby Monocline 106
auritus Subzone *81*, 84
Autissiodorensis Zone *58*, *59*, *64*, *146*

Bagmoor ironstone reserves *132*, **133**
Bagmoor Farm 130P
Bajocian Stage *29*, *44*, *53*, 54, 106
Barmby Barrage 137
Barnetby 84
Barnetby Gap *2*, *110*, 118, 119, 123
Barnetby le Wold 66, 76, 123
Barnetby Pit 87
Barnetby Station Borehole *60*, 61, 141
Barnetby Wold Farm 118
Barnsley Coal *10*, 13, 104
  structure-contour map *104*, 105
Barremian Stage *68*, 73, 74, 76
Barrow eastern and western pits *91*, 143
Barrow Flints *94*, 98, 99, 143
Barrow Haven 125, 126, 129
Barrow upon Humber *91*, 98, 116, 125, 138
  boreholes 137
  quarries 96, 98
Barrow Vale Farm pit *91*
Barton Cliff 129
Barton Hill Farm pit *91*
Barton marl seams *79*, 93
Barton Marl 1 90, *92*, 143
Barton Marl 2 90, *92*
Barton Marl 3 *92*
Barton Marl 4 *92*
Barton pit *91*
Barton School Pit *91*
Barton Vale pit *91*
Barton-upon-Humber *91*, 116, 118, 129, 138
Basal Permian Sands **16**, 17, *18*
Basal Spilsby Nodule Bed 67, 69
Basement Clays 60
Bathonian Stage *29*, *44*, 54, 55, 56, 57, 106
Baylei Zone *58*, *59*, *64*, 66, *146*, *147*
Beacon Hill Marl Seam *79*, *92*
Beaulah Wood 45
The Beck 125
Beck Farm 57, 61, 137
Beeston coals 12
berthierine 30, 34, 35, 36, 39, 40, 41, 45, 71, 72, 131
Better Bed 12
Bielsbeck 117
Bigby *91*, 128
Bigby Quarry 82, 84, 86, 88, *91*, 143
Billingham Main Anhydrite **16**, *18*, *19*, 21
birchi Subzone 36, *42*
'biscuit stone' 34
Black Band *79*, *85*, 87-88, *92*
Black Band (Crow) Coal *9*
Black Bed Coal 12
Blacktoft 126, 127
Blisworth Clay Member *44*, *53*, 55-56
  channel fills 55
Blisworth Limestone 52, 55
?Blocking Coal *9*
blow wells 138
Blown Sand **111**, 122, 124, 136-137
Blue Anchor Formation 25, *28*
Blyborough Borehole 26, *27*, 27, 28, 32, *33*, 34, 36, *37*, 38, 39, 140
Blyton 106, 113, 115
Blyton Borehole *6*, 7, 8, *9*, *10*, *11*, 12-15, 17, *18*, 19, *24*, 129, 140
Blyton School 26
Bonby 122
Bonby Carrs 126
Boreal marine transgression 30
Boreal taxa 30
Boswell Farm Pit 97
Bottesford 32, 34, 35, 106
Bottesford Beck 122, 124, 125
Bottesford Fault 106
'Bottom Blue' 51
Bouguer gravity anomalies *102*, 102, 103
Boulby Halite **16**, *19*, 21-22, 129
boxstones 39, 41, 45, 48, 77
Brantingham 121, 122, 127
Brantingham Dale 124
Brantingham Formation 3, 57, *58*, 58, 62-66, *63*, *64*, 106, 139, *145*
  aquifer 138
  fauna 63, 65
Brat Hill 35, 130
brevispina Subzone 39, *42*

brevispina/polymorphus Subzone *37*
Brickhills 129
Bridge Farm 122
Brierley Coal *11*, 14–15
Brierley Division 15
Brierley Rock *11*, 15
Brigantian Stage **5**, 5
Brigg *1*, *2*, 2, 41, 43, *44*, 45, 52, 54, 55, 56, 106, 109, 112, 113, 122, 123, 125, 129, 139
Brigg Borehole *18*, 20, 22, 23, *24*, 129, 140
Brigg Bridge Farm Borehole 61
Brincliffe Edge Rock 12
Brinsley Coal *10*, 13
British Steel Corporation 131
Bröckelschiefer Member 23
Brocklesby 117
Brocklesby No. 14 Borehole 35
Bronze Age 2, 124, 125
brooki Subzone *42*
Broomfleet 126, 129
Broomfleet Borehole 125, 126
Broomfleet Hope 126
Brough 52, 56, 62, 106, 118, 119, 122, 128, 138, 139
Brough Airfield 125, 126
Brough fault complex *65*, 106
Brough No. 1 Borehole 43, 46, *47*, 52, *53*, 54, 61, 118, 140
Brough No. 2 Borehole *60*, 122
Brough 2
Broughton 125, *132*
Broughton Common 106
ironstone reserves 131, *132*
Broughton Crossroads 50
Broughton Borehole 129, 140
bucklandi Zone *31*, *33*, 34, *42*
building stone 128
Bull Inn, South Kelsey 112
Bunter **16**
Bunter Pebble Beds 23
Bunter Sandstone 23
Bunter Shale Formation 23
Buntsandstein 24
Buntsandstein-Keuper sequence 24
Burnham 87, *91*, 93, 119
Burnham Chalk Formation *79*, 77, **78**, 78, 79, 81, *91*, 93–101
detailed section *94–95*
stratigraphy of *89*
Burnham Lodge pit *91*
Burnham Lodge Quarry 93, 96
Burnham pit *91*, 93
burrow-fills
phosphatised *81*, 82, 83
Burton Salmon 126
Burton upon Stather *2*, 2, 34, 115
Burton upon Stather Borehole (hydrocarbon) 5, *6*, 7, 8, *9*, *10*, *11*, 13, 14, 15, *18*, 20, 21, 22, *24*, *27*, 129, 140
Burton upon Stather Borehole (water) *27*, 141
Butterwick Common 125
Butterwick Borehole 5, *6*, 7, 8, *9*, *10*, *11*, 12, 14, 15, *18*, 19, 20, 22, 23, *24*, 129, 140

Cadney 112, 113, 124, 125, 137
Cadney Common 113
Caistor 2, 67, 86, *91*, 106, 115, 124, 128, 137
Caistor Monocline 77
'Calcareous Beds' 56
calcareous tufa **111**, 125
calcite 38
Callipholites auritus Subzone *81*, 84
Callovian Stage *29*, 29–30, *44*, 56, 57, *58*, 60, *63*, *64*, 106, *145*
Calloviense Subzone 61
Calloviense Zone *58*, *59*, 60, 61, *145*
Calycoceras guerangeri Zone *85*, 87
cambering 45, 73, 106, 107, 108, 109, 127
Cambriense (Top) Marine Band *11*, 14
Campanian Stage **78**
Cancellatum Marine Band *6*, 7, 8
Candley Beck 122
'cannonshot' flints 122
'Cannonshot Gravel' 115, 116
Capper Pass C Borehole 62
Capper Pass Works 118, 139
capricornus Subzone 40, *42*
carbon isotope excursion 90
Carboniferous 3, 5–15, *18*
Dinantian 2, 5, *6*
Namurian 2, 5, *6*, 7–8
Westphalian 8–15
Carboniferous Limestone 5
borehole sections *6*
Carboniferous rocks, structure 104–105
carcitanense Subzone 86
Carnallitic Marl **16**, *18*, 22
Carstone Formation 3, *68*, 71, 76–77, 138
Carstone Grit Member *68*, 76
Castlethorpe 45, 48, 106
Castlethorpe Hall 122, 125
Cathedral Beds 49, 50
Cave Oolite Member *44*, 46, *47*, 49, 50, 51–52, 54, 128, 138
aquifer 139
cement manufacture 2
cementstone 50, 58
Cenomanian Stage **78**, *79*, 84, *85*, 86, 90, 143
Cererithyris intermedia Zone 57
chalcedony 69
Chalk (Group) 3, 77–101, 106, 107, 137
aquifer 138
correlation of Northern Province with Southern Province *79*
outcrops of formations *91*
hydrogeological properties 138
Northern Province, classification of **78**
structure-contours on base 106
Chalk Hill 143
Chalk Hill Marls 88, 90, *92*, 143
chalk
compostion of 78
flaser-bedded 98, 143
laminated 90, 93, *95*, 97, 98, 143
nodular 143
as a raw material 128
'chalk bearings' 138
'Chalky Boulder Clay' 112, 113
Chapel Farm pit *91*
Chatsworth Grit 7
Cherry Farm 55
chert 67, 69, 76
chlorite 32, 69
chloritoid 54
Chokierian Stage **5**
Clactonian Industry 116
Clapgate Farm 56
Clapgate No. 8 Borehole 61
Clapgate No. 9 Borehole *53*, 54
Clarborough Beds 25
Claxby 71, 106, 124, *135*, 136
Claxby Ironstone Formation 1, 3, *68*, 71–73, 129, 136
depositional environment 71
derived fossils 72
fauna 72
mineralogy 71–72
workings *134–135*
Claxby Moor 129
clay
brick, tile and cement making 129
Clay-with-Flints **111**, 117
Cleatham 40, 41, 49, 50
Cleatham Limestone *44*, 46, *47*, 48
Cleveland Basin 30, 106
clevelandicum Subzone *37*, 41, *42*, 43
Cliff, The 25
Clixby 76
Cloughton Formation 52
Clowne Coal 13
Clowne Marine Band 13
coal 2, 54, 129
Coal Dyke End 125
Coal Measures (Westphalian) 3, 8–15, *18*, 129
Cockle Pits Borehole 25, 26, *27*, 27, 28, 32, *33*, 34, 35, *37*, 38, 39, 40, 41, 43, 129, 140
Coleby Opencast Mine 36, 48
Coleby Mudstones Formation 30, *31*, 36–44, 129, 130P, 136
calcareous concretions 43
fauna 38–40, 43, 44
isopachytes *31*
sections *37*
Coleby opencast mine 36
Coleby (Winterton) Mine 35
collophone 67
coloured glass sand **137**
Comb Coal *10*
Common Plantation ironstone reserves *132*, **133**
commune Subzone 44
complanata Subzone 34, *42*
concinnus Subzone 74
concretions 58
calcareous 43

clay-ironstone 36, 38
septarian *64*, 66
Conesby ironstone reserves *132*, **133**
Conesby Bottom 115
Coniacian Stage *79*, **78**, *89*
'Conoco' Marls 100, 144
Contorta Shales 26
conybeari Subzone *33*, 34, *42*
Corallian 62, 63, *64*
coralline knolls 50
Cordatum Zone *58*, *59*, *63*, *64*, *145*
Cornbrash Formation 3, *44*, *53*, 55, 56–57, 60, 61, 128
aquifer 138
fauna 57
sections *53*
Coronatum Zone *58*, *59*, 62, *63*, *145*
Corringham 2
Corringham boreholes 129
No. 1 Borehole 5, *6*, 7, 8, *9*, 12
No. 2 Borehole 7
No. 3 Borehole 5, 7, 140
No. 4 Borehole 5, 12, 141
No. 6 Borehole 8, 12
No. 7 Borehole 5, *6*, 7, 8, *9*, *10*, *11*, 12, 13, 14, 15, 141
No. 8 Borehole 5
No. 9 Borehole 5, 12, 141
No. 10 Borehole 5, 18, 19, 23, *24*, 141
costatus Subzone 86
Cotham Member 26, *27*, 27, *28*, 28, 32
Crawshaw Sandstone 8
Cremnoceramus? rotundatus-erectus lineage *89*
Cretaceous *1*, 3, 71–101
aquifers 138
Lower Cretaceous 71–77
Upper Cretaceous 71, 77–101
cristatum Subzone 83
crog-balls 50
Cromerian Stage 109, **111**
Crosby Borehole 5, *10*, *11*, 13, 14, 16, *18*, 22, 23, *24*, 141
Crosby Warren 35, 39, 40, 132
ironstone reserves *132*, **133**
Crosby Warren Mine *37*, 38, 39, 40, 41
borehole near *37*
Crossi Bed 51
Crow Coal *9*, 12
Crowle No. 1 Borehole 8
Croxton 117, 143
Croxton Marl *92*, 93, 143
Cumbriense Marine Band 8
cyclic sedimentation 3
Cymodoce Zone *58*, *59*, *64*, 66, *146*, *147*
Cythereis luermannae hannoverana ostracod Zone *81*
Cythereis luermannae wermannae ostracod Zone *81*

Dam Road (Barton) Borehole *64*, 140, *148*
davoei Zone *31*, *37*, 40, *42*
De La Pole Flint *95*, 100, 101, 144
De-La-Pole Farm 144
Deepdale 87, 117
Deepdale Flint 90, *92*, 93
Deepdale Lower Marl *79*, 90, *92*, 93
Deepdale Marls 93
Deepdale pit *91*
Deepdale Upper Marl *79*, *92*, 93
denotatus Subzone 36, *42*
densinodulum Subzone *37*, 38, *42*
Densiplicatum Zone *58*, *59*, *63*, *64*, *145*
dentatus Zone *68*, 83
Derwent, River 137
desiccation cracks *27*, *53*, 56
Devensian deposits 117–124
beneath till 117–118
glacial deposits *110*, *120*
in the Humber Gap *120*
lacustrine deposits *120*
in Vale of York *110*
Devensian Stage 109, **111**, 117, 125
Devensian till, stone orientation *110*
diagenesis, calcite 32
Dinantian 2, **5**, 5, *6*
borehole sections *6*
'dip-and-fault' structure 107
discites Zone *44*, 46, 49, 52
discus Zone *44*, *53*, 55, 57
dispar Zone *68*, 84, 86
dolomite 17, 19, 20, 21, 22, 25, 67
Don Group **16**, 16, 17, *18*, 19, 20
Don, River 126
Dowsing Dolomite Formation 25
Dragonby 106, 125, 127, 133
Romano-British site 39
Dragonby Mine 39, 132
ironstone reserves *132*, **133**
drainage evolution, preglacial 3
Dry Valley Deposits **111**, 123
Dunsil Coals *10*, 13

East Butterwick 137
East Clough Borehole *47*, 51, 52, 57, *60*, 61, 62, *63*, *64*, 104, 118, *120*, 122, *148*
East Field 55
East Halton 96
East Halton-Skitter Beck *2*, 119, 126
East Lutton Plantation Pit 143
East Midlands Shelf 30, 46, 106
Eastfield Quarry 50, 51, 52, 54, 115
Easthorpe 143
Easthorpe Tabular Flints *94*, 98, 143
Eastlecottensis Subzone 67
Eau River *2*, *110*, 122, 123
economic products 128–137
Edmondia Marine Band *11*, 14
Elegans Zone *58*, *59*, *64*, *146*
Elland Flags 8
Ellerker 41, 44, 48, 122, 124
Ellerker Beck 48
Ellerker Limestone *44*, 46, *47*, 48
Elloughton Beck 118, 125
stone orientations
Elsham 66, 88, *91*, 115, 137
Elsham Interchange cutting 83, 84
Elsham lower pit 87, 88, *91*, 143
Elsham Quarry Borehole *64*, 66, 140, 146, *148*
Elsham Sandstone Formation 3, 57, *58*, 58, *64*, 66–67, 138, *146*
aquifer 138
Elsham Top 113
Elsham upper pit 90, 143
Elsham Wolds Borehole 36, 41, 43
Endemoceras amblygonium Zone 73
Endemoceras noricum Zone 73
Enodatum Subzone 61
Enthorpe 98
Enthorpe Marls 2, *79*, 98
Enthorpe Marl 1 *94*, 98
Enthorpe Marl 2 *94*
Enthorpe Marl 3 *94*
Enthorpe Oyster Bed *94*, 97, 98
Enthorpe Railway cutting 98, 99, 143, 144
Entolium Assemblage 69
Eodoxus Zone *58*
Eppleworth Chalk Pit 117
Eppleworth Farm cutting 144
Eppleworth Flint *95*, 100, 101
Eppleworth pit 96, 100
Eppleworth Quarry *91*, 144
Eppleworth Wood Farm cutting 100
Ermine Street 2
Esk Evaporate Formation 25
Eskdale Group **16**, 16, 17, *18*, 22–23
Estuarine
Alluvium **111**, 126, 129
Eudoxus Zone 58, *59*, *64*, *146*
evaporites 129
exaratum Subzone *37*, *42*, 43
extranodosa Subzone 34, *42*
EZ1–EZ5 cycles **16**, 17
EZ1 cycle 17, 20
EZ2 cycle 20, 21
EZ3 cycle 21
EZ5 cycle 22
falciferum Subzone 30, 32, *37*, *42*, 43, 44
falciferum Zone *31*, *37*, *42*, 43, 44
Faxfleet 126
feldspar 69
Ferriby Chalk Formation *68*, 77, **78**, 78, *79*, 79, 81–87, *91*, 128
generalised section *84*, *85*
Ferriby Sluice 125, 126
Ferruginous Flint 90, *92*, 93, 143
fibulatum Subzone 44
Field House 117
figulinum Subzone *37*, 40, *42*
fining-upwards cycles 32, 35
First Main Flint 90, *92*
First Terrace **111**, 123
First Waterloo Coal *10*
fissicostatum Zone 74, 75
fissicostatum-rude Zone *68*
Fittoni Zone *58*
Flamborough Chalk Formation 77, **78**, 79, *91*
Flamborough Head 93
Flandrian deposits *110*, 124–127
Flandrian Stage 109, **111**, 117, 123
flaser bedding *95*, 98, 143
flint implements 116
flints 79

burrow-form 77, 79, 87, 90, *95*, 99, 100, 143, 144
carious 79, 93, *95*, 97, 99, 100, 144
lenticular 93, 98, 99, 100, 101, 143, 144
nodular 88, *89*, 90, *95*, 97, 98
paramoudra 79, 93, 96, 100
potstone 144
semitabular 90, *95*, 99, 100, 101, 143, 144
tabular 77, 79, *89*, 93, *95*, 96, 97, 99, 100, 143, 144
Flixborough 106, 115, 122, 124, 127
Flixborough faults 106, 125
?Flockton Thin Coal *9*
Fluvioglacial Sand and Gravel 119–120
The Follies 56, 125
Fonaby 124, 137
Fordon Evaporites **16**, *18*, *19*, 20–21, 129
Forty-Yards Marine Band 8
Foulness, River 126
foundry sand **137**
Fourth Cherry Tree Marker 15
francolite 79
Frodingham Ironstone Member 1, 2, 29, 30, *31*, 32, *33*, 34–36, 38, 41, *42*, 106, 107, 128, 129–133, 130P, 136, 139
annual production *130*
chemical composition 35–36
contours on top of 105
depositional environment 36
detailed analysis (Santon) **131**
fining-upwars cycles 35
ironstone beneficiation 133
mineralogy 35
reserves *132*, **133**
tripartite division 35
Frodingham Ironstone opencast mines 40
Fulletby Beds 73

Gadney 112
Gainsborough 139
Gainsborough Trough 2, 5, *6*, 7, 8, *9*, *10*, *11*, 103, 104
Gallows Farm, Wrawby 66
Gander Farm 55, 57
garnet 54
Garthorpe 126
germanica Belemnite Zone 76
gibbosus Subzone 40, *42*
'ginger-bread' 34, 39, 41, 45
Glacial Sand and Gravel 112–113, 118–119, 136
Glasshoughton Rock *11*, 14
glauconite *37*, 40, 41, 43, 45, 66, 67, 69, 71, 72, 73, 74, 75, 76, 81
Glentham Formation 3, *44*, 52–56, 57, 106
fauna 54–56
sections *53*
Glosense Zone *58*, *59*, *64*, *145*
goethite 34, 35, 38, 39, 40, 45, 48, 61, 71, 72, 73, 74, 75, 76, 77, 131, 136
Gokewell Priory ironstone reserves *132*, **133**
Gokewell Strip 48
Gokewell Strip borehole 43
gottschei Zone *68*, 74
Gowthorpe House 126
Goxhill 117, 118
Goxhill Haven 126
Gracilis Marine Band *6*, 7
'Granby Limestones' 30
Grange Farm 48
granite, concealed 2
Grantham Formation 3, *44*, 45, 45–46, 46, 106
aquifer 138
depositional environment 46
fauna and flora 46
section *44*
Grasby 67, 76, 77, *91*, 112, 123, 124, 138
Grasby Marl *79*, 87, 90, *92*, 93, 143
Grasby Quarry 88, *91*
Grauer Salzton *18*, 21
Gravel Hill Farm 112, 113
Grayingham 41, *44*, 45, 46
Great Limber 39, 98
Great Oolite Limestone 52, 55
Great Ponton Gastropod Beds 51
'Green Beds' 25
greenalite 67
Greenmoor Rock 12
Grenoside Sandstone 12
Grenzanhydrite **16**
Grey Bed 86
'Greystone' 74, 75
Grimsby 34
Gristhorpe Member 52
'Gryphaea Bed' 86–87
Gryphaea vesicularis bed 143
'Gryphite gravel' 122
guerangi Zone *85*, 87
gulls 107
Gunness 126
gypsum 20, 22, 23, 25, 129

Haigh Moor Coal *10*, 13
Haigh Moor Rock *10*, 13
halite 17, 21, 22
Hall Farm Marl 90, *92*, 143
haloes 25
halokinesis 21, 22
Hampole Discontinuity 19
hannoverana ostracod Zone 84
Hardegson Disconformity 24
hardgrounds 81, 84, 86
phosphatised *81*, 82, 83
Hardwick Hill 115
Hatcliffe pit 98
Hatfield High Hazel Coal *10*, 13
Haughton Marine Band *10*, 13
Hauptanhydrit **16**
Hauptdolomit **16**
Hauterivian Stage *68*, 71, 72–73, 74
Haverholme 46, 57
hawskerense Subzone 41, *42*
Hayton Anhydrite **16**, *18*, *19*, 19–20, 20, 129
Head 45, **111**, 118, 124
heavy minerals 54
Hedbergella archaeocretacea Interval Zone 90
hematite 34, 35, 79, 90
hematitisation 36
Hempholme Lock 137
Hemsworth Division 15
Hercynian earth movements 3, 17
Heron Holt ironstone reserves *132*, **133**
Hessle 62, *91*, 109, 116, 117, 118, 138
stone orientations *121*
'Hessle Clay' 118
Hessle Quarry *91*
Hessle Town pit *91*
Hettangian Stage *29*, 30, *31*, 32, *33*, 34, *42*, 106
HF marl 84, 86
Hibaldstow 55, 57, 61, 107, 123
Hibaldstow Airfield 54, 55
Hibaldstow Airfield No. 1 Borehole 48
Hibaldstow Bridge Borehole *60*
Hibaldstow Limestone Member *44*, 46, *47*, 51, 52, 54, 106, 128
aquifer 139
hill wash 45
hodsoni Zone *44*, *53*, 55
hodsoni–aspidoides zonal boundary 55
Holaster subglobosus Zone **78**
Holaster trecensis Zone **78**
Holderness 112
Holkerian Stage **5**, 5
Holton 129
Honley (First Smalley) Marine Band 8, *9*
Horkstow 2
Houghton Thin Coal *11*, 14
Howsham 112, 124
Hoxnian Stage 109, **111**, 112, 115–116
Hudlestoni Zone *58*, *59*, *64*, 67, *146*
Hull *see* Kingston upon Hull
Hull River *2*, 2, 137, 138
Humber boreholes 50, 51, 52, 54, 61, 146
Humber Bridge 2, 108
Humber Bridge Approach Road cutting 87, *91*
Humber Estuary 1, *2*, 2, 34, 35, 43, 44, 50, 51, 57, *91*, 106, *110*, 117, *120*, 137, 138
Humber Gap 4, 109, 118, 121, 126
Devensian deposits in *120*
Humberside 82, 93
humphriesianum Zone *44*, 54
Hundon 76
Hundon Manor 70, 71, 72, 73, 74
Hunstanton 84
Hunstanton Chalk Member (Red Chalk) 3, *68*, 71, **78**, *81*, 81–84
Hunstanton Red Rock 82
Hydraulic Limestone 32, 46, 48
hydrocarbons 129
Corringham area 2
hydrogeology 137–139
Hysteroceras orbigny Subzone *81*, 83
Hysteroceras varicosum Subzone *81*, 84

ibex Zone *31*, *37*, 39, *42*
ice-wedge casts 113, 117
Icehouse Strip ironstone reserves *132*, **133**
illite 79
Immingham 4, 109, 130, 137
LPG storage caverns 93
Immingham Channel Deposits 109, *110*, **111**, 111–112
infiltration, average annual 137–138
inflatum Zone *68*, **78**, *81*, 83
Ings 106
Inoceramus beds 84
Inoceramus lissa Bed 84
Inoceramus lissa bioevent 84
insolubles 36, 39
interglacial deposits 115–117
Interglacial Gravel Beach Deposits 115, 116
Interglacial Silt and Clay 115, 116
intermedius Subzone 83
intertidal conditions 19
inversum Zone *68*, 74
Ipswichian cliff *110*, **111**, 115, 116, 138
Ipswichian deposits 116–117
Ipswichian Stage 109, **111**, 115, 116, 117
iron 35, 39, 41
Iron Age 2, 124
iron ore 129
iron pans 45, 46, 54
ironstone 2, 14, 30, 34, 35, 36, 39, 40, 71, 129–133, 136
Ironstone Junction Bed 54
Island Carr 55, 112, 126, 129

jamesoni Subzone 39, *42*
jamesoni Zone *31*, *37*, 39, *42*
Jason Zone *58*, *59*
johnstoni Subzone 34, *42*
jukesbrownei Zone 87
Jurassic *1*, 3, 29–70
aquifers 138–139
biostratigraphical zonation 29
chronostratigraphical units 29
definition of base 30
depositional environments summarized 29–30

Kamptus Subzone 60
kaolinite 38
kaolinite-illite 30
Kazanian-Tartarian Stage 16
Keelby 131
Kellaways Beds Formation 57, *58*, 58, 60–62, 106, 137, *145*
sections *60*
Kellaways Clay Member *60*, 60, 60–61, 106, *145*
Kellaways Rock Member *60*, 60, 61–62, 128, 138, *145*
Kellaways Sand Member *60*, 60, 61, 139, *145*
aquifer 138
Kelsey New Mill 112
Kent's Thick coals *10*, 13
Kent's Thin Coal *10*
Kerberus Zone *68*, 69
kerogen 136
Kettleby 123, 137
Kettleby Gravel Pit 123
Kettleby Thorpe 122
Keuper **16**
Keuper Anhydritic Member 25
'Keuper Marl' 24
Keuper Sandstone 23
Kilburn Coal 8, *9*, 12
Kilburn Marker 12
Kimmeridge Clay 57, 67, 145, *146*
Kimmeridgian sequences, standard bed numbers 145
Kimmerdigian Stage *29*, 57, *58*, 60, *64*, 69, *146*, *147*
Kinderscout Grit *6*, 7
Kinderscoutian Stage **5**, 7
Kingerby 112
Kingston upon Hull *1*, *2*, 2, *91*, 115, 125, 126, 129, 137, 138
Kiplingcotes 93, 98
Kiplingcotes Flints *94*, 98, 99
Kiplingcotes Marls *79*, 93, *94*, 97–98, 98, 100
Kiplingcotes Marl 1 98
Kiplingcotes Marl 2 98
Kiplingcotes Marl 3 98
Kirk Ella 144
Kirk Ella Marl *79*, *94*, 98, 99, 144
Kirk Ella railway cutting 117
Kirkdale Cave 117
Kirkham Abbey Formation **16**, *18*, *19*, 20
Kirmington 2, 4, 76, 109, 118, 123, 138
Hoxnian deposits 115–116
interglacial deposits 109, *110*
Kirmington Airport 120
Kirmond Le Mire 75
Kirton 129
Kirton Cement Shale 49, 129
Kirton Cementstones Member *44*, 46, *47*, 48, 49P, 49–51, 52, 54, 129
coralline knolls 50
Kirton in Lindsey *2*, 2, 29, 46, 113, 124, 125
Kirton in Lindsey Airfield 48
Kirton in Lindsey railway tunnel 43
Kirton in Lindsey Station 39
Kirton Quarries *47*, 48, 49, 50, 51
Lincolnshire Limestone Formation 49P
Kirton Tunnel Cutting 46, 48
'*koeneni* regression' 100
Koenigi Subzone 60
Kupferscheifer **16**

labiatus Zone **78**, 88, 93
Ladinian Stage 24
laeviuscula Subzone 51
laeviuscula Zone *44*, 50, 51, 52
Lake Humber 4, 109, 121
water levels 123
"Lamberti Limestone" *145*
Lamberti Zone *58*, *59*, 62, *63*, *64*, *145*
Lamplughi Zone *68*, 69
landslips 67, 73, **111**, 127
Langport Member 26
laqueus Subzone *33*, *42*
lata Zone *78*, *89*, 93, 97
Laughton 113, 115
lautus Zone *68*, 83
Leadenham Member 49
Leggat's Quarry, South Ferriby 143
Leine Halit **16**
Lenton Sandstone Formation 23
levesquei Zone *31*
Lias Group 3, **16**, 16, *28*, 28, *29*, 29, 30–44, 129
biostratigraphical zonation 30
graded cycles 30
lithostratigraphical divisions 30
liasicus Zone *31*, *33*, 34, *42*
Lidget Coal *10*, 13
Liesegang rings 77
lignite 23
Lilstock Formation **16**, 25, 26–28, *27*, *28*
Limber Borehole 40
lime 36, 39, 41
limestone, for aggregate 136
limonite 90, 93
Lincoln 45, 50, 51
Lincoln Edge *2*, 2, 48, *110*
Lincoln Marsh *2*, 2
Lincolnshire Limestone Formation 3, *44*, 46–52, 54, 107, 128, 136, 137, 138
aquifer 138–139
fauna 48, 50, 52
hydrogeological properties 139
sections *47*
Lincolnshire Wolds *2*, 2, 67, 71, *110*, 118
Listeri (Alton) Marine Band 8, *9*
Little Weighton 93, 99
Little Weighton Marls 98, 99, 100–101, 144
Little Weighton Marl 1 *79*, *95*, 100, 144
Little Weighton Marl 2 *79*, *95*, 100
Little Weighton Marl 3 *95*, 100
Little Weighton Quarry *91*, 96, 98, 99, 100, 144
railway cuttings near *91*
London Platform 67
London–Brabant Massif 17, 29, 30
loricatus Zone *68*, 83
Louth 96
Low 'Estheria' Band *9*, 12
Low Farm (South Ferriby) Borehole 56, 57, *60*, 61, 118, *120*, 126, 140
Low Field Lane Borehole *60*, 62, *63*, *120*, 122, 140, *148*
?Low Fenton Coal *9*
Low Santon 48
Low Silkstone Coal 12
Lower Albian Stage *81*
Lower Bajocian Substage 54
Lower Callovian Substage *59*
Lower Carboniferous rocks 102, 103
Lower Cenomanian *85*, 86
Lower Chalk *79*, **78**

Lower Claxby Ironstone Member *68*, 71–72
Lower Coal Measures (Westphalian A) 2, 8, 12
borehole sections *9*
Lower Cornbrash Member *44*, 56–57, 60
Lower Cretaceous 3, 71–77
generalised section *68*
Lower Estuarine Series 45
Lower Inoceramus Bed *85*, 86
Lower Jurassic *29*
ammonites *42*
Lower Kimmeridgian Substage *59*
Lower Magnesian Limestone **16**, 17, *18*, *19*, 19, 20
Lower Marl 17, *18*, 19
Lower Marsh Till 118
Lower Mottled Sandstone 23
Lower Orbirhynchia Bed *85*, 86
Lower Oxford Clay 62
Lower Oxfordian Substage *59*
Lower Palaeolithic flakes 2
Lower Pink Band 84, *85*, 86
Lower Pliensbachian Substage *31*, *37*, *42*
Lower Roach 76
Lower Sand and Gravel 121
Lower Sinemurian Substage *31*, 32, *42*
Lower Spilsby Sandstone Member 3, 66, *68*, 137
Lower Tealby Clay Member 3, *68*, 73–74
Lower Toarcian Substage 30, *31*, *37*, *42*
Loxley Edge Rock 8
LPG storage caverns 93, 143
Ludborough Flint *94*, 97
luermannae/ventrocostata ostracod Zone 83, 84
luridum Subzone 39, *42*
lyra Subzone *33*, 34, *42*

M180 Motorway, Bottesford 32, 34
M 513 Borehole *33*
macdonnelli Subzone *37*, 38, *42*
Macrocephalus Zone *44*, *58*, *59*, 60, 61, *145*
maculatum Subzone *37*, 40, *42*
magnetic rocks 103
magnetite 102, 133
Main Anhydrite 22
Main Smut Coal *10*
Maltby (Two-Foot) Marine Band 13
Maltonense Subzone *64*
mammillatum Zone *68*, 77
Manby Wood ironstone reserves *132*, **133**
Manby Wood Borehole *47*
manganese 36, 46, 82
'Mansfield' Cank 14
Mansgate Hill, Caistor 128
Mansgate Quarry 86, 87, 88, *91*, 143
mantelli Zone *85*, 86
Mantelliceras dixoni Zone 86
Mantelliceras mantelli Zone *85*, 86
Manton 45, 48–50, 136
Manton 'Estheria' Band 13
Manton Quarry 48
margaritatus Zone *31*, *37*, 40, 41, *42*
Marginal sand and gravel 122–123
marginatus Zone *68*, 74
Mariae Zone *58*, *59*, 62, *63*, *64*, *145*
marine bands 2
'Marine Gravels' 117, 119
Market Place Borehole 125, 126
Market Weighton 2, 30, 32, 46, 51, 103, 138
Market Weighton Canal *2*
'Market Weighton granite' 103
Market Weighton Structure 3, 30, 44, 46, 67, 77–78, 87, 106
Marl Slate **16**, 17, *18*, 19
Marlstone Rock Member 30, *31*, 36, *37*, 40–41, *42*, 106, 130P, 139
depositional environment 41
fauna 41
Marsdenian Stage **5**, 7
Marsupites testudinarius Zone **78**
masseanum Subzone *42*
Melton 62, 77, 128
Melton Bottom 77
Melton Bottom boreholes 44, *64*, 141
Melton Bottom Quarry 84, 87, *91*
Melton 'Clay' Pit 108
Melton Gallows 76, 77
Melton Gallows Farm 112, 113
Melton Pit *64*
Melton Ross 76, 82, *91*, 123, 128
Melton Ross Marl 90, *92*
Melton Ross Quarry 87, 88, 90, *91*, 93, 143
Melton whiting 128
Meltonfield Coal *10*, 13, 14
Mercia Mudstone Group 3, **16**, *24*, 24–25, 26, *28*, 129, 139
isopachytes *24*
units A–G *24*
Mere Dyke 127
Mesozoic rocks, structure 105–106
Messingham 124, 125, 137
Metheringham 50
Metoicoceras Zone **78**
Mexborough Rock *11*, 14
Mickleholme 56, 57
Micraster coranguinum Zone **78**, *79*, *89*, 100, 101
Micraster cortestudinarium Zone **78**, *79*, *89*, 98, 99
microcline 67, 76
microstromatolites 35
Mid Spilsby Nodule Bed 69, 70
Middle Albian Substage *81*, 83
Middle Callovian Substage *59*
Middle Cenomanian *85*, 86, 87
Middle Chalk **78**
Middle Coal Measures (Westphalian B and C) 12–14
borehole sections *10*
Middle Jurassic *29*, *58*
Middle Manton 39
Middle Marl **16**, *18*, 20, 21, 22
Middle Oxfordian Substage *59*
?Middleton Eleven Yards *9*
Mill Hill, Elloughton 61, 117
Millstone Grit 2, 3, 5, 7–8, 104
borehole sections *6*
montmorillonite *64*, 79
Moortown Hill 129
Mortoniceras inflatum Zone *68*, **78**, *81*, 83
Mount Pleasant 129
Mount Pleasant clay pit 136
mudflows 67, 73, 127
mudstone
brick, tile and cement making 129
murchisonae Zone *44*, 45
Muschelkalk **16**, 25
Muschelkalk Halite Member 25
muscovite 74
Musgrave's Farm 32
Mutabilis Zone *58*, *59*, *64*, 66, *146*

Namurian 2, **5**, 5, *6*, 7–8
borehole sections *6*
Neap House 127
Neocardioceras Zone **78**
Neocythere ventrocostata ostracod zone *81*
neptunean dykes 48, 57
Nettleton 51, 71, 73, 74, 106, 123, 127, 129, *135*, 136
Nettleton Beck *135*
Nettleton Borehole 17, *18*, 21, 22, 23, *24*, 25, 26, 141
Nettleton Bottom 46, 75, 76, *135*
Nettleton Bottom Borehole 32, *33*, 34, 36, *37*, 38, 39, 40, 41, 43, *44*, 44, 45, 46, *47*, 48, 49, *53*, 54, 55, *60*, 60, 61, 62, *63*, *64*, 66, 67, 136, 141, *145–148*
Callovian and Oxfordian sequences, Ancholme Clay Group *145*
correlation with South Ferriby Pit *147*
Kimmeridgian sequence, Ancholme Clay Group *146*
Nettleton Bottom Limestone Member *44*, *53*, 54
Nettleton Bottom Mine *135*, 136
Nettleton Bottom Quarry 76, 77, 84, *91*, 143
Nettleton Grange *135*
Nettleton Grange Barns 76
Nettleton Hill 67, 71, 72, 73, 75
Nettleton Hill Quarry 74
'Nettleton Member' 143
Nettleton Pycnodonte Bed *85*, 86–87, 143
Nettleton Stone beds 84, *85*, 86–87, 143
Nettleton Top 69, 76, *135*
Nettleton Top Farm 71
Nettleton Top Mine 71, *134*, 136
New Cliff Farm 44
New Farm *134*
New Forest Plantation 106
New Holland 129, 138
Newark Gypsum 25, 129
Newbald Wold Pit 97
'Newer Drift' 118
Newhill Coal *10*, 13
Newport 121, 123
Newport Borehole 117

Newstead 127
Newstead Rock *11*, 14
nodules
 clay-ironstone 40
 phosphatic 58, 69, 71, 76, 82
 septarian 40
 septarian ironstone 73
 sideritic 32
noricum Zone *68*
Normanby 73, 75
Normanby le Wold 71, 75, 76, *135*
Normanby Lodge 71, 74, 75
North Ewster Borehole 5, 8, *9*, *10*, *11*, 12, 13, 14, 15, 141
North Ferriby 2, 106, 124, 125
 stone orientations *121*
North Hall 121
North Kelsey 112, 113, 123, 124, 125, 129
North Killingholme boreholes 96
North Landing 93
North Newbald 138
North Ormsby Marl *79*, 93, *94*, 97
North Owersby 112
North Sea 29
Northampton Sand Formation 3, *44*, 44–45
 aquifer 138
 depositional environment 45
 fauna 45
 section *44*
Northampton Sand Ironstone 45
Northern Province 77, 80, 81, 86, 87, 88, 90, 93, 99, 100
Northern Province Chalk *79*
 marker horizons 143–144
Northorpe railway station 115
Norton Coal 8
Norton Mussel Band 8
Nottingham Bight 19
Nottingham Castle Formation 23

Oaks Rock Sandstone *10*, 14
Obovothyris obovata Zone 57
obtusum Subzone 36, *42*
obtusum Zone *31*, *33*, 36, *42*
Oceanic Anoxic Event 88
oil shales 67, 136
'Old Gravels' 122
Older River Gravel 136
oncoliths 21, 35, 51
opalinum Subzone 45
opalinum Zone 45
Oppressus Zone *68*, 69
orbignyi Subzone *81*, 83
orthoclase 67, 76
Osgodby Borehole *63*, *64*, 140, 146, *148*
Otby *135*, 136
Otby House 136
Ouse River *2*, 2, 3
ovalis Subzone 50, 51
Oxford Clay 57, *145*
Oxfordian sequences, standard bed numbers 145
Oxfordian Stage *29*, 57, *58*, 60, *63*, *64*, 106, *145*, *147*
Oxfordian–Kimmeridgian stage boundary 66
Oxgangs 74, 75, *134*
oxynotum Subzone *37*, 38, *42*
oxynotum Zone *31*, *33*, 36, *37*, *42*
Oxytoma Cementstone *147*

Pallasioides Zone *58*
paltum Subzone *42*, 43
paper shales 32, 41, 43, 62, *64*, 66, 67
 bituminous 136
Paradise Lane Borehole *53*, 54
Paradoxica/Sponge Bed 82, 83, 84, *85*, 86
paramoudras 79, 93, 96, 100
Paratollia Zone 72
Park House Farm 113
Parkgate *9*
Parkgate Rock *9*
Passage Beds 55
peat **111**, 125–126
pebbles, phosphatised *37*, 39, 41, 43, 63, *64*, 67, 69, 72, *85*
Pecten Ironstone Member 30, *31*, 36, *37*, 39–40, *42*, 106, 130P, 139
 fauna 39
Pectinatus Zone 67
Pegmatitanhydrit **16**
Penarth Group *3*, **16**, 25–28, *28*, *29*, 32, 129, 139
Pendleian Stage **5**
Pennines 17, 29
perinflatum Subzone 84, 86
Permian *1*, 3, 16–23
 borehole sections *18*
 isopachytes *19*
 structure 105–106
Peturia *see* Brough
phosphate 72, 75, 77, 81
phosphorus 36
Pilham 106, 113
Pilham seismic boreholes *27*
pisoliths 21
plagioclase 76
Planicerclus Subzone 61
planorbis Subzone *33*, *42*
planorbis Zone 30, *31*, *33*, *42*
planus Zone 97, 98
Plattendolomit **16**
playas 21, 22, 24, 25
Pleuromya–Grammatodon Assemblage 69
Pliensbachian Stage *29*, *31*, 106
polita ostracod Biozone 54
polymorphus Subzone 39, *42*
Polyptychites Zone *68*, 72
Portlandian Stage *29*, 30, *68*, 69, 77, *146*
portlocki Subzone *33*, 34, *42*
pot lids 46
potato stones 73, 76
Praeglobotrunca helvetica Partial Range Zone 90
Praeoxyteuthis pugio–Aulacoteuthis belemnite zones boundary 75
preplicomphalus Zone *68*, 69
Priestland Clay Member *44*, 52, *53*, 54–55, 56
 cycles in 54–55
Priestland Covert 54
Primitivus Zone *68*, 69
Protocythere albae/Dolocythereidea vinculum ostracod zone *81*
Protocythere albae/Dolocythereidea bosquetiana ostracod zone *81*
Providence 126
pseudomorphs after halite 20, 23
Pycnodonte Event 87, 143
Pyewipe 55
pyrite 26, 27, 34, 35, 38, 40, 43, 45, 46, 48, 50, *64*, 66, 72, 73, 74, 75, *81*, 82, 90

quasi-hargrounds 40
Quaternary 3–4, 109–127
 Devensian 117–124
Quaternary deposits
 approximately limits 110
 correlation of **111**
 Flandrian 124–127
 hydrogeological significance 138
 interglacial 115–117
 pre-Devensian glacial 109–115

raricostatoides Subzone *37*, 38, *42*
raricostatum Zone *31*, *37*, 38, *42*
rarocinctum Zone *68*, 75
Ravendale Flint 93, *94*, 96, 97
Raventhorpe Beds Member *44*, 46, *47*, 48, 107, 136
Raywell Sanatorium 98
Raywell Sanatorium pit *91*, 98
Red Chalk 3, *68*, 71, **78**, 79, *81*, 81–84
Red Cliff 109. 118, 119P, 122, 125
 borehole at *120*
Red Cliff section *120*
Redbourne 46, 57, 60
Redbourne Church 54
Redbourne Group 3, *29*, 29, 32, 44–57, *58*, 106
 lithostratigraphical and chronostratigraphical correlation *44*
Redbourne Hayes 57, 113, 125, 137
Redbourne Park 54
reduction haloes 22, 23
Reedness 103
Reed's Island 126
regale Zone 73, 74
Regulare Zone *58*, *59*, *64*, 66, *145*
resupinatum Subzone *33*, 34, 36, *42*
Rhaetian Stage 26, *29*, 30, *31*, 32, *33*
rhotomagense Zone *85*, 86
Rhythmic Sequence 55
Riby Marl *92*
Riplingham Grange 144
Riplingham Tabular Flints *94*, 98, 99, 144
Risby ironstone reserves **133**
Risby Borehole 16, *18*, 20, 21, 22, 23, *24*, 26
Risby Opencast Resource *132*
Risby Warren 54, 106, 124, 125
Risby Warren Borehole 41, 43
River Head 57

Roach Formation 3, *68*, 71, 73
Roach Ironstone 136
Roach Stone 75, 76
Rock Abbey Borehole 5, 8, *9*, *10*, *11*, 12, 13, 14, 141
Roman Villa, near Winterton 128
Romano-British remains 125
Romano-British site, Dragonby 39
Rosenkrantzi Zone *58*, *59*, *64*, 66, *145*, *147*
rostratum Subzone 84
Röt Halite Member 25
Rotalipora cushmani Total Range Zone 90
Roter Salzton **16**
rotiforme Subzone *33*, 34, *42*
Rotliegendes **16**, 17
Rotunda Zone *58*
Rough Rock *6*, 8
Rowland Plantation ironstone reserves *132*, **133**
Rowley Estate 144
Rowley Marls 1–4 *95*, 100, 144
Rowley Marl 1 144
Rowley Marl 3 144
Rowley Marl 4 144
Roxby 40, 49
ironstone reserves *132*, **133**
Roxby Mine 36, 38, 39, 41, 43, 136
Roxton Wood No. 11 Borehole 34–35
Ruby Lodge 52
rutile 54, 69
Ryazanian Belemnite Assemblage 70
Ryazanian Stage *29*, 67, *68*, 70, 77

sabkhas 19, 21, 24, 25
St Helens 113
St Helen's Well *53*, 54
St Helen's Well Borehole 61
Saliferous Marl **16**, 16, *18*, 22–23
salt 129
Sancton 103, 106
sand and gravel resources 136–137
Sand House Farm, East Butterwick 137
Sandringham Sands 67, 70
Santon 35, 133
Santon East ironstone reserves **133**
Santon Mine 39, 132
Santon Oolite Member *44*, 46, *47*, 48, 48–49, 49P, 50, 136
Santon West ironstone reserves *132*, **133**
Santonian Stage **78**
sauzei Zone 54
Saxby All Saints 56
Scabcroft 54
Scabcroft Borehole *60*
Scalby Formation 52
Scarborough Formation 54
Scawby 45, 46, 50
Scawby Limestone *47*, 49P, 49, 50, 51, 52, 128, 136
Scawby Park 106
scipionianum Subzone *33*, 34, *42*
scissum Subzone 45
Scitulus Zone *58*, *59*, *64*, *146*
Scofton Coal 15
Scotney Farm No. 2 Borehole 41, 43
Scotter 26, 109, 122, 123, 124
Scotter Moor Drain *27*
Scotter seismic boreholes *27*
Scotter Wood Farm 113
Scotterthorpe 121, 129
Scottlethorpe Beds 50
Scotton Common 115
Scrawby Limestone *44*
Scunthorpe *1*, 1, *2*, 2, 34, 45, 128, 129, 130, 131, *132*, 137
Scunthorpe Bypass 44, 46
Scunthorpe Mudstones Formation **16**, *28*, 30, *31*, 32–36, 36
borehole sections *33*
calcite diagenesis 32
fauna 34
fining-upwards cycles 32
isopachytes *31*
Scythian Stage 24
Searby 76, 84, 115
Seaton Carew Formation 25
Second Ell Coal *10*, 13
Second Wales Coal 14
seismic reflection surveys 104
selenite 58, 73
semiceletum Subzone 32, *37*, *42*, 43
semicostatum Zone *31*, *33*, 34, 36, *42*
Serratum Zone *58*, *59*, *64*, *145*
Severn-Trent Water Authority 137
Sewerby 116, 117
Shafton Coal *11*, 14
Shafton Marine Band *11*, 14
'Shales of the Cornbrash' 60
Sharlston Low Coal *11*, 14
Sharlston Top Coal *11*, 14
Sharlston Yard Coal *11*, 14
Sheepcote Hill 112, 113, 137
'Sheffield' ironstone reserves *132*, **133**
Sheffield's Hill 124
shell marl **111**, 124–125
Sherwood Sandstone Group 3, **16**, 16, *18*, 22, 23–24
aquifer 139
isopachytes *24*
'shrinkage' vugs 20
siddingtonensis Zone 57
siderite 35, 38, 40, 48, 69, 71, 72, 73, 74, 75
Silesian **5**
silica 35, 36
silica sand 137
physico-chemical properties **137**
simpsoni Subzone 36, *37*, 38, *42*
Sinemurian Stage *29*, *31*, 32, *33*, 106
Skipsea (formerly Drab) Till 118
Skitter, River 138
?Slack Bank Rock *9*
Sleaford Member 51
Sleights Siltstone **16**, *18*, 22
South Ferriby 107
smectite 32
smelting sites 39
'snap-bands' 36
Snitterby 54, 55
Snitterby Carr Lane Borehole 60
Snitterby Limestone Member *44*, 52, *53*, 55, 56, 106, 128
aquifer 138
Snitterby, trench near *60*
'Soft Bed' sequence 8
soil creep 45
Sole Pit Trough 30
solution hollows, occupied by Thorncroft Sands 52, 54
Somerby 122
Somerby Pit 86
South Cave *2*, 2, 50, 52, 57, 61, 115, 122, 123, 124, 127, 128, 137
South Cave Borehole 32, *47*, 51, 52, *53*, *60*, 61, 62, *63*, *64*, *148*
South Cliffe Borehole 16, *18*, 21, 22, 23, *24*, 141
South End 119
South Ferriby 66, 83, 84, 87, 90, *91*, 122, 127, 128, 129, 146
stone orientations *121*
South Ferriby Cliff 107–108, 116–117, 118
section *120*
South Ferriby Pit *64*
correlation with Nettleton Bottom Borehole *147*
South Ferriby Quarry 77, 82P, 88, *91*, 143
South Hills 113
South Kelsey 112, 129
South Kelsey Gravel 113
South Killingholme, LPG storage caverns 143
Southern North Sea Basin 17, 21, 22, 23, 24, 29, 30
Southern Province 77, 97
Southern Province Chalk *79*
speetonensis Subzone 74
speetonensis Zone *68*
Sphenoceramus lingua Zone **78**
Spilsby Sandstone Formation *29*, 29, 30, 66, 67–70, *68*, 137, *146*
spinatum Zone *31*, *37*, 41, *42*
Spital Borehole 5, *6*, 7, *9*, *10*, *11*, 12, 16, 17, *18*, 22, *27*
Sponge Bed *see* Paradoxica Bed
Spring Wood ironstone reserves *132*, **133**
Springfield 122
springs 125, 139
Staddlethorpe 126
Stainby Shale *44*, 45
Staintondale Group **16**, 17, *18*, 22
Stassfurt Salze **16**
Station Farm 51
staurolite 54
steghausi/bemerodensis ostracod Zone 84
stellare Subzone 42
Sternotaxis planus Zone **78**, *79*, *89*, 96
stokesi Subzone *37*, 40, *42*
Stoliczkaia dispar Zone **78**, *81*
Stone Age 2
stone orientation measurements, Devensian till 118, *121*
Stoneferry 118, 125, 126
stromatolitic banding 20
structure 102–108
Sturton 125

stylolites 20, 80, 90, 100
subcrenatum (Pot Clay) Marine Band *6*, 8, *9*
subdelaruei Subzone 83
'submerged forest' 125
subnodosus Subzone *37*, 40, *42*
sulphur 36, 39
Superbilinguis Marine Band 7
superficial structures 29, 106–108
Supracoralline Bed *146*
Sverdrup Basin 27
Swallow Vale Farm 143
Swallow Wood Coal *10*, 12–13, 13
Swanland 118
Swinton Pottery Coal *10*, 13
syneresis movements 34
syneresis cracks 20

tardefurcata Zone *68*, 77
taylori Subzone *37*, 38, 39, *42*
Tea Green Marl 25
Tealby 75, *135*
Tealby Clay 66
Tealby Formation *68*, 71, 73–75
Tealby Limestone Member 3, 73, 74–75, 128
Teesside Group **16**, 17, *18*, 21–22
tenuicostatum Subzone *37*, 41, *42*, 43
tenuicostatum Zone *31*, *37*, 41, *42*, 43
Tenuiserratum Zone *58*, *59*, *63*, *64*, 66, *145*
Terebratulina lata Zone *79*, *89*, 93, 97
'*Terebratulina ornata* Subzone' 86
Terrace Deposits 123, 136
undifferentiated **111**
Tertiary 3
Tethyan Sea 29
Tethyan taxa 30
Tetney Lock Borehole 25
Thealby ironstone reserves *132*
Thealby Beck 34
Thealby Mine 36
The Thistles 129
Thoresway 76
Thoresway Borehole 73, 74, 75, 76
Thoresway Sand *see* Carstone Grit
Thorncliffe *9*
Thorncroft 55
Thorncroft Sands Member *44*, 51, 52, *53*, 54, 56, 61, 115
aquifer 138
Thorne Colliery 129
Thornholme 35, 57, 129
Thornton Abbey 119
Thornton Curtis 118
Thornton Curtis Marl *94*
Thornton Curtis pit *91*, 96
Thornton Station 120
Thornton Valley 124
thouarsense Zone *31*
Thoughton Thin Coal *11*
Tidal Flat Deposits **111**, 127
till **111**, 113
Devensian 115, 118, *120*
Toarcian Stage *29*, 30, *31*
Top Anhydrite **16**, *18*, *19*, 22
Totternhoe Stone 84, *85*, 86
tourmaline 54, 69
trecensis Zone 87
Trent Cliff *110*
Trent Opencast Mine 36
Trent River *2*, 2, 3, 137
Trent Remine ironstone reserves *132*, **133**
Trent Valley 109, *110*, 113, 115, 121, 122, 123, 124, 125, 126
Triassic *1*, 3, 23–29
aquifers 139
base of 23
lowest Lias strata 30
Triple Tabular Flints *94*, 97
Triton Anhydritic Member 25
tufa, calcareous 46
turneri Zone *31*, *33*, 34, 36, *42*
Turonian Stage **78**, *79*, 88, *89*, 90, 97
Twigmoor Hall 39
Two-Foot Coal *10*, 13

Uintacrinus socialis Zone **78**
Ulceby *91*, 96, 98, 116, 128
Ulceby Marl *79*, 93, *94*, 96, 97
Ulceby Oyster Bed *94*, 97, 98
Ulceby parish pit *91*, 97
Unterer Lias **16**
Upper Albian Substage *81*
Upper Albian Subzone 83
Upper Anhydrite **16**, *18*, 22
Upper Bajocian Substage 54
Upper Band Coal 8
Upper Callovian Substage *59*
Upper Carboniferous rocks 102
Upper Cenomanian Substage *85*, 87
Upper Chalk **78**, *79*
Upper Claxby Ironstone Member *68*, 72–73
Upper Coal Measures (Westphalian C and D) 14–15
borehole sections *11*
Upper Cornbrash Member *44*, 56–57, 106
Upper Cretaceous 71, 77–101
Upper Estuarine Series 52
Upper Inoceramus Bed *85*, 86
Upper Jurassic *29*, *58*, *68*
Upper Kimmeridgian Substage *59*
Upper Lincolnshire Limestone 51
Upper Magnesian Limestone **16**, *18*, *19*, 21
Upper Marl 22
Upper Marsh Till 118
Upper Orbirhynchia Bed *85*
Upper Oxfordian Substage *59*
Upper Permian 16
Upper Pink Band 84, *85*, 87
Upper Pliensbachian Substage *31*, *37*, *42*
Upper Roach 75, 76
Upper Sand 123
'Upper Sandy Beds' 56
Upper Sinemurian Substage *31*, *37*, *42*
Upper Spilsby Sandstone Member 3, 67, *68*, 69–70, 71
Upper Tealby Clay Member *68*, 73, 75, 76
Upper Toarcian Substage *31*
Upper Turonian Substage 93
?Upper Band Coal *9*

Valanginian ostracod zone, early 72
Valanginian Stage *68*, 71, 73
valdani Subzone *42*
Vale House Quarry *91*, 96, 97, 98
Vale of York 2, 109, *110*, 112, 121
Glacial Lake Deposits *110*, **111**, 120–121, 129, 136, 137
valley bulges 106, 107, 108
Vanderbeckei (Clay Cross) Marine Band 8, *9*, *10*
variabilis Zone *31*, *68*, 73, 74
varicosum Subzone *81*, 84
ventifacts 121, 124
Vertebrale Subzone *64*
vivianite 126
Volgian Stage *68*
Volviceramus involutus Zone 100
Volviceramus koeneni Zone 100

Waddingham 55, 106, 123
Waddingham Beck 125
Waddington 54
Waddington Beck 125
Waddington Holmes 126
Walesby 75, *135*, 136
Walling Fen *2*, *110*, 122, 125, 126
warp **111**, 126–127
Waterloo Marker *10*, 13
wave-cut platform, Ipswichian 116–117
Weedley Dale cutting 87
Weighton Lock 125, 126
Welton *91*
Welton Chalk Formation 77–78, **78**, *79*, 79, 87–93, 128, 143
detailed section *92*
stratigraphy of *89*
Welton Crossing Borehole 56, 122, 140
Welton Waters A Borehole 54, 56, 125, 126, 140
Welton Waters B Borehole 61, *120*, 140
Werraanhydrit **16**, 19
Werradolomit and Zechsteinkalk **16**
West Halton 36, 122, 129, 130, *132*
West Walton Beds 57, *64*, *145*, 145
West Wood ironstone reserves *132*, **133**
Westbury Formation **16**, 25, 26, *27*, 27, *28*
Westfield House pit *91*
Westphalian (Coal Measures) 8–15, 129
Westphalian A (Lower Coal Measures) **5**, 8, *9*, 12
Westphalian B (Middle Coal Measures) **5**, 8, *10*, 12–14
Westphalian C and D (Middle Coal Measures (part), Upper Coal Measures) **5**, 8, *11*, 14–15
Wheatleyensis Zone *58*, *59*, *64*, *146*
Wheatworth Coal *10*, 13–14
White Sands 45
Whitton 121, 122, 124
Whitton Ness Borehole 34, 35, *120*, 126, 140

Willerby 93, 118, 128
  stone orientations *121*
Willerby Flints *94*, 98, 99, 144
Willerby Flint 1 99, 144
Willerby Flint 2 99, 144
Willerby Flint 3 99, 144
Willerby Quarry *91*, 93, 96, 98, 99, 100, 143, 144
Winceby 70
Wing Mine 39
Wingfield Flags 8
Winter Coal *10*, 13
Winteringham 45, 48, 49, 118, 119, 124
Winteringham Haven Borehole 40, 126, 140
Winteringham – Horkstow Moraine 118
Winterton 44, 48, 106, 118, 119, 128, *132*, 137
Winterton Beck *2*, *110*, 122, 125, 126
Winterton Beck Valley 107
Winterton Carrs Borehole 61
Winterton Cliff 132
Winterton Grange *44*, 46, 48
Winterton Grange Borehole *47*
Winterton Holmes 129, 139
Winterton North ironstone reserves *132*
Winterton South ironstone reserves *132*, **133**
Wolds Syncline 106
Wolstonian Stage 109, **111**, 116
Wood Farm 137
Woodshill Quarry, Nettleton 137
Woofham Farm 57
Wootton Hall pit *91*
Wootton Marls *94*, 97
Wootton Wold 117
Worlaby 62, 129, 131
Worlaby boreholes 146
Worlaby E Borehole 36, *37*, 38, 39, 40, 41, 43, 45, *47*, 48, 49, 50, 51, *53*, 54, 55, 56, 57, *60*, 60, 61, 62, *63*, *64*, 140, *148*
Worlaby G Borehole *64*, 66, 140, *148*
Wrawby 113, 122

Yaddlethorpe 121, 122
Yarborough 36
  ironstone reserves *132*, **133**
Yarborough Camp 90, 119, 143
Yarborough Marl 90, *92*, 143
Yarborough Mine 39
Yeadonian Stage **5**, 7
Yellow Clints pit *91*
Yorkshire ironstone reserves *132*, **133**
Yorkshire Coalfield 106
Yorkshire Water Authority 137
Yorkshire Wolds *2*, 2, 71, 87, *110*, 118

Zechstein EZ1 – EZ5 cycles **16**, 17
Zechstein Sea 3, 17, 21, 22
Zechstein transgression, initial 17
Zechsteinletten **16**
zircon 54, 69

BRITISH GEOLOGICAL SURVEY

Keyworth, Nottingham NG12 5GG
0602-363100

Murchison House, West Mains Road,
Edinburgh EH9 3LA 031-667 1000

London Information Office, Natural History Museum Earth Galleries, Exhibition Road, London SW7 2DE
071-589 4090

The full range of Survey publications is available through the Sales Desks at Keyworth and at Murchison House, Edinburgh, and in the BGS London Information Office in the Natural History Museum Earth Galleries. The adjacent bookshop stocks the more popular books for sale over the counter. Most BGS books and reports are listed in HMSO's Sectional List 45, and can be bought from HMSO and through HMSO agents and retailers. Maps are listed in the BGS Map Catalogue, and can be bought from Ordnance Survey agents as well as from BGS.

The British Geological Survey carries out the geological survey of Great Britain and Northern Ireland (the latter as an agency service for the government of Northern Ireland), and of the surrounding continental shelf, as well as its basic research projects. It also undertakes programmes of British technical aid in geology in developing countries as arranged by the Overseas Development Administration.

The British Geological Survey is a component body of the Natural Environment Research Council.

Maps and diagrams in this book use topography based on Ordnance Survey mapping

HMSO publications are available from:

HMSO Publications Centre
(Mail and telephone orders)
PO Box 276, London SW8 5DT
Telephone orders 071-873 9090
General enquiries 071-873 0011
Queueing system in operation for both numbers

HMSO Bookshops
49 High Holborn, London WC1V 6HB
071-873 0011 (Counter service only)
258 Broad Street, Birmingham B1 2HE
021-643 3740
Southey House, 33 Wine Street, Bristol BS1 2BQ
(0272) 264306
9 Princess Street, Manchester M60 8AS
061-834 7201
80 Chichester Street, Belfast BT1 4JY
(0232) 238451
71 Lothian Road, Edinburgh EH3 9AZ
031-228 4181

HMSO's Accredited Agents
(see Yellow Pages)

And through good booksellers